AF411314

Macrophytes in Inland Waters: From Knowledge to Management

Macrophytes in Inland Waters: From Knowledge to Management

Editor

Angelo Troia

MDPI • Basel • Beijing • Wuhan • Barcelona • Belgrade • Manchester • Tokyo • Cluj • Tianjin

Editor
Angelo Troia
Department of Biological,
Chemical and Pharmaceutical
Sciences and Technologies
(STeBiCeF)
University of Palermo
Palermo
Italy

Editorial Office
MDPI
St. Alban-Anlage 66
4052 Basel, Switzerland

This is a reprint of articles from the Special Issue published online in the open access journal *Plants* (ISSN 2223-7747) (available at: www.mdpi.com/journal/plants/special_issues/Macrophytes_Management).

For citation purposes, cite each article independently as indicated on the article page online and as indicated below:

LastName, A.A.; LastName, B.B.; LastName, C.C. Article Title. *Journal Name* **Year**, *Volume Number*, Page Range.

ISBN 978-3-0365-6699-3 (Hbk)
ISBN 978-3-0365-6698-6 (PDF)

Cover image courtesy of Angelo Troia

Contents

About the Editor

Angelo Troia

Angelo Troia is currently Assistant Professor in Botany at the Department of Biological, Chemical and Pharmaceutical Sciences and Technologies (STEBICEF), University of Palermo, Italy. After graduating in Natural Sciences in Palermo, he made his Ph.D. in Plant Biosystematics and Ecology at the University of Bologna, Italy. His ongoing research is focused on the diversity, ecology and biogeography of plants, with a particular focus on the genus Isoetes, freshwater plants and habitats, charophytes, the role of islands in the Mediterranean, and conservation issues.

Editorial

Macrophytes in Inland Waters: From Knowledge to Management

Angelo Troia

Dipartimento di Scienze e Tecnologie Biologiche, Chimiche e Farmaceutiche (STeBiCeF), Università degli Studi di Palermo, 90123 Palermo, Italy; angelo.troia@unipa.it

Abstract: The huge biodiversity of inland waters and the many different aquatic habitats or ecosystems occurring there are particularly threatened by human impacts. In this Special Issue, ten articles have been collected that show new data on the distribution and ecology of some rare aquatic macrophytes, including both vascular plants and charophytes, but also on the use of these organisms for the monitoring, management, and restoration of wetlands.

Keywords: aquatic plants; hydrophytes; charophytes; wetlands; alien invasive species

Citation: Troia, A. Macrophytes in Inland Waters: From Knowledge to Management. *Plants* **2023**, *12*, 582. https://doi.org/10.3390/plants12030582

Received: 18 January 2023
Accepted: 19 January 2023
Published: 29 January 2023

1. Introduction

The importance of freshwater on a global scale as a strategic element for the life of our species and for life in general is more clear today than ever before [1].

Inland waters [2], including not only freshwaters but also brackish or saline waters, host huge biodiversity and many different aquatic habitats or ecosystems that have unfortunately been increasingly threatened, disturbed, and damaged by human impacts in recent decades [1,3,4]. Indeed, we sometimes witness the destruction of important wetlands or aquatic biotopes before we are able to know their inhabitants (including their flora, fauna, microbiota, etc.).

Focusing on the "green" component, the strictly aquatic flora of inland waters (i.e., the plants living in the water or on its surface) are generally poorly known, probably because most "classic" botanists often stop at the edge of the land/water border. In addition, the evolutionarily and ecologically key group of charophytes is often not studied because phycologists focus on marine species (and charophytes usually occur only in inland waters) while botanists think that charophytes are not "plants" but just "algae", so outside their skills or kingdom.

The aim of this Special Issue was to invite people studying aquatic plants (and charophytes) in inland waters to contribute to furthering the knowledge on these organisms, with basic (taxonomical or ecological) or applied research (involving the management of aquatic plants, or their use in assessing the quality of waters or the quality of environments), with a special focus on floating and submersed macrophytes (i.e., the so-called hydrophytes). Invasive alien aquatic species, of course, were also possible subjects.

2. Special Issue Contents

As mentioned in the introduction, aquatic macrophytes are often underknown, so even reports on new populations are often interesting to better understand, for example, the biology, biogeography, or ecology of species. Sciandrello et al. [5] report a new population of the "marsh fern" *Thelypteris palustris* in the well-known island of Sicily, at the southern border of its distribution area. Trbojević et al. [6] report a new population of *Chara baueri* in Serbia: this is an important record, since the species is very rare, with few known populations in Europe and one in Asia. In both articles, the authors supply significant information on the morphology and ecology of the new and isolated populations.

A different case is the report on *Chara zeylanica* in Sardinia by Becker et al. [7]: the new population, in fact, is not only the first one in Europe, but—according to the scenario presented by the authors—also a possible case of "range-shifting"; in other words, the species could be shifting its distribution (in this case northwards), probably due to climate change [8].

Turner et al. [9] focused on *Stratiotes aloides* L., a vascular macrophyte with a wide distribution (from northern Central Europe in the west to Siberia in the east) but that is becoming rare in some areas; in their article, they show the importance of assessing the genetic structure of a species in order to manage it, both for preserving the diversity of the species as a whole and for possible reintroductions. As in other species, the authors verified that, in European populations of this aquatic plant, there is low genetic diversity within each population but high genetic diversity between populations.

Millozza and Abdelahad [10] present a different aspect of the basic research, more connected to the taxonomy. In detail, this paper provides an example of how historical herbarium collections can be used to assist ecology, biogeography, and conservation biology research, in this case supplying information to better define a species described at the end of the nineteenth century.

The articles of Peternel et al. [11] and Panzeca et al. [12] present two different examples in which macrophytes are used to characterize aquatic environments, in a river in Slovenia and in farm ponds of a district of Sicily, respectively. The species composition of the macrophyte community revealed significant changes over the years in a riverine ecosystem in the first case, whereas in the second one it was shown that, although farm ponds are artificial and relatively poor habitats, they seem to be important for aquatic flora and the conservation of local biodiversity.

The article by Ribaudo et al. [13] introduces more proper management aspects. This paper is the only one (in this Special Issue) dealing with alien invasive species, one of the main threats today to the conservation of species and ecosystems on a global scale, but its approach is original and invasive species are not even mentioned in the title or the abstract. The study aims at linking the role of wind action and water oxygenation within dense hydrophyte stands in two shallow lakes located in the southern Atlantic coast of France. Its results highlight the need to consider local hydrodynamics in lake management decisions, and show that mapping hypoxia risk in densely vegetated stands is a promising tool for the management of invasive hydrophytes in shallow lakes. Furthermore, the two invasive alien aquatic species are two submerged rooted Hydrocharitaceae, *Egeria densa* Planch. and *Lagarosiphon major* (Ridl.) Moss.

The re-establishment of submerged macrophytes, and especially charophyte vegetation, is a common aim in wetland management. The contribution of Blindow et al. [14] reviews the knowledge on the life forms, dispersal, establishment, and transplantations of submerged macrophytes, focusing on charophytes, and provides recommendations for an ambitious Swedish project that aims to protect threatened macrophyte species. Rodrigo [15] reviews the available knowledge in wetland restoration based on revegetation with hydrophytes and stresses common challenges as well as potential solutions; the clear negative factors which prevent revegetation success are considered, and useful final suggestions are provided.

Acknowledgments: I would like to thank all colleagues that contributed to this Special Issue by sending in their manuscripts or reviewing them, in addition to the *Plants* Editorial Office for their helpful support.

Conflicts of Interest: The author declares no conflict of interest.

References

1. Strayer, D.L.; Dudgeon, D. Freshwater biodiversity conservation: Recent progress and future challenges. *J. N. Am. Benthol. Soc.* **2010**, *29*, 344–358. [CrossRef]
2. Revenga, C.; Kura, Y. *Status and Trends of Biodiversity of Inland Water Ecosystems*; Secretariat of the Convention on Biological Diversity: Montreal, QC, Canada, 2003; Available online: https://www.cbd.int/doc/publications/cbd-ts-11.pdf (accessed on 18 January 2023).
3. Reis, V.; Hermoso, V.; Hamilton, S.K.; Ward, D.; Fluet-Chouinard, E.; Lehner, B.; Linke, S. A Global Assessment of Inland Wetland Conservation Status. *BioScience* **2017**, *67*, 523–533. [CrossRef]
4. Flitcroft, R.; Cooperman, M.S.; Harrison, I.J.; Juffe-Bignoli, D.; Boon, P.J. Theory and practice to conserve freshwater biodiversity in the Anthropocene. *Aquat. Conserv. Mar. Freshw. Ecosyst.* **2019**, *29*, 1013–1021. [CrossRef]
5. Sciandrello, S.; Cambria, S.; Giusso del Galdo, G.; Tavilla, G.; Minissale, P. Unexpected Discovery of *Thelypteris palustris* (Thelypteridaceae) in Sicily (Italy): Morphological, Ecological Analysis and Habitat Characterization. *Plants* **2021**, *10*, 2448. [CrossRef] [PubMed]
6. Trbojević, I.; Milovanović, V.; Subakov Simić, G. The Discovery of the Rare *Chara baueri* (Charales, Charophyceae) in Serbia. *Plants* **2020**, *9*, 1606. [CrossRef] [PubMed]
7. Becker, R.; Schubert, H.; Nowak, P. *Chara zeylanica* J.G. Klein ex Willd. (Charophyceae, Charales, Characeae): First European Record from the Island of Sardinia, Italy. *Plants* **2021**, *10*, 2069; [CrossRef] [PubMed]
8. Troia, A. The unnoticed northward expansion of *Najas marina* subsp. *armata* (Hydrocharitaceae) in the Mediterranean area: An effect of climate change? *Willdenowia* **2022**, *52*, 91–101. [CrossRef]
9. Turner, B.; Hameister, S.; Hudler, A.; Bernhardt, K.-G. Genetic Diversity of *Stratiotes aloides* L. (Hydrocharitaceae) Stands across Europe. *Plants* **2021**, *10*, 863. [CrossRef] [PubMed]
10. Millozza, A.; Abdelahad, N. The Contribution of Historical and Morphological Studies on Herbarium Specimens to a Better Definition of *Chara pelosiana* Avetta (Charales, Charophyceae). *Plants* **2021**, *10*, 2488. [CrossRef] [PubMed]
11. Peternel, A.; Gaberščik, A.; Zelnik, I.; Holcar, M.; Germ, M. Long-Term Changes in Macrophyte Distribution and Abundance in a Lowland River. *Plants* **2022**, *11*, 401. [CrossRef] [PubMed]
12. Panzeca, P.; Troia, A.; Madonia, P. Aquatic Macrophytes Occurrence in Mediterranean Farm Ponds: Preliminary Investigations in North-Western Sicily (Italy). *Plants* **2021**, *10*, 1292. [CrossRef] [PubMed]
13. Ribaudo, C.; Tison-Rosebery, J.; Eon, M.; Jan, G.; Bertrin, V. Wind Exposure Regulates Water Oxygenation in Densely Vegetated Shallow Lakes. *Plants* **2021**, *10*, 1269. [CrossRef]
14. Blindow, O.; Carlsson, M.; van de Weyer, K. Re-Establishment Techniques and Transplantations of Charophytes to Support Threatened Species. *Plants* **2021**, *10*, 1830. [CrossRef] [PubMed]
15. Rodrigo, M.A. Wetland Restoration with Hydrophytes: A Review. *Plants* **2021**, *10*, 1035. [CrossRef] [PubMed]

Article

Unexpected Discovery of *Thelypteris palustris* (*Thelypteridaceae*) in Sicily (Italy): Morphological, Ecological Analysis and Habitat Characterization

Saverio Sciandrello *, Salvatore Cambria, Gianpietro Giusso del Galdo, Gianmarco Tavilla and Pietro Minissale

Department of Biological, Geological and Environmental Sciences, University of Catania, Via A. Longo 19, I-95125 Catania, Italy; cambria_salvatore@yahoo.it (S.C.); g.giusso@unict.it (G.G.d.G.); gianmarco.tavilla@phd.unict.it (G.T.); p.minissale@unict.it (P.M.)
* Correspondence: s.sciandrello@unict.it

Abstract: *Thelypteris palustris* Schott (*Thelypteridaceae*), known as "marsh fern", is infrequent in the Mediterranean area. The occurrence of this species is known for almost all the Italian regions (except for Sardinia and Sicily), but with rare and declining populations. During floristic fieldwork on the Sicilian wetlands, a new unknown population was found. The aim of this paper is to analyze the morphological traits of the species, as well as its ecological features and the floristic composition of the plant communities where it lives. According to IUCN guidelines, here we provide the regional assessment (Sicily) of *T. palustris*. To analyze its morphological features, many living plants were examined, with particular attention to the spore structure. A total of **179** plots (110 species) and **34** pools were sampled. Our results highlight the relic character of the species which is at the southernmost border of its distribution range. The micro-morphological investigations on the spores show that the Sicilian population belongs to the subsp. *palustris*. The floristic analysis confirms the clear dominance of perennial temperate-cold zones Eurasian taxa. Finally, a new association, *Thelypterido palustris-Caricetum paniculatae*, within the *Caricion gracilis* alliance (*Phragmito-Magnocaricetea* class) is described.

Keywords: distribution; ecology; relic fern; Mediterranean wetlands; conservation status; pteridophytes; vegetation

Citation: Sciandrello, S.; Cambria, S.; del Galdo, G.G.; Tavilla, G.; Minissale, P. Unexpected Discovery of *Thelypteris palustris* (*Thelypteridaceae*) in Sicily (Italy): Morphological, Ecological Analysis and Habitat Characterization. *Plants* **2021**, *10*, 2448. https://doi.org/10.3390/plants10112448

Academic Editor: Angelo Troia

Received: 2 September 2021
Accepted: 21 October 2021
Published: 12 November 2021

Publisher's Note: MDPI stays neutral with regard to jurisdictional claims in published maps and institutional affiliations.

1. Introduction

Thelypteris palustris Schott (*Thelypteridaceae*), known as "marsh fern", is a deciduous species that represents one of the most complex species in the pteridophytes. Fernald [1] recognized four varieties of *Thelypteris palustris*: var. *palustris* of Eurasia (from Europe and NW Africa to eastern Himalayas and southern China); var. *pubescens* (G. Lawson) Fernald of northeastern United States, Canada, and eastern Asia; var. *haleana* Fernald of the southeastern United States and Bermuda; and var. *squamigera* (Schltdl.) Weath. of Africa, southern India, northern New Zealand [2]. Afterwards Tryon et al. [3], mainly analyzing the spore structure, recognized two species, one largely of the southern hemisphere (*T. confluens* (Thunb.) C.V.Morton = *T. palustris* var. *squamigera*) and the other, *T. palustris*, including two varieties (*T. palustris* var. *palustris* and *T. p.* var. *pubescens*), in the northern hemisphere.

This taxonomic view has been confirmed in recent times, so currently the genus *Thelypteris* includes two species, *T. palustris* in the northern hemisphere and *T. confluens* in the southern hemisphere, the only change regarding the rank of the two taxa within *T. palustris* that are now considered subspecies [4].

In Europe *Thelypteris palustris* (subsp. *palustris*) is known from several countries [5].

In Italy, this species is known for almost all regions, except for Sardinia and Sicily, although it is reported as an extinct or doubtful taxon for many territories [6–8].

The species has undergone a considerable decline throughout its distribution range, mainly due to habitat loss and reduction. Despite this significant decrease in area at the European level, it was recently classified as Least Concern (IUCN category) [9].

In northern Europe *Thelypteris palustris* has been found in several plant communities, for example in open habitats and in clear woodland: *Juncus subnodulosus* Schrank fen-meadow, *Salix cinerea* L. woodland, *Alnus glutinosa* (L.) Gaertn. woodland, *Betula pubescens* Ehrh. woods [10]. In Italy, the species has been found in the plant communities of the *Phragmito-Magnocaricetea* [11–13], and in the swamp forests of the *Alnetea glutinosae* class [14–21].

Our finding, during a survey in the Nebrodi Mountains (northern Sicily), is very interesting because the species is quite rare in the Italian territory since the habitats where it grows are in strong reduction, and even more because the Sicilian population represents the southernmost limit of its distribution range. The new finding is certainly unexpected because the flora of Sicily is one of the best studied in Italy and probably in Europe: consider that the start of a "modern" botanical exploration of the island date back to 1664 at least [22]

Marginal habitats in the Mediterranean area represent sites of high ecological importance and a refuge for threatened plants (e.g., hygrophytes) like the case of *Thelypteris palustris*. In fact, these hydrophytic species are linked to peculiar ecological requirements and are highly susceptible to climate changes, and this could be led to their disappearance in the next years. The correct identification of *T. palustris*, as well as the floristic composition of the plant community where it grows and its ecology, are relevant issues for future conservation measures and monitoring actions of this species.

The main objectives of this research are split up into two parts. One is to examine the morphological and ecological features of the new population of *T. palustris*, as well as to assess its conservation status in Sicily. The second one is to provide data about the habitat where *T. palustris* grows and to analyze the floristic composition of the plant community.

2. Results and Discussion

2.1. Description of the Species (Based on the New Population)

Thelypteris palustris Schott, Gen. Fil. [Schott] ad t. 10 (1834) subsp. *palustris* (Figures 1 and 2) Plant terrestrial. **Rhizomes** long creeping, black, glabrous, with more or less solitary leaves. **Fronds** monomorphic, 40–60 cm long; **petioles** 20–36 cm long, bases black, polished, usually glabrous, or rarely with sparse scales, 2.7 × 1.6 mm, irregular to elliptic-lanceolate, light-brown to yellowish, adpressed to patent; **laminae** lanceolate 20–28 cm long, 8–12 cm wide, 1-pinnate-pinnatifid, apices shortly acuminate and pinnatifid; **Rachises** with sparse whitish hairs, 0.1–0.4 mm long; **pinnae** 18–20 pairs, subopposite, flat- or obliquely spreading, usually slightly reflexed, short-petiolulate 0.45–0.50 mm wide, 0.7–0.9 mm long; proximal pair slightly shortened, **middle pinnae** lanceolate, 4.5–5.8 × 0.8–1.4 cm, bases truncate, pinnatifid nearly to costae, apices shortly acuminate; **segments** 4.2–6.8 × 2.1–2.8 mm, rounded-obtuse or obtuse-pointed at apices, fertile segments usually recurved to forming points along margin. **Veins** pinnate in segments, lateral veins 6–8(9) pairs, forked and reaching margins, proximal pair arising from base of costa. **Laminae papery**, grass-green or yellowish green when dry, glabrous on both surfaces, rachises and costae grooved adaxially, raised abaxially, glabrous on both sides or with acicular long hairs abaxially. **Sori** orbicular, dorsifixed at middle of veinlets, located between costa and margins; **indusia** small, orbicular-reniform, membranous, deciduous when mature. **Spores** ca. 45 × 30 μm, with papillate surfaces, papillae 3–5 μm high, perforated at the base. Terrestrial in swamps, bogs, and marshes, also along riverbanks and in wet woods; 0–1400 m.

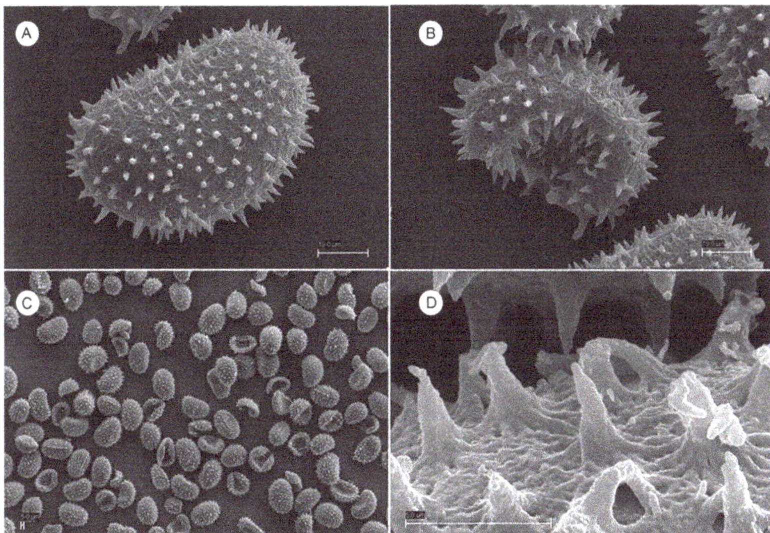

Figure 1. *Thelypteris palustris* Schott subsp. *palustris* spores from the new Sicilian population: (**A**) Spores with diffuse echinate elements; (**B**) Lower spore structure; (**C**) Echinate spores in group; (**D**) Echinate sculpture.

Figure 2. Some views of the new Sicilian population of *Thelypteris palustris* Schott subsp. *palustris*; (**A**) Growth environment (Nebrodi Mountains); (**B–E**) Habit; (**F**) Sporangia (Photos of the Authors).

2.2. Distribution and Conservation Status in Italy

In Italy, the species is reported for almost all regions. Probably, it has never been found in some territories due to the reduction or disappearance of its natural habitat, or in some cases, also owing to incorrect reports. It is reported as extinct in Marche, as a doubtful record in Molise and Campania, and not found in recent times in Umbria, Valle d'Aosta and Abruzzo [7–9]. In southern Italy the species is highly localized, with an altitude range between 0 and 1000 m a.s.l., from the coast to the mountain, occurring in Puglia at Laghi Alimini, Otranto [23,24], in Calabria at Lago dell'Aquila, Reggio Calabria [13,25] and in Sicily at Serra della Testa (Nebrodi) (Figure 3).

Figure 3. Current distribution of *Thelypteris palustris* in southern Italy.

This population recently discovered in Sicily has extended its distribution range and represents the southernmost population of Italy.

According to the European Red List of Vascular Plants [26], the species is classified as Least Concern (**LC**). Currently, in Italy, *Thelypteris palustris* has been recently evaluated as vulnerable (**VU**) by Orsenigo et al. [27] based on the criterion B [28]. In Sicily, the total area occupied by *Thelypteris palustris* is about 0.62 ha. Despite its very small distribution area, it was not possible to carry out a detailed count of the individuals of the population due to the stoloniferous vegetative development of the species. Therefore, thanks to our data and according to the IUCN criterion B, we recommend considering *Thelypteris palustris* as Critically Endangered (**CR B2abii, iii, iv**) for Sicily, due to a very small AOO (4 km^2), the occurrence on one location, and possible decline of the population especially because of the grazing practices and water flow reduction due to climate change.

2.3. Plant Communities with T. palustris in Italy

Thelypteris palustris is indicated as a characteristic/diagnostic species of the swamp forests of the *Alnion glutinosae* alliance (*Alnetea glutinosae* class). In Italy, especially in the northern sector, several plant communities of the *Alnetea glutinosae* class include *Thelypteris palustris*, such as *Carici acutiformis-Alnetum glutinosae* Scamoni 1935 [16]; *Carici elatae-Alnetum glutinosae* Franz ex Sburlino, Poldini, Venanzoni et Ghirelli 2011 [17–20]; *Carici elongatae-Alnetum glutinosae* Tüxen 1931 [29]; *Thelypterido-Alnetum glutinosae* Klika 1940 [14,15]; *Rhamno catharticae-Ulmetum minoris* Poldini, Vidali, Castello, Sburlino [30]; *Hydrocotylo vulgaris-Alnetum glutinosae* Gellini, Pedrotti ex Venanzoni, 1986 [21,31]; *Limnirido pseudacori-Fraxinetum oxycarpae* Gennai, Gabellini, Viciani, Venanzoni, Dell'Olmo, Giunti, Lucchesi, Monacci, Mugnai et Foggi 2021 [21]; *Cladio marisci–Fraxinetum oxycarpae* Piccoli, Gerdol & Ferrari ex Piccoli 1995 [21]; *Valeriano dioicae-Fraxinetum oxycarpae* Poldini et Sburlino 2018 [21].

Thelypteris palustris is reported, also, for Lake Massaciuccoli (northern Tuscany), in peculiar reed-beds (*Thelypterido palustris-Phragmitetum australis*) developing on floating islands rich in decaying organic matter [12]. This community was included in the *Carici pseudocyperi-Rumicion hydrolapathi* alliance (*Magnocaricetalia elatae, Phragmito-Magnocaricetea*). Probably, also the community of Lake Alimini (Otranto, Lecce) [23,24], which hosts *Thelypteris palustris*, is to be referred to *Thelypterido palustris-Phragmitetum australis*.

Moreover, the species is reported for Lake Aquila (Calabria), in tall sedges marsh vegetation (*Cladietum marisci* Allorge 1921) included in the *Magnocaricion elatae* alliance (*Magnocaricetalia elatae* Pignatti 1954) [13].

In Sicily, *Thelypteris palustris* falls within the sedges of the *Caricion gracilis* alliance (*Magnocaricion elatae*). This alliance, until now never reported in Sicily, groups plants communities growing on eutrophic clayey soils flooded for long time with a temperate Europe distribution.

2.4. Vegetation Ecology and Habitat

Overall, **15** different plant communities, each one with specific floristic compositions, were identified (Appendix A). Most of these plant communities were investigated by Brullo et al. [32] for Nebrodi Mounts. Therefore, we avoid a detailed description of the investigated communities. The wide sampling and cluster analysis allowed us to highlight the uniqueness and rarity of *T. palustris* in Sicily and define objectively the correct syntaxonomic framework. The cluster analysis of all relevés carried out on the Nebrodi Mounts showed 2 main groups (Figure 4). The first group (**cluster A**) includes mainly the helophytic perennial vegetation of the *Phragmito-Magnocaricetea* class, while the second group (cluster B) includes the aquatic vegetation of the *Lemnetea* and *Potametea* classes [31]. Within the *Phragmito-Magnocaricetea* four alliances can be distinguished: the first one (**A11**) *Phragmition communis* includes the vegetation dominated by tall graminoid species subjected to regular, prolonged periods of flooding that grow on mineral meso-eutrophic, often muddy, soils; the second one, *Magnocaricion elatae* (**A121**) consist of plant communities of mesotrophic to dystrophic soils, often peaty and flooded for prolonged periods; the third alliance *Caricion gracilis* (**A122**) groups communities of eutrophic soils, flooded for prolonged periods; the fourth alliance *Alopecuro-Glycerion spicatae* (**A2**), that includes the vegetation of hygrophilous herblands of shallow montane pools characterized by large water-depth fluctuations at high altitudes of Sicily. This last alliance is grouped with a peculiar annual amphibious vegetation dominated by *Lythrum portula* which falls within the *Nanocyperetalia* order (*Isoeto-Nanojuncetea*). Within the second group (**cluster B**) two subclusters can be distinguished: the first one (**B1**) (*Potametea pectinati*) delimits the perennial macrophytic communities of fresh, mesotrophic to eutrophic, waters; while the second one subcluster (**B2**) includes (*Lemnetea minoris*) the floating pleustophyte communities eutrophic to hypertrophic waters.

Bray-Curtis ordination shows a marked correspondence with cluster analysis (Figure 5). The highest data dispersion is obtained with axes 1 and 2. On the positive side of axis 1 there are the helophytic perennial vegetation of the *Phragmito-Magnocaricetea* class with high floristic diversity values, while on the negative side of axis 1 are distributed the aquatic vegetation of the *Lemnetea* and *Potametea* classes, with low values of floristic diversity. The *Carex paniculata* community (cluster 7) is very isolated from the other associations, probably due to the peculiar ecological and floristic conditions of the wet habitat.

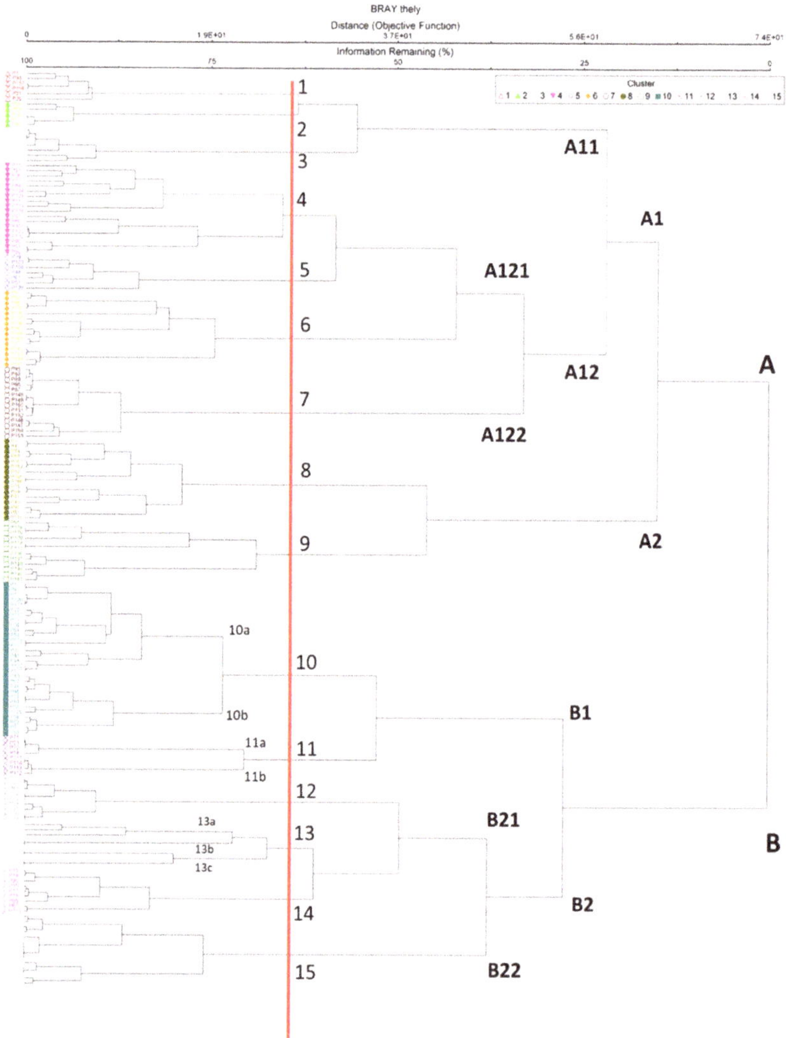

Figure 4. Cluster analysis of 179 unpublished phytosociological relevés. Plant communities: 1. *Sparganietum erecti*; 2. *Scirpetum lacustris*; 3. *Typhetum domingensis*; 4. *Galio palustris-Juncetum inflexi*; 5. *Eleocharitetum palustris*; 6. *Iridetum pseudacori*; 7. *Thelypterido palustris-Caricetum paniculatae*; 8. *Lythrum portula* comm.; 9. *Glycerio spicatae-Oenanthetum aquaticae*; 10a. *Potametum natantis*; 10b. *Utricularietum australis*; 11a. *Potamogetono natantis-Polygonetum natantis*; 11b. *Potametum pusilli*; 12. *Myriophylletum verticillati*; 13a. *Lemnetum minoris*; 13b. *Potamogetono-Ceratophylletum submersi*; 13c. *Ranunculetum omiophylli*; 14. *Ranunculetum aquatilis*; 15. *Wolffietum arrhizae*.

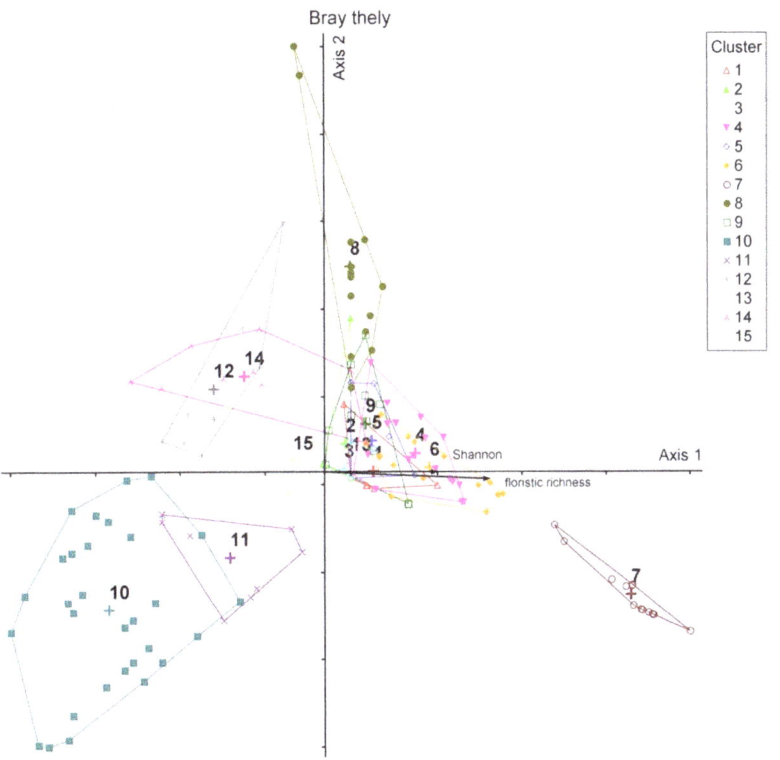

Figure 5. Bray-Curtis ordination. Axis 1 extracted 9.97% of the original distance matrix Cumulative: 9.97%; Axis 2 extracted 4.89% of the original distance matrix. Cumulative: 14.86%. Plant communities according to Figure 3.

2.5. Floristic Composition and Phytosociological Insights of the Thelypteris palustris Population in Sicily

In the study area, *Thelypteris palustris* was found exclusively in a perennial wetland characterized by *Carex paniculata* L. and *Juncus subnodulosus*. This perennial vegetation grows on flat or slightly sloping surfaces, on clayey-silty acid soils, permanently wet and rich in organic matter. The structure is determined mainly by *Carex paniculata*, the dominant species in terms of biomass and number of individuals, joined to several hygrophilous species, as *Galium palustre* L. subsp. *elongatum*, *Mentha aquatica* L., *Cirsium creticum* (Lam.) d'Urv. subsp. *triumfettii* (Lacaita) K.Werner, *Juncus subnodulosus*, *Carex distans* L., *Cyperus longus* L., *Hypericum tetrapterum* Fr., *Phragmites australis* (Cav.) Trin. ex Steud., *Epilobium parviflorum* Schreb., *Lolium arundinaceum* (Schreb.) Darbysh., *Eupatorium cannabinum* L., *Lotus rectus* L., *Eleocharis palustris* (L.) Roem. & Schult., *Heloscidium nodiflorum* (L.) W.D.J. Koch, *Rumex conglomeratus* Murray. The constant presence of *Thelypteris palustris* highlights the mesophilous character of the plant community, clearly differentiating it from the other sedge communities present in central and northern Italy. Therefore, because of its ecological features, *Thelypteris palustris* is proposed as a characteristic species of a new association named *Thelypterido palustris-Caricetum paniculatae* ass. nova hoc loco (Table 1, Rel. 10, cluster 7) included in the *Caricion gracilis* alliance and *Magnocaricetalia elatae* order (*Phragmito-Magnocaricetea* class. The new association is also characterized by a floristic component of the *Holoschoenetalia vulgaris* order, as *Lysimachia nemorum* L., *Holcus lanatus* L., *Dactylorhiza maculata* (L.) Soó subsp. *saccifera* (Brongn.) Diklić, *Juncus effusus* L., *Lythrum junceum* Banks & Sol. This later order includes hygrophilous communities dominated by helophytes (rushes and sedges) that grow in depressions in the supratemperate thermotype subjected to periodic submersions, on soils with low permeability and a rich silty-clayey component [33–35]. In addition, the association hosts floristic elements, very rare in Sicily, of high phytogeographic value (Figure 6), such as *Epipactis palustris* (L.) Crantz, *Equisetum palustre* L., *Rhynchocorys elephas* (L.) Griseb., *Juncus conglomeratus* L., *Carex flacca* Schreb. subsp. *flacca, C. pallescens* L., etc. The muscinal component also plays an important ecological role, particularly *Calliergonella cuspidata* (Hedw.) Loeske with a high degree of coverage and sociability. From the chorological and structural viewpoint, this vegetation highlights the relevance of the species with an Euroasiatic-Circumboreal distribution (34%), with geophytes (34%) and hemicryptophytes (61%) being the dominant life forms. This new association can be considered a southern vicariant of the *Caricetum paniculatae*, with a central and northern Italian distribution [36,37]. This last association shows structural affinities with *Thelypterido palustris-Caricetum paniculatae* owing to a high cover of *Carex paniculata*. However, the two plant communities can be clearly separated, based on many differential diagnostic species, such as *Epipactis palustris, Equisetum palustre, Juncus subnodulosus,* and *Rhynchocorys elephas.* From a bioclimatic point view the *Thelypterido palustris-Caricetum paniculatae* falls into the lower Supramediterranean belts with lower subhumid ombrotype [38], in contact with deciduous thermophilic *Quercus cerris* oak forests, referable to the *Arrhenathero nebrodensis-Quercetum cerridis* [39].

Figure 6. Photo plate illustration of some rare hygrophilous species of the Nebrodi Mountains: (**A**) *Epipactis palustris;* (**B**) *Carex flacca* subsp. *flacca;* (**C**) *Rhynchocorys elephas;* (**D**) *Equisetum palustre;* (**E**) *Carex paniculata;* (**F**) *Carex pallescens.* (Photos of the Authors).

Table 1. *Thelypterido palustris-Caricetum paniculatae ass. r ova hoc loco*–Sampling plots main features of plant community investigated.

LF	Corology	Pools	18	18	18	18	18	18	18	18	18	18	18	18	18	31	presence
		Relevé number	1	2	3	4	5	6	7	8	9	10	11	12	13	14	
		Altitude (m a.s.l.)	1000	1000	1000	1000	1000	1030	1030	1028	1028	1028	1000	1000	1028	1057	
		Surface (mq)	50	50	50	50	50	50	50	50	50	50	50	50	50	50	
		Coverage(%)	100	100	100	100	100	100	100	100	100	100	100	100	100	100	
		Slope (°)	-	-	-	-	-	-	-	5	5	5	5	5	5	-	
		Aspect	-	-	-	-	-	-	-	W	W	W	W	W	W	-	
		Vegetation height (m)	1.5	1.3	1.3	14	1.5	1.5	1.3	1.3	1.4	1.4	1.4	1.3	1.2	1.5	
		No. species	18	19	20	19	20	25	27	23	24	27	28	24	25	23	
		Simpson_1-D	0.9	0.9	0.9	0.9	0.9	0.95	0.95	0.95	0.95	0.95	0.96	0.95	0.95	0.95	
		Shannon_H	2.8	2.8	2.9	2.9	2.9	3.1	3.2	3.1	3.1	3.2	3.2	3.1	3.1	3.1	
		Evenness_e^H/S	0.9	0.9	0.9	0.9	0.9	0.9	0.9	0.9	0.9	0.89	0.9	0.9	0.9	0.9	
		Equitability_J	0.96	0.96	0.97	0.97	0.97	0.97	0.97	0.97	0.97	0.96	0.97	0.96	0.97	0.97	
LF	**Corology**	**Characteristic species**															
G	Subcosmop	*Thelypteris palustris* Schott subsp. *palustris*	2	1	2	1	2	+	+	2	2	3	3	2	1	-	13
G	Circumbor.	*Epipactis palustris* (L.) Crantz	+	+	-	+	-	-	-.	+	-	+	+	-	-	-	
		Char. Magnocaricion elatae and Magnocaricetalia															
H	Europ.-Caucas.	*Carex paniculata* L.	4	3	2	3	3	4	3	3	4	4	4	3	1	4	14
H	Euri-Medit.	*Galium palustre* L. subsp. *elongatum* (C. Presl) Arcang.	1	1	2	1	1	1	2	2	1	2	1	2	1	1	14
G	Paleotemp.	*Cyperus longus* L.	1	2	1	1	+	-	-	1	1	1	+	1	+	+	12
G	Europ.	*Carex flacca* Schreb. subsp. *flacca*	-	-	+	+	-	-	+	+	-	+	+	-	-	-	6
H	Circumbor.	*Carex pallescens* L.	.	.	+	.	.	+	.	.	.	+	.	.	.	-	3
H	Eurasiat.	*Rumex conglomeratus* Murray	.	.	.	.	.	.	.	.	.	.	.	.	.	+	1
		Char. Phragmito-Magnocaricetea															
H	Paleotemp	*Mentha aquatica* L.	1	+	2	1	1	1	2	1	1	+	1	1	1	1	14
H	Orof. NE-Medit.	*Cirsium creticum* (Lam.) d'Urv.subsp. *triumfettii* (Lacaita) K. Werner	+	1	+	1	1	2	1	+	1	+	+	1	+	+	14
G	Europ.-Caucas.	*Juncus subnodulosus* Schrank	2	3	3	2	2	1	2	2	1	1	1	3	1	.	13
H	Eurosiber.	*Angelica sylvestris* L.	+	+	+	+	1	1	+	+	1	+	1	1	+	.	13
G	Circumbor.	*Equisetum palustre* L.	1	2	+	1	+	2	1	1	1	1	+	+	3	.	13
H	Euri-Medit.	*Carex distans* L.	2	.	1	1	+	2	1	1	+	1	1	1	.	2	13
H	Paleotemp.	*Hypericum tetrapterum* Fr.	+	-	1	1	1	+	+	.	.	1	1	+	.	1	11

Table 1. *Cont.*

		Pools	18	18	18	18	18	18	18	18	18	18	18	18	18	31	
		Relevé number	1	2	3	4	5	6	7	8	9	10	11	12	13	14	
G	Subcosmop.	*Phragmites australis* (Cav.) Trin. ex Steud.	.	+	+	.	+	+	+	+	+	+	+	.	.	.	9
H	Paleotemp.	*Epilobium parviflorum* Schreb.	+	1	+	+	+	.	.	.	.	+	+	+	.	.	8
H	Paleotemp.	*Lolium arundinaceum* (Schreb.) Darbysh.	.	.	.	.	.	+	1	+	+	+	.	+	.	+	7
H	Paleotemp.	*Eupatorium cannabinum* L.	1	1	1	+	1	.	.	.	.	.	+	+	.	.	7
Ch	Medit.	*Lotus rectus* L.	.	.	+	.	.	.	.	.	+	+	1	+	.	.	5
G	Subcosmop.	*Eleocharis palustris* (L.) Roem. & Schult.	.	.	.	.	.	.	.	+	+	+	+	.	.	1	5
H	Euri-Medit.	*Helosciadium nodiflorum* (L.) W.D.J. Koch	.	.	.	.	.	+	+	.	.	.	.	.	1	1	4
G	Subcosmop	*Glyceria spicata* Guss.	.	.	.	.	.	+	+	.	.	.	.	.	+	1	4
		Trasgr. Holoschoenetalia vulgaris															
H	Europ.-Caucas. Subatl.	*Lysimachia nemorum* L.	2	+	1	1	1	1	1	1	+	1	1	2	2	.	13
H	Circumbor.	*Holcus lanatus* L.	1	+	1	+	+	+	+	+	+	+	+	+	+	.	13
G	Medit.	*Dactylorhiza maculata* subsp. *saccifera* (Brongn.) Diklić	+	+	.	.	.	+	1	+	+	+	1	+	+	1	11
G	Cosmop.	*Juncus effusus* L.	.	.	.	+	+	1	2	+	+	.	+	1	1	1	10
G	Eurosiber.	*Juncus conglomeratus* L.	+	.	+	+	+	+	+	.	+	+	+	.	.	.	9
H	NE-Medit.	*Rhynchocorys elephas* (L.) Griseb.	.	.	.	.	.	+	+	1	1	+	1	+	1	.	8
H	Medit.	*Lythrum junceum* Banks & Sol.	.	.	.	.	.	+	+	+	+	+	+	+	+	1	9
		Other species															
		Calliergonella cuspidata (Hedw.) Loeske	1	1	1	1	1	1	1	1	1	1	1	1	1	1	14
H	Paleotemp.	*Lathyrus pratensis* L.	.	.	.	.	.	.	+	1	2	+	+	+	.	+	7
H/T	Euri-Medit.-Sett.	*Myosotis sicula* Guss.	.	.	.	.	.	+	+	.	.	+	+	+	.	1	6
H	Circumbor.	*Prunella vulgaris* L.	.	.	.	.	.	.	+	+	+	.	+	.	+	.	5
G	Circumbor.	*Juncus articulatus* L.	.	.	.	+	+	.	.	.	.	.	.	.	2	2	4
H	Subcosmop.	*Samolus valerandi* L.	.	.	.	.	.	+	+	.	.	.	.	.	1	+	4
H	Subcosmop.	*Isolepis cernua* (Vahl) Roem. & Schult.	.	.	.	.	.	+	+	.	.	.	.	.	1	+	4
H	Europ.-Caucas.	*Carex remota* L.	.	.	.	.	.	.	.	.	.	.	.	.	3	.	1
H	Paleotemp.	*Trifolium repens* L.	.	+	+	.	.	.	.	.	.	.	.	.	.	.	2
T	Euri-Medit.	*Ranunculus ophioglossifolius* Vill.	.	.	.	.	.	.	.	.	.	.	.	.	+	+	2
G	NE-Medit	*Geranium versicolor* L.	.	.	.	.	.	.	.	.	.	.	.	.	+	.	1
H	Eurasiat.	*Ajuga reptans* L.	.	.	.	.	.	.	.	.	.	.	.	.	.	+	1

Localities and dates of relevés. Rel. 1–5, Serra della Testa 3 (inf.); Rel. 6–7, Serra della Testa 3 (sup.) 15.06.2021; Rel. 8–12, Serra della Testa 3 (inf.) 15.06.2021; Rel. 13, Serra della Testa 3 (rigagnolo, tra inf. e sup.) 15.06.2021; Rel. 14, Serra della testa 2 15 June 2021.

3. Materials and Methods

3.1. Study Area

The study area is situated in the Nebrodi Mounts, Sicily's largest mountain complex (Figure 7). They are located in the N-E part of the island, between the west side of the Peloritani Mountains and the east side of the Madonie Mounts, constituting the extension of the Apennine ridge on the island. They are a mountain range without major roughness that reaches its maximum altitude at Monte Soro (1847 m a.s.l.). From a geological point of view, this territory is mainly made up of sedimentary successions belonging to different periods. The dominance of Flysch is mostly noted, the oldest sediments belonging to the Alpine Tethys Units [40], they are Cretaceous in age and are represented by deep-water flyschs and scaly clays. Most of the outcropping rocks are part of the so-called Flysch of Monte Soro (upper Tithonian, lower Cretaceous) and Numidian (lower Oligocene Miocene) [41].

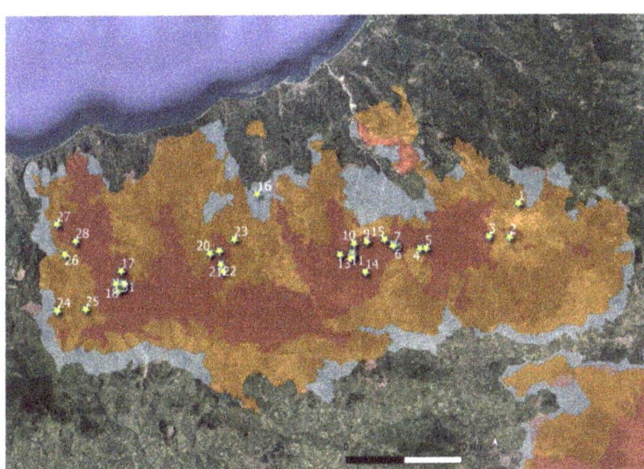

Figure 7. Wetlands investigated in Nebrodi Mounts (localities are given in Table S1), with zonation of the regional park (zone A: red; zone B: orange; zone C: blue; zone A is the most protected zone).

The outcrop of clayey layers favors the formation of humid environments and ponds, and lakes originate where the orographic conditions allow it. The existence of humid environments on the Nebrodi is possible because of favourable climatic conditions that characterize this mountain area that is the most mesic and rainy in Sicily, being affected by average annual rainfall between 1000 and 1400 mm. According to Rivas Martínez et al. [42], the bioclimate of this area is supra-Mediterranean lower middle-humid bioclimatic conditions [38]. Although these small wetlands can have a relatively short lifespan due to landfills, climate changes, etc., for some of these areas in the Nebrodi Mounts an existence has been documented since the end of the last glaciation (about 10,000 years ago) when it seems that the climate had become wetter in Sicily [43]. The climatic and geomorphological conditions of the Nebrodi Mounts make it the area with the greatest wooded coverage and with the highest values of biodiversity in Sicily [44]. In particular, this territory is characterized by very extensive oak forests (*Quercus cerris* L.) at medium altitudes, and beech woods (*Fagus sylvatica* L.) at higher altitudes. However, grazing meadows and small wetlands (mostly natural) are the main discontinuities in the forest cover of this territory.

3.2. Data Sets and Data Processing

The morphological study regarding *Thelypteris palustris* was carried out on living material (15 specimens), all coming from Nebrodi Mounts territory. The collected samples were kept at the Catania Herbarium (CAT). For scanning electron microscopy (SEM) images, samples of spores were transferred from herbarium specimens to aluminum SEM stubs coated with double sided carbon tape. The stubs were then sputter-coated with gold and imaged digitally using a Zeiss EVO LS10, with an accelerating voltage of 30 kV, in the Center for Microscopy at the University of Catania. The morphological terminology used in the description follows Lellinger [45], while to spore nomenclature follows Tryon & Lugardon [46].

To analyze the structure and floristic composition of the marsh vegetation in the Nebrodi Mounts, **34** pools were examined. A total of **179** unpublished phytosociological relevés (**110** species) were collected, personally sampled in the period April 2018-June 2021. The floristic composition and cover of species in each plot were determined by using the standard method of relevés [47]. All the relevés were classified using classification and ordination methods. Numerical analysis was performed using the software package "PC-ORD", 6.08 software. A multivariate analysis (Linkage method: Ward's, Distance measure: Sorensen (Bray-Curtis) was applied. Bray-Curtis ordination (Distance measure: Jaccard) takes into account different quantitative data, such as vegetation coverage (%), altitude, number of species (N. sp.), Altitude (m a.s.l.), Slope (°), Aspect, and Simpson/Shannon index. Quantum GIS software version 3.6 and GPS Garmin Montana was used to geolocate the surveyed wetlands.

For the risk assessment at the regional scale (Sicily), we followed the IUCN protocol and the most recent guidelines for its application [28]. In particular, we applied the IUCN criterion B by estimating trends in the Area of Occupancy (AOO), that is, the area covered by a taxon. AOO was assessed by using a 2×2 km grid [48]. Syntaxa classification follows Biondi et al. [33], and Mucina et al. [49]. Taxonomic nomenclature follows Bartolucci et al. [9] and Pignatti [50–53].

4. Conclusions

The species we found in Sicily seems to be very rare in the island, unless new discoveries that might be made in the future. It is localized in a microrefuge area that means, according to Rull [54], a small area with local favorable environmental features, in which small populations can survive outside their main distribution area, protected from the unfavorable regional environmental conditions. *T. palustris* grows in contexts which, even if somewhat subject to disturbing factors such as grazing, maintain good natural characteristics. However, these places are vulnerable to further disturbances, such as drainage, and above all to the decrease in rainfall triggered by climate change as detected for Sicily [55,56] that could jeopardize its precarious survival. Therefore, even the microrefuge area may not be enough to guarantee the existence in Sicily of this species; in any case, it will need to be monitored over time. This is a general trend that can undermine a species that, although with a large distribution range, is linked to peculiar environmental conditions. These circumstances could fail especially in semi-arid areas such as around the Mediterranean basin where climate change overlaps the usual intense anthropogenic disturbance that particularly affects wetlands [57]. In the Mediterranean area, the populations of *T. palustris* are likely to be declining following the general trend of destruction and degradation of shallow wetlands. It is not considered common anywhere in its Mediterranean range. It is very rare in Morocco and in Algeria. In Morocco it is known from two localities only: Bou Charen and near to Açilah in the western Rif. In Algeria, *T. palustris* is known from three localities, including Senhadja in Numidie. It is widespread in Turkey, but its habitats are under threat and because of this in the future this taxon may be threatened [58]. In Italy, as we have shown, many reports are old and no longer reconfirmed. The species, going south, is highly localized, occurring only in Puglia, Calabria, and Sicily (Figure 3), with only one sub-population for each region. Therefore, this recently discovered in Sicily extend its distribution area and represents the southernmost population of Italy.

Plants characterizing wet environments represent one of the most threatened groups of the Mediterranean flora [59–65]. For the reasons quoted above, these areas require urgent and effective conservation policies not only to safeguard the biodiversity but also for the important ecosystem services they perform [66]. An example can be our study that illustrates the environmental context of the *T. palustris* populations and the morphological features of the species. Moreover, it clarifies some ecological requirements which are relevant issues for future conservation measures for this species, especially in the Mediterranean areas where the extremely scattered distribution with isolated populations make it vulnerable to disappearance. Unlike the northern European populations, quite widespread, the risk of local-scale extinction is really high in all the Mediterranean populations.

The floristic composition of the Sicilian *Thelypteris palustris* plant community shows a clear affinity with the common sedge communities of the northern Europe. Perhaps, this plant community is a relict vegetation type of the last glacial stage, which currently is localized exclusively in the humid stands of the Nebrodi Mounts. These microrefuges were originated by peculiar geological characteristics of the territory, with a humid supramediter-ranean bioclimate that facilitate the growth of these hygrophilous species [43].

In conclusion, our study has made possible to highlight the unexpected occurrence in Sicily of the marsh fern *Thelypteris palustris*, growing together with some floristic elements of the highest nature value, such as *Equisetum palustre, Epipactis palustris, Utricularia australis* R. Br., *Rhynchocorys elephas, Juncus conglomeratus, J. subnodulosus, Carex paniculata, C. flacca* subsp. *flacca, C. pallescens,* and *Hypericum tetrapterum*. These vascular species, linked to wetlands, show in Sicily a narrow distribution range due to a strong reduction of their habitat in recent decades. Although they are included in the "A" zone of the Nebrodi Park (Figure 7) and within the Natura2000 site SAC ITA030014, targeted conservation and monitoring actions would be desirable, aimed at the long-term conservation of the floristic component and especially their humid habitats.

Supplementary Materials: The following are available online at https://www.mdpi.com/article/10.3 390/plants10112448/s1, Table S1: Unpublished phytosociological relevés.

Author Contributions: S.S. conceived the project and organised the research group and developed the first theoretical framework; S.S., S.C., G.T., P.M. carried out the field work; S.S., G.T., P.M. contributed to setting the theoretical framework and data processing; S.S., S.C., G.G.d.G., G.T., P.M. prepared and revised the data; S.S., S.C., P.M. processed the data and prepared the graphical outputs; S.S., G.G.d.G., G.T., P.M. prepared the first draft of the manuscript; all the authors contributed to the finalization of the paper and agreed to submit its last draft. All authors have read and agreed to the published version of the manuscript.

Funding: This research was financially supported by the research programme (PIA.CE.RI. 2020-2022 Line 2, cod. 22722132149) funded by the University of Catania and by Convention with PIM in the frame of the project MedIsWet funded by MAVA Foundation.

Data Availability Statement: Data is contained within the article or Supplementary Material.

Acknowledgments: The authors wish to thank G. Siracusa for his helpful assistance with the Scanning Electron Microscopy performed in the Plant Biology Laboratory of the University of Catania.

Conflicts of Interest: The authors declare no conflict of interest.

Appendix A. Phytosociological Survey of Aquatic and Marsh Vegetation

Syntaxonomical Scheme

LEMNETEA O. de Bolos et Masclans 1955
LEMNETALIA MINORIS O. de Bolos et Masclans 1955
LEMNION MINORIS O. de Bolos et Masclans 1955
Lemnetum minoris von Soó 1927
Wolffietum arrhizae Myawaki & J.Tx. 1960

POTAMOGETONETEA Klika in Klika et Novak 1941
POTAMOGETONETALIA Koch 1926
POTAMION (Koch 1926) Libbert 1931
Myriophylletum verticillati Lemnée 1937
NINPHAEION ALBAE Oberd. 1957
Potamogetono natantis-Polygonetum natantis Knapp et Stoffers 1962
Potametum pusilli von Soó 1927
Potametum natantis von Soó 1927
UTRICULARIETALIA Den Hartog & Segal 1964

UTRICULARION VULGARIS Passarge 1964
Utricularietum australis Müller & Görs 1960
CERATOPHYLLION DEMERSI Den Hartog & Segal ex Passarge 1996
Potamogetono-Ceratophylletum submersi Pop 1962
RANUNCULION AQUATILIS Géhu 1961
Ranunculetum aquatilis Géhu 1961
RANUNCULION OMIOPHYLLO-HEDERACEI Rivas-Martínez et al. 2002
Ranunculetum omiophylli Br.-Bl. & Tüxen ex Pizarro 1995

PHRAGMITO-MAGNOCARICETEA Klika in Klika et Novak 1941
PHRAGMITETALIA Koch 1926
PHRAGMITION COMMUNIS Koch 1926
Sparganietum erecti Philippi 1973
Scirpetum lacustris Schmale 1939
Typhetum domingensis Brullo, Minissale & Spamp. 1994
MAGNOCARICETALIA Pignatti 1953
MAGNOCARICION ELATAE Koch 1926
Galio palustris-Juncetum inflexi Venanzoni et Gigante 2000
Eleocharitetum palustris Savic 1926
Iridetum pseudacori Krywanski 1974
CARICION GRACILIS Géhu 1961
Thelypterido palustris-Caricetum paniculatae ass. nova hoc loco
OENANTHETALIA AQUATICAE Hejny ex Balatova-Tulackova et al. 1993
ALOPECURO-GLYCERION SPICATAE Brullo, Minissale, Spamp. 1994
Glycerio spicatae-Oenanthetum aquaticae Brullo, Minissale & Spamp. 1994

ISOËTO-NANOJUNCETEA Br.-Bl. et Tx. in Br.-Bl. et al. 1952
Nanocyperetalia Klika 1935
Nanocyperionflavescentis W. Koch ex Libbert 1932
Lythrum portula comm.

References

1. Fernald, M.L. A study of *Thelypteris palustris*. *Rhodora* **1929**, *31*, 27–36.
2. Tryon, A.F. Structure and variation in spores of *Thelypteris palustris*. *Rhodora* **1971**, *73*, 433–460.
3. Tryon, A.F.; Tryon, R.; Badré, F. Classification, spores, and nomenclature of the marsh fern. *Rhodora* **1980**, *82*, 461–474.
4. Fraser-Jenkins, C.R. *New Species Syndrome in Indian Pteridology and the Ferns of Nepal*; International Book Distributors: Dehra Dun, India, 1997.
5. Christenhusz, M.; von Raab-Straube, E. Polypodiopsida. In *Euro+Med Plantbase—The Information Resource for Euro-Mediterranean plant Diversity*. 2013. Available online: http://ww2.bgbm.org/EuroPlusMed/ (accessed on 9 January 2021).
6. Conti, F.; Bartolucci, F.; Iocchi, M.; Tinti, D. Atlas of the pteridological knowledge of Abruzzo (Central Italy). *Webbia* **2011**, *66*, 251–305. [CrossRef]
7. Bovio, M. Flora Vascolare della Valle d'Aosta. In *Repertorio Commentato e Stato Delle Conoscenze*; Testolin Editore: Sarre, Italy, 2014.
8. Bartolucci, F.; Peruzzi, L.; Galasso, G.; Albano, A.; Alessandrini, A.; Ardenghi, N.M.G.; Astuti, G.; Bacchetta, G.; Ballelli, S.; Banfi, E.; et al. An updated checklist of the vascular flora native to Italy. *Plant Biosyst.* **2018**, *152*, 179–303. [CrossRef]
9. García Criado, M.; Väre, H.; Nieto, A.; Bento Elias, R.; Dyer, R.; Ivanenko, Y.; Ivanova, D.; Lansdown, R.; Molina, J.A.; Rouhan, G.; et al. *European Red List of Lycopods and Ferns*; IUCN: Brussels, Belgium, 2017; pp. iv + 59.
10. Rodwell, J.S. British Plant Communities. In *Woodlands and Scrub*; Cambridge University Press: Cambridge, UK, 1991.
11. Angiolini, C.; Ferretti, G.; Foggi, B.; Frignani, F.; Gestri, G.; Landi, M.; Lastrucci, L.; Monacci, F.; Peruzzi, L.; Sani, A.; et al. Segnalazione 51. In *Contributi Alla Flora Vascolare di Toscana I (1–85)*; Peruzzi, L., Viciani, D., Bedini, G., Eds.; Atti. Soc. Tosc. Sci. Nat. Mem. Serie B: Italy, 2009; Volume 116, pp. 33–44. Available online: https://www.researchgate.net/publication/264273119_Contributi_per_una_flora_vascolare_di_Toscana_I_1-85 (accessed on 1 June 2021).
12. Lastrucci, L.; Dell'Olmo, L.; Foggi, B.; Massi, L.; Nuccio, C.; Vicenti, C.; Viciani, D. Contribution to the knowledge of the vegetation of the Lake Massaciuccoli (northern Tuscany, Italy). *Plant Sociol.* **2017**, *54*, 67–87.
13. Spampinato, G.; Sciandrello, S.; Giusso del Galdo, G.; Puglisi, M.; Tomaselli, V.; Cannavò, S.; Musarella, C.M. Contribution to the knowledge of Mediterranean wetland biodiversity: Plant communities of the Aquila Lake (Calabria, Southern Italy). *Plant Sociol.* **2019**, *56*, 53–68.
14. Pedrotti, F. Nota sulla flora e vegetazione del Lago di Madrano (Trentino). *Inf. Bot. Ital.* **1991**, *22*, 182–193.

15. Pedrotti, F. Nota sulla vegetazione degli ambienti umidi della bassa Valsugana (Trentino). *Doc. Phytosoc.* **1995**, *15*, 417–449.
16. Tasinazzo, S.; Fiorentin, R. I relitti boschetti ad *Alnus glutinosa* delle risorgive vicentine (Pianura Veneta). *Ann. Mus. Civ. Rovereto* **2003**, *17*, 125–135.
17. Pedrotti, F. Ricerche geobotaniche al Laghestel di Pin, (1967–2001). *Braun-Blanquetia* **2004**, *35*, 1–55.
18. Lonati, M.; Lonati, S. Le comunità a *Carex elata* All. della torbiera di Vanzone (Piemonte, Vercelli). *Fitosociologia* **2005**, *42*, 15–21.
19. Tisi, A.; Minuzzo, C.; Siniscalco, C.; Caramiello, R. La vegetazione acquatica e palustre della zona dei "Cinque Laghi" di Ivrea. *Riv. Piem. St. Nat.* **2007**, *28*, 87–126.
20. Sburlino, G.; Poldini, L.; Venanzoni, R.; Ghirelli, L. Italian black alder swamps: Their syntaxonomic relationships and originality within the European context. *Plant Biosyst.* **2011**, *145*, 148–171. [CrossRef]
21. Gennai, M.; Gabellini, A.; Viciani, D.; Venanzoni, R.; Dell'Olmo, L.; Giunti, M.; Lucchesi, F.; Monacci, F.; Mugnai, M.; Foggi, B. The floodplain woods of Tuscany: Towards a phytosociological synthesis. *Plant Sociol.* **2021**, *58*, 1–28. [CrossRef]
22. Sciandrello, S.; Cambria, S.; Giusso del Galdo, G.; Guarino, R.; Minissale, P.; Pasta, S.; Tavilla, G.; Cristaudo, A. Floristic and Vegetation Changes on a Small Mediterranean Island over the Last Century. *Plants* **2021**, *10*, 680. [CrossRef]
23. Macchia, F. Vegetazione e flora dei Laghi Alimini. *Atti e Relazioni Acc. Pugliese Scienze ns. Cl. Sc. Fis. Med. Nat.* **1967**, *25*, 221–267.
24. Annese, B.; Beccarisi, L. Thelypteris palustris Schott. In *Notule Pteridologiche Italiche, 3 (64–84)*; Marchetti, D., Ed.; Ann. Mus. Civ. Rovereto Sez. Arch. St. Sc. Nat.: Sardinia, Italy, 2003; Volume 18, pp. 65–81. Available online: http://www.museocivico.rovereto. tn.it/UploadDocs/202_C__WINDOWS_Desktop_annali_pdf_18_art04_marchetti.pdf (accessed on 1 June 2021).
25. Crisafulli, A.; Cannavò, S.; Maiorca, G.; Musarella, C.M.; Signorino, G.; Spampinato, G. Aggiornamenti floristici per la Calabria. *Inform. Bot. Ital.* **2010**, *42*, 431–442.
26. Bilz, M.; Kell, S.P.; Maxted, N.; Lansdown, R.V. *European Red List of Vascular Plants*; Publications Office of the European Union: Luxembourg, 2011; p. 130.
27. Orsenigo, S.; Fenu, G.; Gargano, D.; Montagnani, C.; Abeli, T.; Alessandrini, A.; Bacchetta, G.; Bartolucci, F.; Carta, A.; Castello, M.; et al. Red list of threatened vascular plants in Italy. *Plant Biosyst.* **2021**, *155*, 310–335. [CrossRef]
28. IUCN. *Guidelines for Using the IUCN Red List Categories and Criteria*, 14th ed.; IUCN Standards and Petitions Committee: Gland, Switzerland, 2019; p. 113. Available online: http://cmsdocs.s3.amazonaws.com/RedListGuidelines.pdf (accessed on 1 January 2021).
29. Guglielmetto Mugion, L.; Montacchini, F. La vegetazione del Lago di Viverone. *Allionia* **1994**, *32*, 1–26.
30. Poldini, L.; Vidali, M.; Castello, M.; Sburlino, G. A novel insight into the remnants of hygrophilous forests and scrubs of the Po Plain biogeographical transition area (Northern Italy). *Plant Sociol.* **2020**, *57*, 17–69. [CrossRef]
31. Gellini, R.; Pedrotti, F.; Venanzoni, R. Le associazioni forestali ripariali e palustri della Selva di San Rossore (Pisa). *Doc. Phytosoc. n.s.* **1986**, *10*, 27–41.
32. Brullo, S.; Minissale, P.; Spampinato, G. Studio fitosociologico della vegetazione lacustre dei Monti Nebrodi (Sicilia settentrionale). *Fitosociologia* **1994**, *27*, 5–50.
33. Biondi, E.; Blasi, C.; Allegrezza, M.; Anzellotti, I.; Azzella, M.M.; Carli, E.; Casavecchia, S.; Copiz, R.; Del Vico, E.; Facioni, L.; et al. Plant communities of Italy: The Vegetation Prodrome. *Plant Biosyst.* **2014**, *148*, 728–814. [CrossRef]
34. Brullo, S.; Grillo, M. Ricerche fitosociologiche sui pascoli dei monti Nebrodi (Sicilia settentrionale). *Not. Fitosoc.* **1978**, *13*, 23–61.
35. Brullo, S.; Scelsi, F.; Spampinato, G. La vegetazione dell'Aspromonte. In *Studio Fitosociologico*; Laruffa: Reggio Calabria, Italy, 2001; p. 370.
36. Landucci, F.; Gigante, D.; Venanzoni, R.; Chytrý, M. Wetland vegetation of the class Phragmito-Magno-Caricetea in central Italy. *Phytocoenologia* **2013**, *43*, 67–100. [CrossRef]
37. Ciaschetti, G.; Pirone, G.; Venanzoni, R. Sedge vegetation of the 'Major Highlands of Abruzzo' (Central Italy): Updated knowledge after new discoveries. *Plant Biosyst.* **2021**, *155*, 647–662. [CrossRef]
38. Bazan, G.; Marino, P.; Guarino, R.; Domina, G.; Schicchi, R. Bioclimatology and vegetation series in Sicily: A geostatistical approach. *Ann. Bot. Fenn.* **2015**, *52*, 1–18. [CrossRef]
39. Brullo, C.; Brullo, S.; Giusso del Galdo, G.; Guarino, R.; Siracusa, G.; Sciandrello, S. The class Querco-Fagetea sylvaticae in Sicily: An example of boreal-temperate vegetation in the central mediterranean area. *Ann. Bot.* **2012**, *2*, 19–38.
40. Henriquet, M.; Dominguez, S.; Barreca, G.; Malavieille, J.; Monaco, C. Structural and tectono-stratigraphic review of the Sicilian orogen and new insights from analogue modeling. *Earth-Sci. Rev.* **2020**, *208*, 103257. [CrossRef]
41. Lentini, F.; Carbone, S. Geologia della Sicilia—Geology of Sicily. *Mem. Descr. Carta Geol. D'italia* **2014**, *95*, 7–414.
42. Rivas-Martínez, S.; Penas, A.; Diaz, T.E. *Biogeographic Map of Europe*; Cartographic Service, University of Léon: León, Spain, 2004.
43. Bisculm, M.; Colombaroli, D.; Vescovi, E.; van Leeuwen, J.F.N.; Henne, P.D.; Rothen, J.; Procacci, G.; Pasta, S.; La Mantia, T.; Tinner, W. Holocene vegetation and fire dynamics in the supra-mediterranean belt of the Nebrodi Mountains (Sicily, Italy). *J. Quat. Sci.* **2012**, *27*, 687–698. [CrossRef]
44. Signorello, G.; Prato, C.; Marzo, A.; Ientile, R.; Cucuzza, G.; Sciandrello, S.; Martínez-López, J.; Balbi, S.; Villa, F. Are protected areas covering important biodiversity sites? An assessment of the nature protection network in Sicily (Italy). *Land Use Policy* **2018**, *78*, 593–602. [CrossRef]
45. Lellinger, D.B. A Modern Multilingual Glossary for Taxonomic Pteridology. *Pteridologia* **2002**, *3*, 1–263.
46. Tryon, A.F.; Lugardon, B. *Spores of Pteridophyta: Surface, Wall Structure and Diversity Based on Electron Microscope Studies*; Springer: New York, NY, USA, 1991; p. 648.

47. Westhoff, V.; van der Maarel, E. The Braun-Blanquet approach. In *Classification of Plant Communities*, 2nd ed.; Whittaker, R.H., Ed.; Junk: The Hague, The Netherlands, 1978; pp. 287–297.

48. Gargano, D. Verso la redazione di nuove Liste Rosse della flora d'Italia: Una griglia standard per la misura dell'Area of Occupancy (AOO). *Inform. Bot. Ital.* **2011**, *43*, 35–38.

49. Mucina, L.; Bültmann, H.; Dierßen, K.; Theurillat, J.-P.; Raus, T.; Čarni, A.; Šumberová, K.; Willner, W.; Dengler, J.; García, R.G.; et al. Vegetation of Europe: Hierarchical floristic classification system of vascular plant, bryophyte, lichen, and algal communities. *Appl. Veg. Sci.* **2016**, *19*, 3–264. [CrossRef]

50. Pignatti, S. Volume 1: Flora d'Italia & Flora Digitale. In *Flora d'Italia: In 4 Volumi*, 2nd ed.; Edagricole-Edizioni Agricole di New Business Media srl: Milano, Italy, 2017.

51. Pignatti, S. Volume 2: Flora d'Italia & Flora Digitale. In *Flora d'Italia: In 4 Volumi*, 2nd ed.; Edagricole-Edizioni Agricole di New Business Media srl: Milano, Italy, 2017.

52. Pignatti, S. Volume 3: Flora d'Italia & Flora Digitale. In *Flora d'Italia: In 4 Volumi*, 2nd ed.; Edagricole-Edizioni Agricole di New Business Media srl: Milano, Italy, 2018.

53. Pignatti, S.; Guarino, R.; La Rosa, M. Volume 4: Flora d'Italia & Flora Digitale. In *Flora d'Italia: In 4 Volumi*, 2nd ed.; Edagricole-Edizioni Agricole di New Business Media srl: Milano, Italy, 2019.

54. Rull, V. Microrefugia. *J. Biogeogr.* **2009**, *36*, 481–484. [CrossRef]

55. Cannarozzo, M.; Noto, L.V.; Viola, F. Spatial distribution of rainfall trends in Sicily (1921–2000). *Phys. Chem. Earth* **2006**, *31*, 1201–1211. [CrossRef]

56. Arnone, E.; Pumo, D.; Viola, F.; Noto, L.V.; La Loggia, G. Rainfall statistics changes in Sicily. *Hydrol. Earth Syst. Sci.* **2013**, *17*, 2449–2458. [CrossRef]

57. Cuena-Lombraña, A.; Fois, M.; Cogoni, A.; Bacchetta, G. Where we Come from and where to Go: Six Decades of Botanical Studies in the Mediterranean Wetlands, with Sardinia (Italy) as a Case Study. *Wetlands* **2021**, *41*, 69. [CrossRef]

58. Kavak, S. *Thelypteris palustris*. The IUCN Red List of Threatened Species 2014: e.T164136A42331187. Available online: https://dx.doi.org/10.2305/IUCN.UK.2014-1.RLTS.T164136A42331187.en (accessed on 1 January 2021).

59. Daoud-Bouattour, A.; Muller, S.D.; Ferchichi-Ben Jamaa, H.; Ghrabi-Gammar, Z.; Rhazi, L.; Mokhtar Gammar, A.; Raouf Karray, M.; Soulié-Märsche, I.; Zouaïdia, A.; de Bélair, G.; et al. Recent discovery of the small pillwort (*Pilularia minuta* Durieu, Marsileaceae) in Tunisia: Hope for an endangered emblematic species of Mediterranean temporary pools? *Comptes Rendus Biol.* **2009**, *332*, 886–897. [CrossRef]

60. Gianguzzi, L.; D'Amico, A.; Troia, A. Notes on the distribution, ecology and conservation status of two very rare sedges (*Carex*, Cyperaceae) rediscovered in Sicily (Italy). *Bot. Lett.* **2017**, *164*, 339–349. [CrossRef]

61. Minissale, P.; Sciandrello, S. Ecological features affect patterns of plant communities in Mediterranean temporary rock pools. *Plant Biosyst.* **2016**, *150*, 171–179. [CrossRef]

62. Minissale, P.; Santo, A.; Sciandrello, S. Analisi geobotanica del SIC: Capo Murro di Porco, Penisola della Maddalena e Grotta Pellegrino (Siracusa, Sicilia). *Fitosociologia* **2011**, *48*, 77–98.

63. Minissale, P.; Molina, J.A.; Sciandrello, S. *Pilularia minuta* Durieu (Marsileaceae) discovered in south-eastern-Sicily: New insights on its ecology, distribution and conservation status. *Bot. Lett.* **2017**, *164*, 197–208. [CrossRef]

64. Sciandrello, S.; Puglisi, M.; Privitera, M.; Minissale, P. Diversity and spatial patterns of plant communities in volcanic temporary ponds of Sicily (Italy). *Biologia* **2016**, *71*, 793–803. [CrossRef]

65. Gazaix, A.; Klesczewski, M.; Bouchet, M.A.; Cartereau, M.; Molina, J.; Michaud, H.; Muller, S.D.; Pirsoul, L.; Gauthier, P.; Grillas, P.; et al. A history of discoveries and disappearances of the rare annual plant *Lythrum thesioides* M. Bieb.: New insights into its ecology and biology. *Bot. Lett.* **2020**, *167*, 201–211. [CrossRef]

66. Taylor, N.G.; Grillas, P.; Al Hreisha, H.; Balkız, Ö.; Borie, M.; Boutron, O.; Catita, A.; Champagnon, J.; Cherif, S.; Çiçek, K.; et al. The future for Mediterranean wetlands: 50 key issues and 50 important conservation research questions. *Reg. Environ. Chang.* **2021**, *21*, 1–17. [CrossRef]

Communication

The Discovery of the Rare *Chara baueri* (Charales, Charophyceae) in Serbia

Ivana Trbojević * , Vanja Milovanović and Gordana Subakov Simić

Faculty of Biology, University of Belgrade, Studentski trg 16, 11000 Beograd, Serbia;
b3026_2019@stud.bio.bg.ac.rs (V.M.); gsubak@bio.bg.ac.rs (G.S.S.)
* Correspondence: itrbojevic@bio.bg.ac.rs

Received: 30 October 2020; Accepted: 11 November 2020; Published: 19 November 2020

Abstract: *Chara baueri* is one of the rarest charophytes worldwide. It had been considered extinct in Europe for more than a century, from the 1870s to 2006, when it was rediscovered in Germany. The current distribution of this species is limited to a few localities in Europe (Germany, Poland and Russia), and one locality in Asia (Kazakhstan). We present a new finding of *Chara baueri*, to be a significant contribution to the species ecology and biogeography, and helping to review and update the current scarce knowledge. *Chara baueri* was discovered in Serbia and monitored for two vegetative seasons in 2018 and 2019, along with the associated macrophyte vegetation and water quality parameters. The morphology and ecology data of the species are presented comparatively with the literature data and the biogeography is critically reviewed. The population in Serbia is the first verified record of *Chara baueri* in southern Europe. Considering the recent findings and the knowledge accumulated in these records, *Chara baueri* was very possibly never extinct at all, but overlooked in Europe for the entire 20th century. We suggest that waterfowl migrating from the northern parts of Europe should be considered as the important spreading agent of *Chara baueri* in southern regions.

Keywords: charophyta; *Chara baueri*; species ecology; species biogeography

1. Introduction

Chara baueri A. Braun is considered to be one of the rarest charophytes worldwide [1]. It was first discovered and often collected in the Berlin area in the first half of the 19th century, until the 1870s. During the 19th century, only a few other localities all over the world could be reliably recognized–in Austria, southern Sweden and near Schwerin in Germany [2], all of them single records from the first half of the 19th century. Since the end of the 19th century, *C. baueri* was thought to be extinct in Europe [2,3]. The first record of *C. baueri* since then was made far away from the formerly known localities, in Kazakhstan in 1994, and this is still the only record in Asia [4]. Recently, it was rediscovered in Germany (Brandenburg) in 2006 [2] and newly discovered for Poland (2008) [5,6] and Russia (2010) [7].

C. baueri is morphologically, on first sight, very similar to *Chara braunii*, which is commonly found all over the world [3,8]. Still, there is a clear difference between these two taxa: a triplostique cortex that develops on the main axis (at least at the upper internodes) of *C. baueri*, and well-developed solitary spines, while *Chara braunii* is completely ecorticated [2,3,6]. Although Krause [3] suggested that these two species could be closely related, this has not yet been proven by molecular phylogenetic analysis. However, it has been confirmed that *C. baueri* could be even more phylogenetically close to the genus *Lamprothamnium* than to the other *Chara* species [9]. Phylogenetic relationships of these rare charophyte remain to be resolved in future studies, and collecting material from a wide geographic range is crucial so that these studies could offer a reliable answer.

The purpose of this study is to report about the first record of *C. baueri* in Serbia (ponds in the locality Štrbac, in the special nature reserve (SNR) "Gornje Podunavlje"). Discovered populations were monitored for two vegetative seasons in 2018 and 2019, along with the associated macrophyte vegetation and water quality parameters. The specimens of *C. baueri* are morphologically described in detail, comparative with the reliable literature sources. Habitat characteristics are discussed in terms of updating knowledge on the ecology of this rare charophyte. This new finding of *C. baueri* is discussed as a significant contribution to the knowledge on this species biogeography.

2. Results

The measured environmental parameters in pond 1 and pond 2 are presented in Table 1. It can be noticed that values of parameters are quite variable, since these are small and very shallow ecosystems. The measured values for total phosphorus (TP) and total nitrogen (TN) point to both ponds being eutrophic water bodies. The other measured parameters were more or less in range of moderate values. *C. baueri* was found in pond 1 in 2018 in July and August, and in 2019, at the end of July (31st), young plants were detected. In pond 2, *C. baueri* was found in August and September 2018. The bottoms of the both ponds were covered with a thick layer of very fine silt.

The following macrophyte and Charophyte species, apart from *C. baueri*, were recorded at the sampling sites: *Eleocharis palustris* agg., *Hydrocharis morsus-ranae* L., *Myriophyllum spicatum* L. (dominant in Pond 1), *Potamogeton gramineus* L., *Potamogeton nodosus* Poir., *Ranunculus aquatilis* agg., *Salvinia natans* (L.) All., *Schoenoplectus lacustris* (L.) Palla, *Spirodela polyrhiza* (L.) Schleid., *Utricularia* L. sp., *Chara braunii* C.C.Gmel., *Chara globularis* Thuill., *Chara tenuispina* A.Braun, *Nitella capillaris* (Krock.) J.Groves & Bull.-Webst., *Nitella mucronata* (A.Braun) Miq. and *Nitella* C.Agardh sp. In general, macrophyte vegetation was better developed (in terms of both diversity and cover) in pond 1 (more or less stable water level) in comparison to pond 2 (more prone to drastic water level changes). When pond 2 was filled with water, usually flotant macrophyte species dominated. In pond 1, *Myriophillum spicatum* was found to be dominant in the deeper parts along with *Potamogeton nodosus*, while the shallower part was completely covered with a meadow of *Chara globularis* and *Chara tenuispina*, and in the most shallow region, *C. baueri* formed patchy groups (*Nitella* specimens were found only sporadically, only a few specimens in total). Although we expected ponds to dry up completely at some point during the summer period, and despite the very dry August and September 2018, that did not happen.

Description of C. Baueri from Serbia

The macro habitus of the discovered specimens of *C. baueri* are presented in Figure 1. Specimens were found in groups in very shallow water (0.1 to 0.3 m), almost on the shoreline, in areas prone to drying up due to water level changes. Triplostichous and isostichous cortex on axis was clearly noticeable throughout the stem, though in some parts irregular (Figure 2c,d). This is an important taxonomic parameter for distinguishing this species from *C. braunii*. Specimens were fructifying, with both female and male gametangia being well developed (Figure 2a,b). Ripe oospores were also abundantly represented. Oospores were large and markedly black, while the oospore membrane color was light brown and showed a finely granulated structure.

Specimens of *C. baueri* found in Serbia are described in detail, and presented comparatively with the data from the relevant literature sources [1,4,6,10–15] (Table 2).

Table 1. Environmental parameters measured in pond 1 and pond 2 in 2018 and 2019. Bold is for the dates when *Chara baueri* was found in ponds. Water temperature (T), oxygen concentration/saturation (O_2), pH, conductivity (Cond.), degree of General Hardness of water ($°dH$), calcium ions concentration (Ca^{2+}), magnesium ions concentration (Mg^{2+}), nitrites (NO_2^-), nitrates (NO_3^-), ammonia (NH_4^+, total nitrogen (TN), Orthophosphates (OP), total phosphates (TP).

	T	O_2	O_2	pH	Cond	$°dH$	Ca^{2+}	Mg^{2+}	NO_2^-	NO_3^-	NH_4^+	TN	OP	TP
	°C	mg/L	%		µS/cm		mg/L	mg/L	mg/L	mg/L	mg/L	mg/L	mg/L	mg/L
POND 1														
May 2018	20.5	9.05	102	8.23	325	-	-	-	-	-	-	-	-	-
June 2018	22	4.42	50.5	8	300	-	-	-	-	-	-	-	-	-
July 2018	**30.7**	**15.45**	**190**	**9.35**	**338**	**1.1**	**3.7**	**2.7**	**0.021**	**<0.5**	**0.24**	**2.5**	**<0.020**	**0.028**
August 2018	**31**	**12**	**162**	**8.76**	**379**	**2.6**	**8.8**	**5.7**	**<0.020**	**<0.5**	**0.23**	**1.9**	**<0.020**	**0.026**
September 2018	14.5	9.94	101.2	8.68	346	6.3	30.2	9.1	0.027	<0.5	<0.05	2.9	<0.020	0.031
May 2019	18	15	152.6	9.61	328	5.2	12	15	<0.005	<0.5	2.2	1.2	<0.010	0.24
June 2019	29.15	11.9	157.5	9.13	344	3.7	11.6	9	<0.005	<0.5	0.45	1.5	<0.010	0.07
July 2019	**28.3**	**2.57**	**32.4**	**8.13**	**445**	**6.1**	**23.2**	**12.4**	**<0.005**	**<0.5**	**0.54**	**1.9**	**<0.010**	**0.07**
POND 2														
May 2018	-	-	-	-	-	-	-	-	-	-	-	-	-	-
June 2018	21	2.21	25.4	7.56	275	-	-	-	-	-	-	-	-	-
July 2018	28.1	7.5	96.5	8.12	342	6.1	31.2	7.5	0.043	<0.5	0.21	1.7	<0.020	0.029
August 2018	**26.1**	**6.24**	**77.1**	**8.44**	**435**	**4.4**	**21.8**	**5.9**	**<0.020**	**<0.5**	**0.18**	**3**	**<0.020**	**0.046**
September 2018	**15**	**12**	**120**	**9.08**	**355**	**6.7**	**33.8**	**8.4**	**0.08**	**0.91**	**0.79**	**5.3**	**<0.020**	**0.04**
May 2019	15.5	13.7	135.2	9.18	352	-	-	-	-	-	-	-	-	-
June 2019	28	3.63	43.6	7.96	354	-	-	-	-	-	-	-	-	-
July 2019	25.6	2.42	28.4	8.13	416	-	-	-	-	-	-	-	-	-

Figure 1. Macro habitus, scale 5 cm.

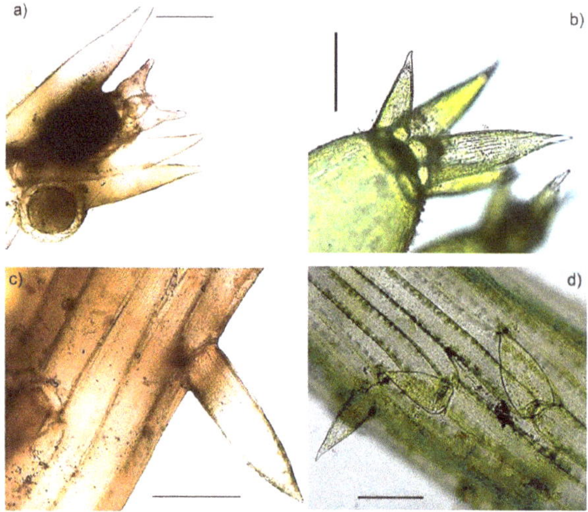

Figure 2. Microscopic taxonomic features: (**a**) Gametangia, (**b**) terminal corona on the branchlet terminal segment, (**c**) triplostichous, isostichous cortex, (**d**) cortex detail showing irregular structure, scale 200 μm.

Table 2. Species traits of *Chara baueri*, as described in the literature, and traits measured in samples of *C. baueri* discovered in Serbia (19 specimens and 127 oospores were measured for obtaining presented values, measured material was previously conserved in formalin).

Literature Source	Reichenbach, 1829 [10]	Braun, 1835 [11] Braun and Nordstedt, 1882 [12]	Ganterer, 1847 [13]	Migula, 1897 [14]	Langangen and Sviridenko, 1995 [4]	Hutorowicz, 2007 [15]	Pukacz, 2012 [1]	Urbaniak and Gąbka, 2014 [6]	Specimens from Serbia
Note					Specimen from Kazakhstan (1994)/ Specimen from Sweden (1849), herbarium specimen examined in dry condition	Oospores from specimens from Kazakhstan [4]	Oospores from specimens from Cedynia, Western Poland (August 2008)/ Oospores from specimens from near Batzlow, Germany (2006)		
Habitus description	Nice green color, turning black towards the bottom. Stiff stature, almost cartilaginous, "shining" similarly to *Ceratophyllum*.	Hard to differentiate from *Ch. coronata* (*Chara braunii*) without the loupe	Richly branched, bushy habitus, mostly light green in color	Habitus is not easy to distinguish from *Ch. coronata* (*Chara braunii*), all parts are proportionally thicker. Light yellow-green or green, rarely brownish-green.	-/-			Rather small plants, richly branched, yellow green to light green, strongly resembling *C.braunii*	Light green to yellowish green. Resembling *C. braunii*, but branchlets more rounded, plump and voluminous—like succulent.
Axes									
1. Incrustation	-	Not incrusted	-	Incrustation is abundant	Not incrusted/ slightly incrusted	-	-	Slightly incrusted	Very slightly incrusted
2. Diameter			More than 1.0975 mm (more than $\frac{1}{2}$ Linie, 1 Linie = 2.195 mm)	Up to 1 mm	Up to 0.57 (0.65 in Figure 2 caption) mm/0.93 mm			0.6–2.1 mm	0.55–0.95 mm
Cortex	-	Finely striped cortex	In the uppermost internodes, or throughout the stem more or less clearly and finely striped, and covered with scattered spines	Triplostichous, isostichous	2–3 corticate, isostichous, young internodes mostly 2 corticate/ Mostly 2 corticate, isostichous (hard to determine in dry material)	-	-	Triplostichous, in some parts irregular	Triplostichous, isostichous cortex developed throughout the stem, in some parts irregular
Spine cells									
1. Description	-	Fine, spiked	-	Solitary, up to $\frac{1}{2}$ of axes-diameter, acuminate–spiky, especially in young segments	Solitary, acute, to 1 × stem diameter, commonly shorter, many on young internodes/ Solitary, acute, many in young internodes, only few in the lower parts of stem	-	-	Solitary, common, acuminate, not exceeding axes diameter	Large, solitary, acuminate, more numerous in the upper section of axes

Table 2. *Cont.*

Literature Source	Reichenbach, 1829 [10]	Braun, 1835 [11] Braun and Nordstedt, 1882 [12]	Ganterer, 1847 [13]	Migula, 1897 [14]	Langangen and Sviridenko, 1995 [4]	Hutorowicz, 2007 [15]	Pukacz, 2012 [1]	Urbaniak and Gąbka, 2014 [6]	Specimens from Serbia
Branchlets									
1. Number in a whorl and description	-	Ecorticated	Ecorticated	8–9	8–9, ecorticated/ 7–8, ecorticated	-	-	7	8–11, ecorticated
Oogonia									
1. Number in node	Usually geminate, rarely 1 or 3	-	Single, geminate or 3, ovoid	-	-	-	-	-	1, mostly 2, sometimes even 3
2. Dimensions (length × width)	-	700–880 μm × 380–480 μm	-	650 × 500 μm	1000 μm/ 600 μm × 250 μm (immature)	-	-	585–950 μm × 365–610 μm	580–710 (775) μm × 370–450 μm
3. Number of convolutions	8–10	10–11	Mostly 10	8–10	-	-	-	-	9–10
4. Coronula width × height	-	140–200 μm × 220–250 μm	-	150–200 μm	-	-	-	-	(120) 170–260 (275) μm × (150) 200–330 (350) μm
Oospore									
1. Color	-	-	-	Black	-/Black	Light brown to black/-	Dark brown or black/-	Black/dark brown	Black
2. Dimensions (length × width)	-	500–560 μm × 310–350 μm	-	500–550 μm × 280–340 μm	-/600 μm	436–574 μm × 281–340 μm	400–667 μm × 183–300 μm/417–550 μm × 216–300 μm	465–740 μm × 280–340 μm	475–620 μm × 230–350 μm
3. Ridges description	-	-	-	Irregular, blunt/sharp edges	-	Unpronounced in young and prominent in mature oospores	Prominent ridges/-	-	Oval, blunt
4. Ridges number	-	-	-	8	-	8–11	8–11 (most often 9)/8–11	-	8–10 (11)
5. Fossa	-	-	-	-	-	-	-	-	50–70 μm
6. Membrane coloration	-	-	-	-	-	-	-	-	Light brown
7. Membrane structure	-	-	-	-	-	Either smooth or finely granulated	-	-	Finely granulated
Antheridia									
Diameter and description	In pairs or individual, brick red color	280–370 μm	Individual or in pairs, beneath the oogonia	250–300 μm	-/200 μm	-	-	250–330 μm, below oogonia	250–300 μm, below oogonia, individual, brick red color

3. Discussion

In the context of biogeography of the species, it is very interesting to review the historical and current distribution of *C. baueri*. Recent records are known from central, southern and eastern Europe, and central Asia (Figure 3). Fully reliable historical findings are known only from central/northern Europe (Germany and Sweden [2]). Other historical records (in Italy [16] and Lithuania [17]) could not be confirmed or verified [2]. As far as the authors' knowledge reaches, the historical finding from Austria [13] also cannot be supported by the herbarium material, but according to Krause [3], Raabe [2] and other authors' opinions, it should be considered valid (Figure 3).

Figure 3. Distribution of *C. baueri* across Eurasian continent: Red circles—recent findings, new discovery or rediscovery. Black circles—reliable historical findings. Empty black circles—historical records which could not be confirmed or verified.

Considering the description of his own material collected in the ponds near St. Andrä in southern Austria, and the drawing of these specimens, Ganterer's [13] finding is likely reliable. The most interesting point is that this locality in southern Austria is relatively close to the locality in Serbia, only about 340 km in a straight line [18]. While comparing his specimens with Bauer's original material, in which only a few upper internodes or even only one internode was corticated, Ganterer [13] commented that the plants he found near St. Andrä were larger and corticated throughout the stem, with long and numerous spine cells. This peculiarity was also seen in the specimens found in Serbia, which were large, and the cortex was present throughout the stems, with long spine cells (Table 2). Habitat specificities, as well as population origin and/or isolation, could be the basis of these specificities. Still, after Ganterer's record, *C. baueri* was never confirmed again in Austria, or anywhere near. The record from Serbia that we are reporting here is the only verified one in southern Europe. The fact is that recent findings (Figure 3) substantially widen the distribution area of *Chara baueri* in comparison to the historical records. In Germany, *C. baueri* is considered highly endangered ("stark gefährdet"), it is threatened and under the risk of extinction in Poland, and it is regionally extinct in Sweden [19].

Data on the ecology of the *C. baueri* are very scarce, and authors could only reach data published by Pukacz et al. [1] and Doege et al. [20] (since data presented in Doege et al. [20] are practically the same as in Pukacz et al. [1], we are further referencing the older, but original publication). When the literature data is compared to the data obtained in this study (Table 3), it is clear that the values for almost all comparable parameters are lower, meaning that water in the localities in Serbia were poorer in Ca and Mg content and electrolytes in general (conductivity), as well as softer and a bit more alkaline (Table 3). Considering the nutrient content, lower concentrations were detected in our localities, though according to the total phosphorus (TP) and total nitrogen (TN), both ponds in Serbia were also eutrophic water bodies, as expected [21].

Table 3. Summed available data on the ecology of *C. baueri* according to Pukacz et al. [1], comparatively with the data obtained in this study (mean values presented).

	O_2 (mg/L)	Hardness (°dH)	pH	Conductivity (μS/cm)	Mg^{2+} (mg/L)	Ca^{2+} (mg/L)	TP (mg/L)	PO_4^- (mg/L)	TN (mg/L)	NO_2^- (mg/L)	NO_3^- (mg/L)	NH_4^+ (mg/L)
Pukacz et al., 2012 [1]												
Cediniya (Poland)	6.14	12.9	8.01	611	14.9	71.7	1.04	0.63	3.89	-	0.47	1.31
Batzlow (Germany)	3.25	13.2	7.92	632	16.3	67.6	1.12	0.71	5.15	0.02	0.68	1.45
This study												
Pond 1	10.04	4.17	8.7	350.6	8.98	14.92	0.08	<0.015	1.98	0.02	<0.5	0.73
Pond 2	6.8	5.73	8.3	361.3	7.27	28.93	0.04	<0.02	3.33	0.06	0.91	0.39

Recently added details on the habitat type of the species *C. baueri* [21] correspond to the characteristics of the localities in Serbia—small water bodies in the fields, rich in nutrients; in Serbia, those are ponds, artificially made to serve as a watering place for wild animals, prevalently wild boars and deer (locality Štrbac, where ponds are placed in the special nature reserve (SNR) "Gornje Podunavlje"). Digging the watering places in the fields of Štrbac in the SNR "Gornje Podunavlje", could have activated a diaspore bank with *C. baueri* oospores in it, which could explain the surprising occurrence of this rare charophyte in this area. Sediment characteristics and the depth range where the species occurs matches literature [6,21] descriptions.

Among the most often associated macrophyte species that Gregor [21] emphasized, in our study, only *Nitella mucronata* and *Chara globularis* were recorded along with *C. baueri*. Nevertheless, *Elatine alsinastrum* which was recorded in almost all literature sources as being associated with *C. baueri* [21], was not detected in the localities in Serbia. Considering the phenology of the *C. baueri* species, it is clear that its yearly occurrence depends on the existence of ephemere habitats where it typically grows [21].

When considering the taxonomically relevant morphological characteristics of *C. baueri*, the triplostichous and isostichouse cortication is a distinctive feature, clearly separating this taxon from *C. braunii*, which is completely ecorticated [21] (see Table 2). Still, Langangen and Sviridenko [4] ascertained that the specimens found in Kazakhstan had a diplo- to triplo-stichous cortex, while young internodes were mostly diplostichous. These authors also described the old Swedish specimen from 1849 as mostly 2 corticate, although they say it was difficult to determine, probably because the material was old and the herbarium specimen was studied in dry conditions [4]. Urbaniak and Gąbka [6] remarked the irregularity of the cortex, which was also noted in this study. It was already shown that the number of cortex cell rows may be variable within a genetically homogeneous *Chara* group [22], which may be the case in *C. baueri* as well.

Finally, the origin of the population of *C. baueri* in Serbia is debatable. Langangen and Sviridenko [4] suggested that long-distance dispersal by birds migrating from Europe to the Kazakhsthan area in the spring could explain the occurrence of *C. baueri* in Central Asia. We find this hypothesis plausible, as it could potentially explain the finding in Serbia. The SNR "Gornje Podunavlje", where the locality Štrbac and studied ponds belong, is declared a Ramsar site and Important Bird Area (IBA), where many migratory bird species stop to rest or nest, and this area is the most significant national nesting area of the wild (graylag) goose (*Anser anser*) in Serbia [23]. For the sake of illustration, the distribution range of this waterfowl species covers all recent localities of *C. baueri* (in Europe and Asia) [24]. Also, according to Dick et al. [25], the central European graylag goose population migration routes perfectly link all localities where *C. baueri* was ever found in Europe (including non-reliable findings). Considering the habitat characteristics of *C. baueri*, various waterfowl species migrating from the north of Europe could be dispersal agents in more southern regions, and IBA and nesting areas of these birds should be the first ones surveyed in search for the potential new localities of *C. baueri* in Europe. Also, we suggest that future detailed multidisciplinary research on if and how the migration routes of waterfowl coincide with the distribution of other charophytes across Europe would be valuable input regarding charophyte biogeography, especially rare species.

Summarizing the current knowledge of ecological requirements and habitat characteristics of *C. baueri*, as well as the distribution range, it is quite possible that this species was simply overlooked in Europe for the entire 20th century. Contributing to this knowledge gap is the lack of organized, targeted and continued monitoring of charophytes (i.e., certain type of habitats), which are often completely overlooked in macrophyte studies or at best recognized as *Chara* sp.

4. Materials and Methods

The study area was located in the Special Nature Reserve (SNR) "Gornje Podunavlje", Štrbac area, in Serbia's Northern Province, Vojvodina (Figure 4). This area is a very unique and complex mosaic of meadows, woods, ponds and wetlands, and it includes the river Danube and its meanders.

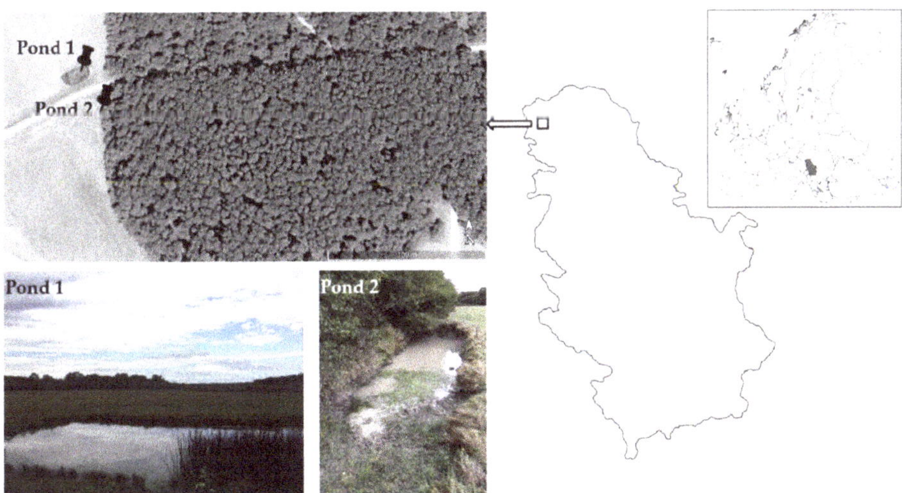

Figure 4. Study area and sampling localities.

Two ponds separated by a dirt road, located in the meadow near the woods (Figure 4), were labeled as pond 1 (N 45.812775, E 18.960583) and pond 2 (N 45.812577, E 18.960785). They were monitored monthly from May until September in 2018 and May until July in 2019. According to the nature reserve rangers, these ponds were man-made with the purpose of forming a watering place for wild animals.

Each time the ponds were sampled for charophytes, the environmental parameters: temperature (T), pH, oxygen concentration (O_2 mg/L) and saturation (O_2%), and conductivity, were measured in situ in the ponds, using digital field instruments made by Eutech Instruments Oakton® and YSI ProODDO

Optical Dissolved Oxygen Meter. Simultaneously, water samples from both ponds were also taken for further laboratory analyses of chemical water properties, which were conducted in the accredited laboratories of the Institute of Public Health of Serbia "dr Milan Jovanović Batut", using standard analytical methods.

Charophytes were collected by wading and using rakes and grapnels. Material was stored in plastic bags and transported to the laboratory where it was identified using a STEMI DV4 stereomicroscope and a Nikon YS100 microscope and standard literature [3,6,14,16,17,26].

Micrographs were made using a Carl Zeiss AxioImager M1 microscope and a digital camera AxioCam MRc5, with AxioVision 4.8 software. Part of the identified material was herbarized and part was stored in 4% formalin (final concentration) in the collection of wet specimens of the Department of Algology, Mycology and Lichenology, Faculty of Biology, University of Belgrade (BEOU, Belgrade, Serbia).

5. Conclusions

This study reported the first record of *C. baueri* in Serbia, but also the first reliable record of the species in southern Europe. Results of our study supplement the knowledge on the habitat characteristics and overall ecology of this rare charophyte. Our finding significantly contributes to the species biogeography, which is reviewed and discussed, thus concluding that the distribution of *C. baueri* should be observed across the Eurasian continent. Waterfowl species migrating from the north of Europe are suggested as the most probable dispersal agent of *C. baueri* in more southern regions, where IBA and nesting areas of these birds could be considered the potential new localities of *C. baueri* in Europe. Considering recent findings and knowledge accumulated in these records, *C. baueri* has very possibly never been extinct, but overlooked in Europe for the entire 20th century.

Author Contributions: Conceptualization, I.T.; methodology, I.T. and V.M.; investigation, I.T., V.M. and G.S.S.; resources, I.T. and G.S.S.; data curation, I.T. and G.S.S.; writing—original draft preparation, I.T.; writing—review and editing, I.T., V.M. and G.S.S.; visualization, I.T. and V.M.; supervision, G.S.S.; funding acquisition, I.T. and G.S.S. All authors have read and agreed to the published version of the manuscript.

Funding: This research was funded by RUFFORD FOUNDATION, grant number 25789-1, and the APC was funded by Ministry of Education, Science and Technological Development, Republic of Serbia.

Acknowledgments: The authors are sincerely grateful to Milica Petrović Đurić for the technical support, expertise and commitment in material processing. We thank two anonymous reviewers for their careful reading of our manuscript and their insightful comments and suggestions.

Conflicts of Interest: The authors declare no conflict of interest. The funders had no role in the design of the study; in the collection, analyses, or interpretation of data; in the writing of the manuscript, or in the decision to publish the results.

References

1. Pukacz, A.; Boszke, P.; Pelechaty, M.J.; Raabe, U. Comparative study of the oospore morphology of two populations of a rare species *Chara baueri* A. Braun in Cedynia (Poland) and Batzlow (Germany). *Acta Soc. Bot. Pol.* **2012**, *81*, 131–136. [CrossRef]
2. Raabe, U. *Chara baueri* rediscovered in Germany—Plus additional notes on Gustav Heinrich Bauer (1794–1888) and his herbarium. *Int. Res. Group Charophytes News* **2009**, *20*, 13–15.
3. Krause, W. *Charales (Charophyceae) Süßwasserflora von Mitteleuropa*; Gustav Fischer Verlag: Jena, Germany, 1997.
4. Langangen, A.; Sviridenko, B.F. *Chara baueri* A. Br., a charophyte with a disjunct distribution. *Cryptogam. Algol.* **1995**, *16*, 125–132.
5. Pukacz, A.; Pelechaty, M.; Raabe, U. Pierwsze stanowisko *Chara baueri* (Characeae) w Polsce. *Fragm Florist Geobot Pol.* **2009**, *16*, 425–429.
6. Urbaniak, J.; Gąbka, M. *Polish Charophytes. An Illustrated Guide to Identification*; Universytet Przyrodiczy We Wroclawiu: Wroclaw, Poland, 2014; ISBN 978-83-7717-166-0.
7. Klinkova, G.Y.; Zhakova, L.V. New and rare species Charales in the flora of the lower Volga region. *Byull. Moskovsk. Obshch. Isp. Prir. Otd. Biol.* **2014**, *119*, 61–66.

8. Noedoost, F.; Riahi, H.; Sheidai, M.; Ahmadi, A. Distribution of Charophytes from Iran with three new records of Characeae (Charales, Chlorophyta). *Cryptogam. Algol.* **2015**, *36*, 389–405. [CrossRef]

9. Schneider, S.C.; Nowak, P.; von Ammon, U.; Ballot, A. Species differentiation in the genus *Chara* (Charophyceae): Considerable phenotypic plasticity occurs within homogenous genetic groups. *Eur. J. Phycol.* **2016**, *51*, 282–293. [CrossRef]

10. Reichenbach, H.G.L. *Dr. Joh. Christ. Mössler's Handbuch der Gewächskunde, Enthaltend eine Flora von Deutschland mit Hinzufügung der Wichtigsten Ausländischen Cultur-Pflanzen*, 2nd ed.; Johann Friedrich Hammerich: Altona, Germany, 1829; Volume 3, pp. 1592–1601.

11. Braun, A. Uebersicht der genauer bekannten Chara-Arten. *Flora Allg. Bot. Ztg.* **1835**, *18*, 49–73.

12. Braun, A.; Nordstedt, C.F.O. *Fragmente einer Monographie der Characeen*; Königlichen Akademie der Wissenschaften: Berlin, Germany, 1882; pp. 1–211.

13. Ganterer, U. *Die Bisher Bekannten Österreichischen Charen: Vom Morphologischen Standpunkte Bearbeitet*; Carl Haas'schen Verlag: Vienna, Austria, 1847.

14. Migula, W. *Die Characeen Deutschlands, Oesterreichs und der Schweiz: Unter Berücksichtigung aller Arten Europas*; E. Kummer: Leipzig, Germany, 1897.

15. Hutorowicz, A. Morphological variability of oospores of *Chara baueri* A. Braun [Characeae]. *Acta Soc. Bot. Pol.* **2007**, *76*, 235–237. [CrossRef]

16. Wood, R.D.; Imahori, K. *A Revision of the Characeae. First Part:Monograph of the Characeae*; J. Cramer Verlag: Weinheim, Germany, 1965.

17. оа, ..; ааа, .. *аоВ000. ПооВ000ССС*; Nauka: Leningrad, Russia, 1983.

18. Google Earth. Available online: https://earth.google.com/web/search/Sankt+Andr%c3%a4,+Austria/@45. 81280243,18.96052335,83.72340015a,114.48633148d,35y,0h,0t,0r/data=CigiJgokCQAz16vOpUdAEdpVE1p BRUdAGQnvaQoHsjBAIfzZ_8Vy3CxA (accessed on 26 October 2020).

19. Becker, R. Gefährdung und Schutz von Characeen. In *Armleuchteralgen. Die Characeen Deutschlands*; Deutschlands, A.C., Ed.; Springer: Heidelberg, Germany, 2016; pp. 149–191. ISBN 978-3-662-47796-0.

20. Doege, A.; van de Weyer, K.; Becker, R.; Schubert, H. Bioindikation mit Characeen. In *Armleuchteralgen. Die Characeen Deutschlands*; Deutschlands, A.C., Ed.; Springer: Heidelberg, Germany, 2016; pp. 149–191. ISBN 978-3-662-47796-0.

21. Gregor, T. *Chara baueri*. In *Armleuchteralgen. Die Characeen Deutschlands*; Deutschlands, A.C., Ed.; Springer: Heidelberg, Germany, 2016; pp. 247–248. ISBN 978-3-662-47796-0.

22. Trbojević, I.; Marković, A.; Blaženčić, J.; Subakov Simić, G.; Nowak, P.; Ballot, A.; Schneider, S. Genetic and morphological variation in *Chara contraria* and a taxon morphologically resembling *Chara connivens*. *Bot. Lett.* **2020**, *167*, 187–200. [CrossRef]

23. Simić, D.; Puzović, S. *Ptice Srbije i Područja od Međunarodnog Značaja*; Liga za ornitološku akciju Srbije: Belgrade, Serbia, 2008; ISBN 978-86-911303-0-5.

24. BirdLife International (2020) Species factsheet: *Anser Anser*. Available online: http://datazone.birdlife.org/sp ecies/factsheet/greylag-goose-anser-anser/distribution (accessed on 4 September 2020).

25. Dick, G.; Baccetti, N.; Boukhalfa, D.; Darolova, A.; Faragó, S.; Hudec, K.; Leito, A.; Markkola, J.; Witkowski, J. Graylag Goose *Anser anser*: Central Europe/North Africa. In *Goose Populations of the Western Palearctic. A Review of Status and Distribution*; Madsen, J., Cracknell, G., Fux, T., Eds.; Wetlands International Publication: Wageningen, The Netherlands; National Environmental Research Institute: Rønde, Denmark, 1999; pp. 202–213. ISBN 87-7772-437-2.

26. Wood, R.D.; Imahori, K. *A Revision of the Characeae. Second Part: Iconograph of the Characeae*; J. Cramer Verlag: Weinheim, Germany, 1964.

Publisher's Note: MDPI stays neutral with regard to jurisdictional claims in published maps and institutional affiliations.

Communication

Chara zeylanica J.G.Klein ex Willd. (Charophyceae, Charales, Characeae): First European Record from the Island of Sardinia, Italy

Ralf Becker [1], Hendrik Schubert [2] and Petra Nowak [2,*]

[1] Independent Researcher, 26121 Oldenburg, Germany; becker.r@posteo.de
[2] Aquatic Ecology, Institute of Biosciences, University of Rostock, 18059 Rostock, Germany;
     hendrik.schubert@uni-rostock.de
*   Correspondence: petra.nowak@uni-rostock.de

Abstract: The first record of a species belonging to the genus *Chara* L. subgenus *Chara* R.D.Wood section *Grovesia* R.D.Wood subsect. *Willdenowia* R.D.Wood from Europe is presented here, thus challenging the interpretation of its distribution pattern as an intertropical group of charophytes. The morphological characters of the specimens, as well as the results of a phylogenetic analysis, clearly identified them as *Chara zeylanica* J.G.Klein ex Willd. Although the subsection *Willdenowia* has yet to receive a thorough taxonomic treatment, a discussion of its relationship to other taxa of this subsection is provided despite the lack of a commonly agreed upon taxonomic concept. The ecological conditions of the Sardinian site of *C. zeylanica* are presented. Moreover, the status of and threats to this taxon, and hypotheses regarding potential pathways through which it reached Europe, are discussed.

Keywords: charophytes; *Willdenowia*; Sardinia; biogeography; *Chara zeylanica*

Citation: Becker, R.; Schubert, H.; Nowak, P. *Chara zeylanica* J.G.Klein ex Willd. (Charophyceae, Charales, Characeae): First European Record from the Island of Sardinia, Italy. *Plants* **2021**, *10*, 2069. https://doi.org/10.3390/plants10102069

Academic Editor: Angelo Troia

Received: 27 July 2021
Accepted: 24 September 2021
Published: 30 September 2021

Publisher's Note: MDPI stays neutral with regard to jurisdictional claims in published maps and institutional affiliations.

1. Introduction

Charophytes are morphologically complex macrophytic green algae with a worldwide distribution. Because they are close relatives of the earliest land plants [1], they have attracted growing scientific interest in recent decades. However, in addition to becoming a subject of academic interest, charophytes play a major role in bioindication systems due to their species-specific pattern of niche occupation [2,3]. Moreover, Characeae are among the most threatened groups of organisms on earth [4–6], and have thus been targeted by nature conservation actions [7–9]. As charophytes occur in an astonishingly wide variety of habitats, ranging from ultraoligotrophic freshwater to hypersaline and hypertrophic environments, their presence is often measured in water quality assessments and other related fields [10,11].

For the development of such bioindication systems, having comprehensive and reliable knowledge about the habitat preferences and distribution patterns of the individual species is essential, as is the accurate identification of charophyte species, and the formulation of a sound taxonomic concept. In recent decades, a large number of studies have attempted to fulfil these requirements [12–19]. As a result, our knowledge about the biogeography of charophytes has increased substantially. However, whereas in the past site-specific information about the occurrence of the individual species was provided [20], recent treatments have led to the development of large-scale distribution grid maps and detailed descriptions of the species' preferred habitat conditions [21].

For several species, a strong correspondence between the distribution range and the niche structure was found. For example, the strictly circumpolar distribution of *Tolypella normaniana* Nordst. can be explained by its temperature preference (cold-stenothermic). Moreover, it has been shown that species such as *Chara vulgaris* L. or *C. braunii* C.C.Gmelin occur in a broad range of habitats on all continents, except for Antarctica [22].

However, unresolved questions regarding charophytes have hampered the development of general bioindication schemes that are also applicable outside of the reference regions, which have mainly been restricted to specific geographic scales [3]. One of these questions related to the absence of subsection *Willdenowia* R.D.Wood (section *Grovesia* R.D.Wood) in Europe is dealt with here, as we provide the first record of the presence of *Chara zeylanica* J.G.Klein ex Willd. in Europe from the Mediterranean island of Sardinia (Italy).

In a global taxonomic treatment of charophytes by Wood [23,24], the genus *Chara* L. was divided into five sections with a total of eight subsections. Subsection *Willdenowia* comprises diplostephanous species with triplostichous cortication and a completely ecorticated basal branchlet segment. According to Wood [23,24]), this subsection includes just one species, *Chara zeylanica*, which has several varieties and forms. This approach was not universally accepted because it combined A) monoecious and dioecious taxa, B) monoecious taxa with sejoined and conjoined gametangia, and C) taxa with tetra- and octoscutate antheridia [23]. However, several authors used Wood's concept as a basis for investigating the distribution pattern of subsect. *Willdenowia,* and came to the conclusion that it can best be described as an intertropical taxon [25,26]. On the other hand, as distinct patterns of the distribution of subspecies and varieties of *Chara zeylanica sensu* Wood [23] emerged, a fine-resolution taxonomic treatment of subsection *Willdenowia* was clearly needed for biogeographical purposes [25,27]. In an approach designed to overcome the problems caused by Wood's taxonomic concept, van Raam [28,29] presented an alternative view in which subsect. *Willdenowia* was divided into 20 species that were mainly distinguished by the abovementioned criteria of gametangia position, antheridia morphology, and sexuality.

However, irrespective of which concept was applied, neither *Chara zeylanica* nor any other taxon of subsect. *Willdenowia* has previously been recorded anywhere in Europe, even though numerous investigations of charophytes have been performed throughout the Mediterranean area in recent decades [30–36]. *Chara zeylanica* occurs mainly in tropical and subtropical regions of the world [23,37–43]. As it is an "intertropical taxon", the first record of the presence of *C. zeylanica* in Europe could be considered a surprise. On the other hand, Corillion and Guerlesquin [26] and Proctor et al. [25] have reported, taxa of subsect. *Willdenowia* have been found in North America at up to 45° N under climatic conditions comparable to those in Northern Europe. There are historical records of the presence of the species from Egypt and Israel [26,44], as well as reports of extinct occurrences from Algeria [45]. Consequently, limitations other than climatic conditions should be responsible for the failure to observe the presence of taxa of subsect. *Willdenowia* in Europe, which is a well-investigated region that certainly cannot be regarded as undersampled. Recent records of the presence of non-native charophyte species with predominantly tropical and subtropical distributions—such as reports of the presence of *Chara fibrosa* C.Agardh ex Bruzelius ssp. *benthamii* (A.Braun) Zaneveld or *Chara* c.f. *chrysospora* J.Groves and Stevens in rice fields, lakes, and an artificial stormwater retention pond in Southern France, Italy, and Crete, respectively [32,46–48]—indicate that the climatic conditions in the Mediterranean area are suitable for the establishment of intertropical taxa.

The main aim of this study is to document the first record of the presence of *Chara zeylanica* in Europe, and the morphological features of the Sardinian specimens we collected. Moreover, this study contributes to knowledge about the taxonomic classification, the ecological requirements, and the geographic distribution of this mainly tropical and subtropical species. To support our morphological analysis, we used *rbcL* and *matK* barcodes, as previous barcoding studies have shown that a combination of these sequences is suitable for investigating species of the genus *Chara* [49–52].

2. Results

2.1. Ecology

Chara zeylanica can colonize a broad range of both brackish and freshwater habitats throughout the tropical and subtropical zones of the world. These habitats include permanent and temporary bodies of water, such as lakes, ponds, pools, ditches, temporarily flooded wetlands, canals, rice fields, and retention ponds [26,38,40–43]. Few of the existing hydrochemical datasets cover a spectrum ranging from low-impacted waterbodies with total P-concentrations below 20 µg L^{-1} and total N-concentrations between 0.425 and 1.9 mg L^{-1} [43] to eutrophic habitats [37]. According to Muller et al. [45], *C. zeylanica* needs temperatures of approximately 25 °C for fructification.

The only European site (reported here for the first time) where the presence of *C. zeylanica* has been detected is at Cala Fuili, which is located north of Orosei on the east coast of Sardinia, Italy (coordinates: 40°25′03″ N, 9°46′13″ E; coordinate system WGS 84) (Figure 1). The specimens were found in September 2019 at a depth of about 1 m, mainly in sandy to silty places with stony substrate, in a shallow and probably permanent small stream located close to the beach, or 110 m from the Mediterranean Sea. The specific site where the *C. zeylanica* specimens were found was situated directly next to a bridge (Figure 1, left image below), and was therefore disturbed by the structure. By contrast, the neighbouring stream sections and landscape areas can be considered semi-natural habitats. The small population of *C. zeylanica* was observed to have high fertility, with ripe antheridia, oogonia, and oospores. The nutrient conditions at the sampling date were as follows: NH_4-N 0.108 mg L^{-1}, NO_3-N 0.279 mg L^{-1}, total N 1.143 mg L^{-1}, PO_4-P 0.073 mg L^{-1}, and total P 0.137 mg L^{-1}. The water hardness was 26.4 °dH (Ca 62.2 mg L^{-1}, Mg 76.8 mg L^{-1}), pH 8.3. Although the salinity at the sampling date was 1.9, the salinity of the site probably varies because it is close to the coast. At the same site in May 2016, a salinity level of 4.4 was recorded and the Cl concentration was found to be 2819 mg L^{-1}, instead of 1290 mg L^{-1}, as measured in September 2019.

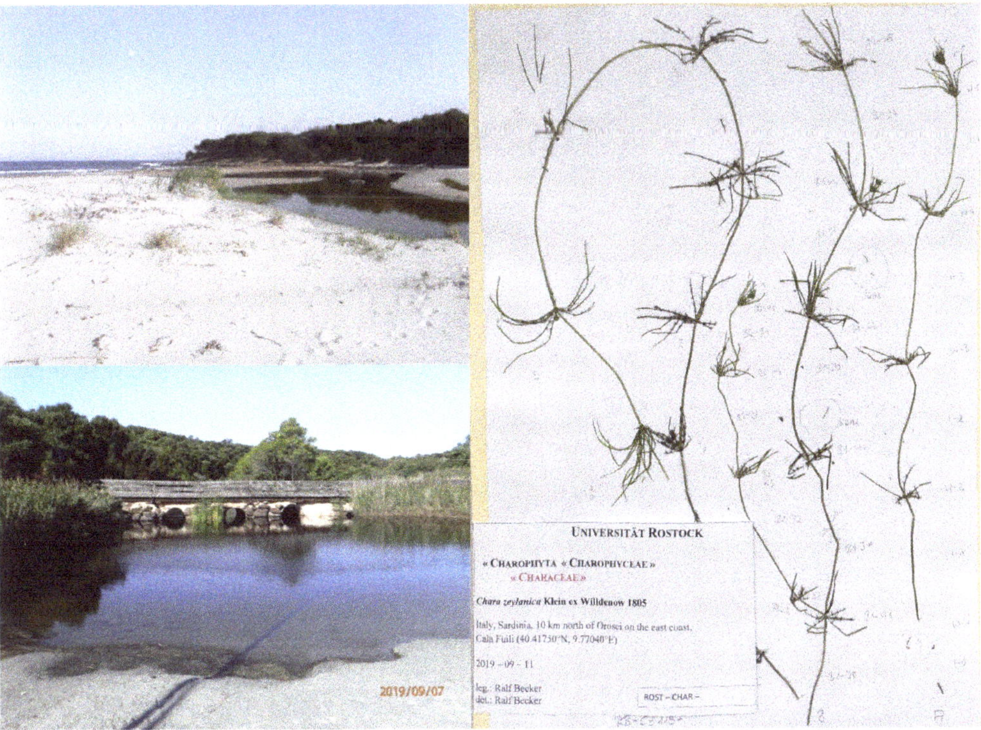

Figure 1. Habitat photographs (**left panels**) and habitus (**right panel**) of specimens collected at Cala Fuili, Sardinia. Plants were collected in the widening of the small stream just in front of the bridge.

2.2. Morphological Description

The specimens are 30–60 cm long, erect and straight, stout, fresh to greyish green, and slightly incrusted (Figure 2). The main axis diameter is 589–1076 μm with a mean value of 844 μm, slightly (0–4) branched. Most of the internodes are 4.4–8.0 cm long, and are usually much longer (up to 4 ×) than the branchlets. The uppermost 1–2 internodes are only 0.5–2.2 cm long, and are generally shorter than the adjacent branchlets. The cortex is usually triplostichous, and is rarely (partly) diplostichous and tylacanthous to isostichous (Figure 2D). Single, acute, thin, and needle-like spines are observed on the young internodes, and rarely on the older internodes. These spines can vary in length (182–1468 μm long) even on the same plant, and mainly point downwards (Figure 2D). The stipulodes are acute, elongated, and well developed. They are arranged in two regular tiers with two pairs per branchlet (Figure 2E). The upper stipulodes are longer than the lower ones. As the upper stipulodes are 515–1045 μm long (a mean value of 760 μm), they are usually longer than the diameter of the axes, and are much longer than the lowermost branchlet segment. The lower stipulodes are sometimes of unequal lengths, at 161–475 μm long, with a mean value of 293 μm. The branchlets are 9–12 in whorl and generally much shorter than the internodes, at 2.0–4.3 cm long. The branchlets of the uppermost 1–2 youngest whorls are even shorter, at just 0.2–2.0 cm long. The lowermost basal segments of the branchlets are ecorticated, and are very short at 208–479 μm long (mean value 343 μm) and 189–470 μm wide (mean value 318 μm). These segments are hidden behind the upper stipulodes (Figure 2A). The branchlets consist of 7–10 segments, with the lowermost segments always being ecorticated, followed by 4–6 corticated segments and 2–5 ecorticated distal segments with a tiny acute end cell on top, surrounded by a ring of bract cells (Figure 2C,F). The bract cells (5–8) are well developed (220–843 μm long), slender, and acute, and are shorter than the bracteoles. The two bracteoles are very long (990–1948 μm), at 1–2.5 × longer than the oogonia and oospores (Figure 2B). All of the fertile specimens are monoecious with conjoined gametangia (Figure 2B). Gametangia usually occur only at the nodes of corticated segments, and are rarely observed at the lowest nodes just above the ecorticated segment. The gametangia are mainly solitary, and very rarely geminate. The oogonia are elliptical to elongated oval in shape, are yellow or greenish in colour, and generally have constricted coronulae. The length of the oogonia (without coronula) is (600) 700–850 (900) μm, and the width of the oogonia is 417–575 (600) μm. The length of the coronula is 69–125 (150) μm, and the width of the coronula is (127) 160–200 (250) μm. The oospores are elliptical in shape and black in colour, with a length of (539) 600–685 μm, a width of 375–475 (500) μm, and 10–13 striae. The antheridiae are tetrascutate with a diameter of (300) 350–400 (450) μm. The dried specimens are stored at the herbarium of the University of Rostock (ROST).

2.3. Phylogenetic Analysis

The three individuals collected on Sardinia had identical *rbcL* and *matK* sequences. The BLAST of the GenBank nucleotide collection under default settings with *rbcL* from the Sardinian samples as query sequences matched the individuals to *C. zeylanica* from New Caledonia (AB440257) with 100% identity. One basepair (bp) substitution (99% identity) was detected for two further *C. zeylanica* (HQ380481: Sri Lanka, AY720934: Taiwan), but also for a sequence belonging to *C. hydropitys* Rchb. (HQ380464: Puerto Rico).

The BLAST of the GenBank nucleotide collection using *matK* from the Sardinian samples as query sequences matched the individuals with 99% identity (1 bp substitution) to *C. zeylanica* from Myanmar (MT739758). *Chara guairensis* R.M.T.Bicudo (KY656924) and *C. hydropitys* (KY656921) differed from the Sardinian samples by 15 bp substitutions (98% identity), respectively.

Phylogenetic analyses were performed for *rbcL* and *matK* separately to confirm the species identified through the BLAST search. The final *rbcL* alignment was trimmed to 1051 bp. Within the subsect. *Willdenowia*, 30 variable sites were identified. In the *rbcL* tree (Figure 3), the relationships within the subsect. *Willdenowia* were ambiguous, because several nodes did not have significant supports. The specimens from Sardinia formed a cluster together with *C. zeylanica*, but only with a low level of support (BS: 50%, PP: 0.6). The phylogeny based on the *rbcL* gene sequences only could not be resolved, and relationships of *C. zeylanica* to other species of subsect. *Willdenowia* were ambiguous.

Figure 2. Detailed photographs of *C. zeylanica* collected at Cala Fuili, Sardinia. (**A**) branchlet whorl with ecorticated basal segments; (**B**) conjoined gametangia; (**C**) ecorticated end segments with bract cells; (**D**) triplostichous main axis cortication with single spines; (**E**) diplostephanous stipulodes; (**F**) branchlet tip cell, surrounded by bract cells.

The final *matK* alignment was trimmed to 970 bp. Within the subsect. *Willdenowia*, 64 variable sites were identified. Phylogenetic analysis of the *matK* alignment provided strong bootstrap support for the sequences from the Sardinian samples forming a monophyletic clade with *C. zeylanica* sequence: MT739758 (BS: 100% and BP: 1, Figure 4). The *matK* phylogeny assigned the Sardinian specimens to *C. zeylanica*, and differentiated them from other species of subsect. *Willdenowia* (*C. guarensis*, *C. rusbyana* M.Howe, *C. haitensis* Turpin, *C. foliolosa* Muhl. ex Willd.) and sect. *Imahoria* (*C. hydropitys*).

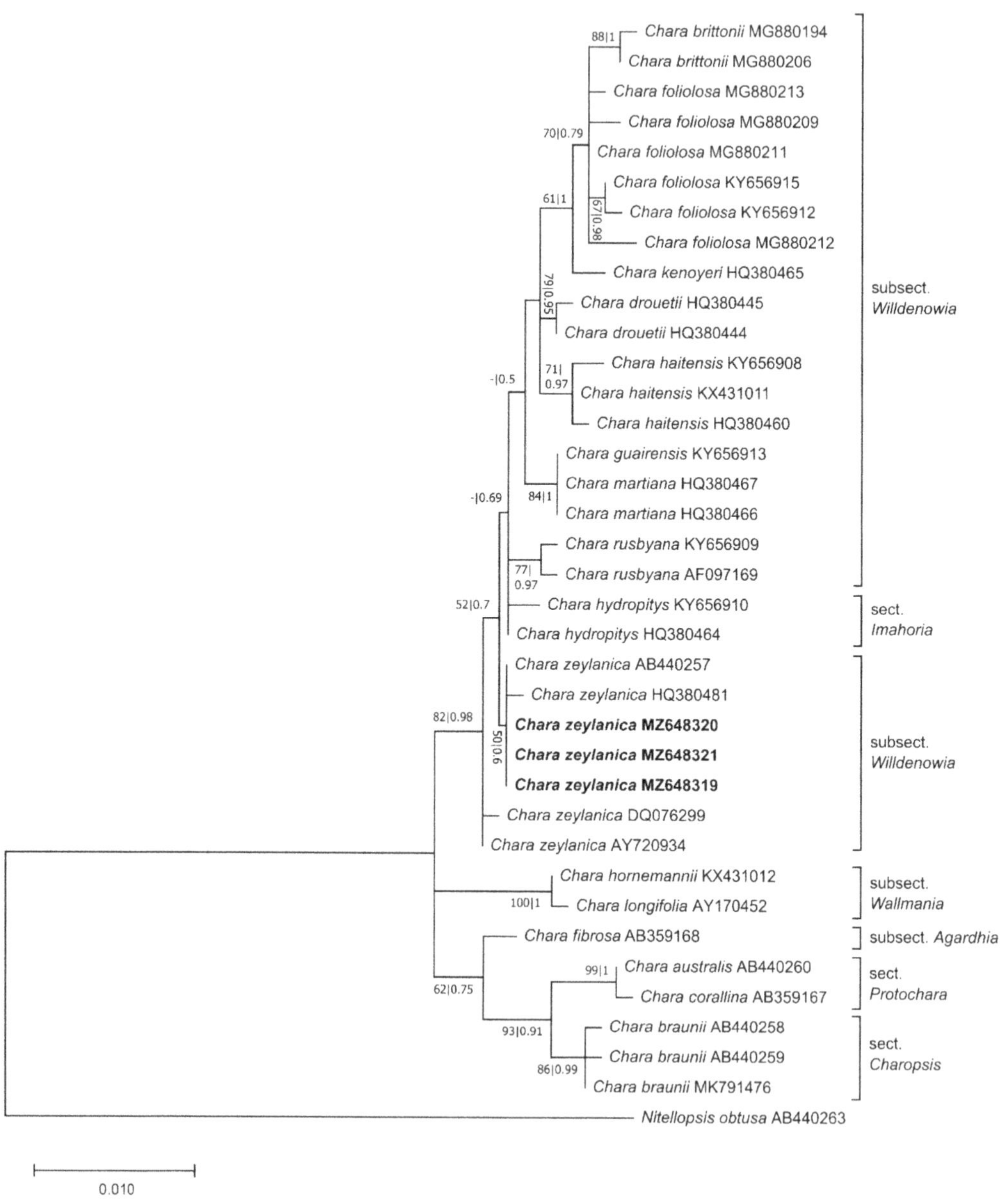

Figure 3. Phylogenetic tree based on 1051 base pairs (bp) of the *rbcL* sequences. The evolutionary history was inferred using the GTR+G + I model. Maximum likelihood bootstrap values (>50%, ML, **left**) and Bayesian posterior probabilities (>0.5, BI, **right**) are given at the branches. The scale indicates sequence divergence in percent. The specimens collected in Sardinia are marked bold. Information about the section and the subsection of the species is given at the tree [24,29,53].

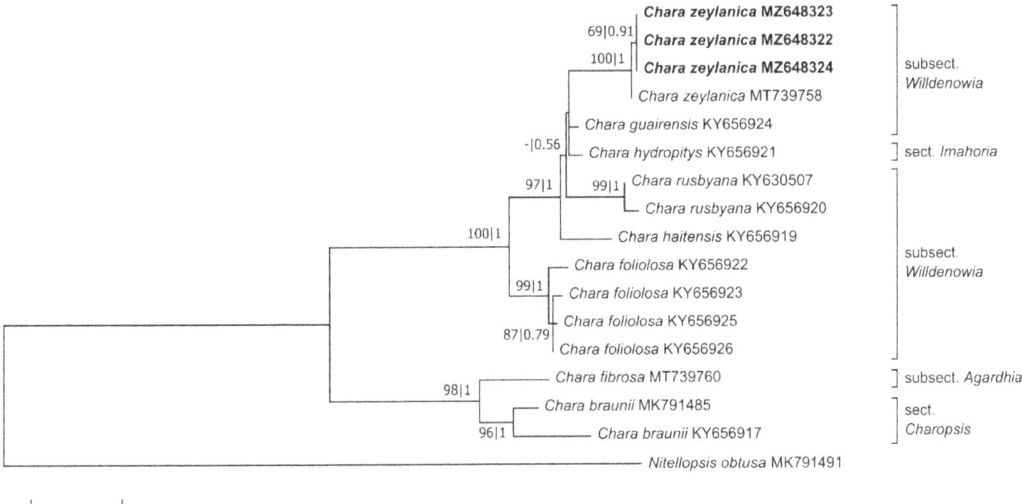

Figure 4. Phylogenetic tree based on 970 base pairs (bp) of the *matK* sequences. The evolutionary history was inferred using the GTR+G+I model. Maximum likelihood bootstrap values (>50%, ML, **left**) and Bayesian posterior probabilities (>0.5, BI, **right**) are given at the branches. The scale indicates sequence divergence in percent. The specimens collected in Sardinia are marked bold. Information about the section and the subsection of the species is given at the tree [24,29,53].

3. Discussion

3.1. Taxonomical Remarks

The specimens collected at Cala Fuili (Sardinia) were shown to qualify, based on their morphological characters, as a taxon belonging to subsect. *Willdenowia* because they are diplostephaneous with triplostichous cortication and have ecorticated basal segments of otherwise corticated branchlets [23]. Following the approach of van Raam [28], who distinguished 20 species within subsect. *Willdenowia*—in contrast to Wood [23], who identified a monospecific subsection—the question of to which species the specimens belong is discussed in detail below.

Van Raam [28] analyzed systematically the problem of gymnopodial (ecorticated first branchlet segment) taxa of the genus *Chara* L. using a stepwise approach. A total of 37 taxa of the genus *Chara* were found to share the character of an ecorticated basal branchlet segment. We recall that a taxon is a taxonomic unit of any rank, which can be species, but also varieties and forms. Eight taxa from subsect. *Willdenowia* can be excluded because they are haplostephaneous (and can thus be assigned to sect. *Imahoria* J. van Raam). Of the remaining 29 diplostephaneous taxa, *C. kenoyeri* M.Howe and *C. rusbyana* can be excluded here because they are dioecious. As a haplostichous species, *Chara pseudohydropitys* Imahori belongs to section *Aghardia* R.D.Wood, and can also be excluded here. Similarly, *C. foliolosa*, *C. tenuifolia* (Allen ex R.D.Wood) R.D.Wood, *C. guairensis*, *C. haitensis*, *C. indica* Bertero ex Spreng., *C. martiana* Wallman, and *C. paucicorticata* Cáceres can be excluded because they have octoscutate antheridia. Unlike the specimens described here, *Chara drouetii* (R.D.Wood) R.D.Wood, *C. michauxii* (A.Braun) Kütz., and *C. formosa* C.B.Rob. are characterised by a sejoined gametangia arrangement. *Chara cubensis* Allen, *C. depauperata* Allen, *C. oerstediana* A.Braun, and *C. diaphana* (Meyen) R.D.Wood have fewer than four corticated branchlet segments, whereas all the specimens found in Sardinia have at least four corticated segments. According to van Raam [28], the remaining 12 taxa belong to *C. zeylanica* as varieties or forms based on quantitative characters, such as spine length relative to axis diameter and the length of the stipulodes.

At this stage, we can conclude that the Sardinian specimens fit the character combination of *C. zeylanica*. Because van Raam [28] failed to provide an adequate description of infraspecific taxa beyond offering a series of tables, the specimens discussed here will not be related to varieties or forms. In any case, the specimens clearly do not belong to *C. foliolosa*, which can be found in the northernmost distribution range of subsect. *Willdenowia* in North America [25].

However, a sound comparison between our specimens and specimens described by other authors [39,40,43,54] is still impossible because of the different taxonomic concepts applied. Taking Wood [23] as a basis, many authors [27,45,55] did not take antheridia morphology into account. Thus, it is extremely difficult to compare their data with the recent concept proposed by van Raam [28].

To obtain independent proof of the morphology-based determination, phylogenetic analyses were performed with the regularly used barcode markers *rbcL* and *matK*. The analyses classified independently the individuals from Sardinia along with other *C. zeylanica*. Morphologically similar species such as *C. foliolosa*, *C. haitensis*, and *C. rusbyana* (previously considered to be varieties or forms of *C. zeylanica sensu* Wood) could be excluded through alignment with GenBank sequences. The phylogenetic analyses showed that *C. hydropitys*, a haplostephaneous species belonging to sect. *Imahoria*, is closely related to the abovementioned *Willdenowia* species, consistent with the findings of previous studies [43,56,57]. The phylogenetic relationships between *C. zeylanica* and *C. hydropitis* were not evident based on *rbcL* sequence data. However, the Sardinian samples were shown to have *rbcL* gene sequences identical to those of a *C. zeylanica* individual from GenBank (HQ380481), which made the categorisation unambiguous. The assignment of the specimens to this taxon was supported by the results of a *matK* analysis, which showed that *C. zeylanica* obtained from GenBank (MT739758) formed a monophyletic clade together with the Sardinian specimens [43,56,57]. Thus, the genetic classification based on the *rbcL* and *matK* sequences clearly supported the morphological determination of the individuals collected at Cala Fuili. The phylogeny of the subsect. *Willdenowia* was not the main focus of this study. Nevertheless, in order to test the phylogeny of *Willdenowia* species in future studies, the taxonomic and geographical basis for an analysis should be broadened, and additional molecular data should be gathered.

3.2. Status and Threats

Many charophyte species and their habitats are threatened throughout Europe, and are mentioned in several national Red Lists [4]. Sardinia has a key role to play in the conservation of Characeae in the Mediterranean region [30,58,59]. Becker [30] identified numerous Sardinian hotspots for the conservation of charophytes, and proposed specific action plans that mainly focused on Characeae in brackish habitats. The Sardinian site where *C. zeylanica* has been found is in the hotspot area between Orosei and Capo Comino.

In contrast to rare and threatened taxa, introduced non-native species can become invasive and cause ecological damage, as the example of *Nitellopsis obtusa* (Desv.) J.Groves in North America shows [60]. However, the examples of two alien charophyte species with mainly intertropical distribution that were previously introduced into Europe have so far not been found to have any serious environmental impacts. Both species, *Chara fibrosa* (including ssp. *benthamii*) and *Chara* c.f. *chrysospora*, were probably introduced by humans into rice fields in Southern France and Northern Italy through the importation of contaminated rice seeds [32,46,47,61]. Moreover, while the presence of a population of *Chara fibrosa* ssp. *benthamii* was recorded on the Greek island of Crete [48], it appears that it has been extinct since 2010 [62].

Chara zeylanica cannot currently be considered an invasive species among the European charophyte flora. For the moment, the Sardinian population is very small, and is limited to a single and relatively isolated location. Although the species has a high rate of fertility in Sardinia, strong dispersal cannot be expected at this stage. Nevertheless, the development of the Sardinian population of *C. zeylanica* should be monitored.

Although the abovementioned intertropical species *Chara fibrosa* and *C.* c.f. *chrysospora* were probably introduced into Europe by anthropogenic factors [47], this is unlikely to be the case for *C. zeylanica*. The Sardinian site is situated more than 100 km away from the nearest rice fields. The surrounding land is used primarily for grazing sheep and small-scale tourism. Other anthropogenic dispersal pathways (e.g., fishery, bathing, or diving) also appear to be unlikely. On the other hand, Sardinia is an important interim stop for birds migrating between Europe and Africa. As the nearest previous records of the presence of *C. zeylanica* are from a Saharan pond in Algeria at least 88 years ago [45 and literature therein], and from Senegal and Egypt [26,27], we assume that the species was introduced into Sardinia by migrating water birds. However, against the backdrop of climate change, future investigations of *C. zeylanica* and other Characeae should consider whether rice fields in Sardinia and throughout the Mediterranean area play a role in the dispersal of the species.

4. Material and Methods

4.1. Hydrochemical and Morphological Analyses

Hydrochemical analyses were conducted in a laboratory according to standard methods and national DIN norms, as published by Wasserchemische Gesellschaft [63–65]. The nutrient concentrations (NH_4-N, NO_3-N, total N, PO_4-P and total P) were measured using a photometer (CADAS 200 by Dr Lange). The cation concentrations (Ca, Mg) were determined by means of an atomic absorption spectrometer (SpectrAA 55 by Varian). The pH values were analyzed using WTW Multi 3510 IDS. The conductivity, salinity, and chloride levels were determined using WTW Cond 3130, with the specific probe being applied in each case.

The morphological analysis was done by means of a stereo microscope (SZX16; Olympus, Tokyo, Japan) equipped with a digital camera for recording photographs.

4.2. DNA Barcoding

The total genomic DNA was extracted using the DNeasy Plant Mini Kit (Qiagen, Hilden, Germany) following the manufacturer's protocol. Partial sequences of the *rbcL* and *matK* genes were amplified using the primers rbcL-1a (5′-TCG TGT AAC TCC ACA ACC TG-3′) and rbcL-1b (5′-TAC TCG GTT AGC TAC AGC TC-3′), and matK-F2 (5′-GAA TGA GCT TAA ACA AGG ATT C-3′) and matK-R1b (5′-GCA GCC TTA TGA ATT GGA TAG C-3′). The PCR tests were performed in a 30 μL reaction volume with a Taq PCR Master Mix (Qiagen, Hilden, Germany) consisting of 2.5 mM $MgCl_2$ (final concentration), and 0.5 pmol of each primer. The PCR products were extracted from agarose gels following the protocol of the Biometra-innuPrep Gel Extraction Kit (Analytik Jena, Jena, Germany), and were sequenced directly using a 3130×L Genetic Analyser (Applied Biosystems, New York, NY, USA) with sequencing primers identical to the primers that were used for the PCR reaction. The quality of the chromatograms of the generated sequences were checked using the BIOEDIT software [66]. The nucleotide sequences identified in this study have been deposited in the GenBank (MZ648319- MZ648324).

Sequences from three specimens collected in Sardinia were submitted to the National Center for Biotechnology Information's (NCBI) Basic Local Alignment Search Tool (BLAST) [67] to allow them to be checked against the nucleotide collection in the GenBank in order to identify other *Chara* sequences with high scoring similarity pairs (HSP) in the NCBI web server. The phylogenetic analysis was performed with the sequence data from the *Chara* specimens collected in Sardinia, and with data on closely related taxa in the GenBank's nucleotide database (https://www.ncbi.nlm.nih.gov/nuccore) for both the *rbcL* and *matK* sequences separately, because the sequences available in the GenBank were completely different. Alignments were created and trimmed using BIOEDIT software [66]. Identical sequences were merged into one entry. Sequences differing only in length were also reduced to one genotype. If different taxa had identical sequences, they were retained in the alignment (Table 1). The *rbcL* dataset contained 36 sequences belonging to nine species of the subsect. *Willdenowia*, and seven of haplostephanous species belonging to the sect. *Charopsis, Protochara* and *Imahoria*, and to the subsect. *Wallmania* and *Agardhia*. In addition, *Nitellopsis obtusa* was used as the outgroup (Table 1). For the *matK* dataset, the three Sardinien samples of *C. zeylanica* were analysed together with 13 sequences belonging to six species of the subsect. *Willdenowia* and *Agardhia*, and one species of the sect. *Charopsis* and *Imahoria*, respectivly. *Nitellopsis obtusa* was used as the outgroup (Table 1). Phylogenetic trees were created using the Maximum likelihood (ML) method and Bayesian inference (BI) analysis. The best-fit model of sequence evolution was determined using MEGA v. X [68]. The ML method was applied using MEGA v. X [68], with the HKY+G+I model used as the nucleotide substitution model for the *rbcL* dataset, and the GTR+G+I model used for the *matK* dataset. Branch supports were evaluated using 1000 bootstrap replicates (BS). MrBayes 3.2.7 [69] was used for the BI method. Two independent runs with four chains were run for 10 million generations using the MCMC method. Calculations of the consensus tree, including clade posterior probability (PP), were performed based on the trees sampled after the chains converged using Tracer 1.7 [70]. The first 25% were discarded as burn-in.

Table 1. Sample list of specimens used for phylogenetic analyses. Indicated are the accession numbers of the haplotypes (=non-redundant genotypes) downloaded from the GenBank. Information about the section and the subsection of the species is given in the third column [24,29,53] and identical sequences are given in the last column. n ǀ a = not applicable.

Marker	Species	Section/Subsection	Strain Designation and/or Collection Information	Accession	Reference	Redundant Accessions
rbcL	*Chara australis*	*Protochara / -*	S002/Unknown	AB440260	[71]	-
	Chara braunii	*Charopsis / -*	S036/Japan, Lake Ashino	AB440259	[71]	-
			S019/Japan, Lake Haryu-numa	AB440258	[71]	KJ395929
			GR10-UW37/Greece, Etoloakarnania	MK791476	[59]	-
	Chara brittonii	*Grovesia / Willdenowia*	KGK2610/USA, Wisconsin	MG880194	[72]	MG880191, MG880186 MG880205, MG880204, MG880203, MG880202, MG880201, MG880200, MG880199, MG880198, MG880197, MG880196, MG880195, MG880193, MG880192, MG880190, MG880189, MG880188, MG880187, MG880185, MG880184, MG880183
			KGK3120/USA, New Jersey	MG880206	[72]	
	Chara corallina	*Protochara / -*	SK026/Japan, Hiroshima	AB359167	[73]	-
	Chara drouetii	*Grovesia / Willdenowia*	Proctor Loc 36/Guatemala, San Luis	HQ380445	[74]	-
			KGK0467/Mexico, Quintana Roo	HQ380444	[74]	-
	Chara fibrosa	*Agardhia / Agardhia*	SK066/ Japan, Hiroshima	AB359168	[73]	AB440261
	Chara foliolosa	*Grovesia / Willdenowia*	Proctor 138/Mexico	MG880213	[72]	HQ380448
			NY 02146579/USA, Lake Erickson	MG880212	[72]	MG880210, HQ380452, HQ380449
			NY 02145914/USA, Clopper Lake	MG880211	[72]	MG880208, MG880207, KY656911, HQ380451, HQ380447, HQ380446
			NY 00739274/USA, Near New Deal	MG880209	[72]	HQ380450
			SJRP31534/Brazil, Neves Paulista	KY656915	[57]	KY656914, KY656907
			SJRP31529/Brazil, São Paulo	KY656912	[57]	-

Table 1. *Cont.*

Marker	Species	Section/Subsection	Strain Designation and/or Collection Information	Accession	Reference	Redundant Accessions
	Chara guairensis	*Grovesia/Willdenowia*	SJRP31523/Brazil, São Paulo	KY656913	[57]	-
	Chara haitensis	*Grovesia/Willdenowia*	SJRP28306/Brazil, Mato Grosso	KY656908	[57]	-
			NY02145934/Michigan, USA	KX431011	[75]	-
			X-930/USA, Everglades	HQ380460	[74]	HQ380459, HQ380458, HQ380457, HQ380456, HQ380455, HQ380454, HQ380453
	Chara hornemannii	*Agardhia/Wallmania*	NY00739162/Peru, Lima	KX431012	[76]	-
	Chara hydropitys	*Imahoria/-*	SJRP28308/Brazil, Mato Grosso do Sul	KY656910	[57]	-
			KGK0774/Puerto Rico, Lago Carite	HQ380464	[74]	HQ380463, HQ380462, HQ380461
	Chara kenoyeri	*Grovesia/Willdenowia*	TP118/Panama, Gatun Lake	HQ380465	[74]	-
	Chara longifolia	*Agardhia/Wallmania*	MB/Canada, Saskatchewan	AY170452	[77]	-
	Chara martiana	*Grovesia/Willdenowia*	Proctor X-952/Venezuela, Caracas	HQ380467	[74]	-
			Proctor TP097/Brazil, São Paulo	HQ380466	[74]	-
	Chara rusbyana	*Grovesia/Willdenowia*	SJRP28307/Brazil, Mato Grosso do Sul	KY656909	[57]	KY630506
			LG/ Unknown	AF097169	[75]	AF097168
	Chara zeylanica	*Grovesia/Willdenowia*	n l a/Taiwan, Gueishan Island	AY720934	unpubl.	-
			n l a/Australia: Elizabeth Creek	DQ076299	unpubl.	-
			S111/New Caledonia	AB440257	[71]	KT343914, KT343913, AB359169, HQ380480, HQ380479, HQ380477, HQ380475, HQ380474, HQ380473, HQ380472, HQ380471, HQ380469, HQ380468
			Proctor X-574/Sri Lanka, Ceylon	HQ380481	[74]	HQ380478, HQ380476, HQ380470
			RB-CZ119A	MZ648319	this study	-
			RB-CZ119B	MZ648320	this study	-
			RB-CT1119-B	MZ648321	this study	-

Table 1. *Cont.*

Marker	Species	Section/Subsection	Strain Designation and/or Collection Information	Accession	Reference	Redundant Accessions
matK	*Chara braunii*	*Charopsis/-*	GR10-UW37/ Greece, Etoloakarnania	MK791485	[59]	-
			48/ Brazil, São Paulo	KY656917	[57]	-
	Chara fibrosa	*Agardhia/Agardhia*	MY-33/Myanmar, Yezin	MT739760	[43]	MT739765, MT739766, MT739768
	Chara foliolosa	*Grovesia/Willdenowia*	SJRP31527/Brazil, Paulicéia	KY656925	[57]	-
			SJRP31929/Brazil, São Paulo	KY656923	[57]	-
			SJRP31534/Brazil, Neves Paulista	KY656926	[57]	-
			SJRP28309/Brazil, Mato Grosso do Sul	KY656922	[57]	-
	Chara guairensis	*Grovesia/Willdenowia*	SJRP31523/Brazil, São Paulo	KY656924	[57]	-
	Chara haitensis	*Grovesia/Willdenowia*	SJRP28306/Brazil, Mato Grosso	KY656919	[57]	-
	Chara hydropitys	*Imahoria/-*	SJRP23308/Brazil, Mato Grosso do Sul	KY656921	[57]	-
	Chara rusbyana	*Grovesia/Willdenowia*	n l a/B-azil. Mato Grosso do Sul	KY630507	[57]	KY656916
			SJRP28307/Brazil. Mato Grosso do Sul	KY656920	[57]	-
	Chara zeylanica	*Grovesia/Willdenowia*	MMYA-1/Myanmar, Inlay Lake	MT739758	[43]	MT739759, MT739761, MT739762, MT739763, MT739764, MT739767
			RB-CZ119A	MZ648322	this study	-
			RB-CZ119B	MZ648323	this study	-
			RB-CT1119-B	MZ648324	this study	-
	Nitellopsis obtusa		GEC4-1/Poland, Lake Lagowskie	MK791491	[59]	-

Author Contributions: Conceptualization, R.B., H.S. and P.N.; methodology, R.B., H.S. and P.N.; formal analysis, R.B., H.S. and P.N.; investigation, R.B.; writing—original draft preparation, R.B., H.S. and P.N.; writing—review and editing, R.B., H.S. and P.N.; visualization, R.B., H.S. and P.N. All authors have read and agreed to the published version of the manuscript.

Funding: Part of the equipment was funded by the European Regional Development Fund (ERDF, UHRO26).

Acknowledgments: We are very greatful to Melanie Willen (University of Oldenburg) for carrying out the hydrochemical analyses. The acquisition of some of the equipment was supported by a grant of the European Regional Development Fund (ERDF, UHRO26). We thank our anonymous reviewers and Angelo Troia, our scientific editor, for their helpful comments and suggestions, which greatly improved the final version of our paper. We thank Miriam Hils-Cosgrove for thorough language editing.

Conflicts of Interest: No potential conflict of interest was reported by the authors.

References

1. Karol, K.G.; McCourt, R.M.; Cimino, M.T.; Delwiche, C.F. The closest living relatives of land plants. *Science* **2001**, *294*, 2351–2353. [CrossRef] [PubMed]
2. Schubert, H.; Blindow, I.; Bueno, N.C.; Casanova, M.T.; Pełechaty, M.; Pukacz, A. Ecology of charophytes—Permanent pioneers and ecosystem engineers. *Perspect. Phycol.* **2018**, *5*, 61–74. [CrossRef]
3. Doege, A.; van de Weyer, K.; Becker, R.; Schubert, H. Chapter 8 Bioindikation mit Characeen [Bioindication using Characeae]. In *Armleuchteralgen. Die Characeen Deutschlands [Stoneworts. Characeae of Germany]*; Arbeitsgruppe Characeen Deutschlands, Ed.; Springer: Berlin/Heidelberg, Germany, 2016; pp. 97–138.
4. Becker, R. Chapter 10 Gefährdung und Schutz von Characeen [Threats and conservation of charophytes]. In *Armleuchteralgen. Die Characeen Deutschlands [Stoneworts. Characeae of Germany]*; Arbeitsgruppe Characeen Deutschlands, Ed.; Springer: Berlin/Heidelberg, Germany, 2016; pp. 149–191.
5. Azzella, M. Italian volcanic lakes. A diversity hotspot and refuge for European charophytes. *J. Limnol.* **2014**, *73*, 502–510. [CrossRef]
6. Blindow, I.; Garniel, A.; Munsterhjelm, R.; Nielsen, R. Conservation and threats. Proposal for a red data book for charophytes in the Baltic Sea. In *Charophytes of the Baltic Sea*; Schubert, H., Blindow, I., Eds.; Gantner, Ruggel: Königstein, Germany, 2003; pp. 251–260.
7. Schwarzer, A. *Aktionsplan Vielästige Glanzleuchteralge (Nitella hyalina (DC.) C. Agardh). [Action Plan for Many-Branched Stonewort (Nitella hyalina (DC.) C. Agardh)]*; Baudirektion Kanton Zürich, Amt für Landschaft und Natur: Zürich, Switzerland, 2019.
8. Blindow, I. Schwedische Artenschutzprogramme für bedrohte Characeen [Swedish conservation programmes for endangered Characeae]. *Rostock. Meeresbiol. Beitr.* **2008**, *19*, 23–28.
9. Stewart, N. *Important Stonewort Areas. An Assessment of the Best Areas for Stoneworts in the United Kingdom (Summary)*; Plantlife International: Salisbury, UK, 2004.
10. Marquardt, R.; Schubert, H. Photosynthetic characterisation of *Chara vulgaris* in bioremediation ponds. *Charophytes* **2009**, *2*, 1–8.
11. Braun, M.; Foissner, I.; Löhring, H.; Schubert, H.; Thiel, G. Characean algae: Still a valid model system to examine fundamental principles in plants. In *Progress in Botany*; Esser, K., Löttge, U., Beyschlag, W., Murata, J., Eds.; Springer: Berlin/Heidelberg, Germany, 2007; pp. 193–220.
12. Selig, U.; Steinhardt, T.; Schubert, H. Interannual variability of submerged vegetation in a brackish coastal lagoon on the southern Baltic Sea. *Ekológia* **2009**, *28*, 412–423. [CrossRef]
13. Jäger, D. Beiträge zur Characeen-Flora Vorarlbergs (Österreich) [Contributions to the Characeae Flora of Vorarlberg (Austria)]. *Ber. Des Nat.-Med. Ver. Innsbr.* **2000**, *87*, 67–85.
14. Burns, N.M.; Rutherford, J.C.; Clayton, J.S. A monitoring and classification system for New Zealand lakes and reservoirs. *Lake Reserv. Manag.* **1999**, *15*, 255–271. [CrossRef]
15. van Raam, J.C. *Handboek Kranswieren*; Chara Boek: Hilversum, The Netherlands, 1998.
16. Gutowski, A.; Hofman, G.; Leukart, P.; Melzer, A.; Mollenhauer, M.; Schmedtje, U.; Schneider, S.; Tremp, H. Trophiekartierung von aufwuchs- und makrophytendominierten Fließgewässern. Trophic monitoring of emergent and macrophyte-dominated rivers and streams. *Inf. Bayer. Landesamtes Wasserwirtsch.* **1998**, *4/98*, 1–501.
17. Melzer, A. Möglichkeiten einer Bioindikation durch submerse Makrophyten—Beispiele aus Bayern [Bioindication opportunities through submerged macrophytes—Examples from Bavaria]. *Beiträge Angew. Gewässerökologie Norddtschl.* **1994**, *1*, 92–102.
18. Kohler, A. Wasserpflanzen als Belastungsindikatoren [Aquatic plants as stress indicators]. *Dechen. Beih.* **1982**, *26*, 31–42.
19. Krause, W. Characeen als Bioindikatoren für den Gewässerzustand [Characeae as bioindicators of water quality]. *Limnol.-Ecol. Manag. Inland Waters* **1981**, *13*, 399–418.
20. Migula, W. *Die Characeen Deutschlands, Oesterreichs und der Schweiz [The Characeae of Germany, Austria and Switzerland]*; Eduard Kummer: Leipzig, Germany, 1897.

21. Deutschlands, A.C. (Ed.) Chapter 9: Die Armleuchteralgen-Gesellschaften Deutschlands. In *Armleuchteralgen. Die Characeen Deutschlands [Stoneworts. Characeae of Germany]*; Springer: Berlin/Heidelberg, Germany, 2016.

22. Korsch, H. The worldwide range of the Charophyte species native to Germany. *Rostock. Meeresbiol. Beitr.* **2018**, 45–96.

23. Wood, R.D.; Imahori, K. *A revision of the Characeae. First Part: Monograph of the Characeae*; Cramer: Weinheim, Germany, 1965.

24. Wood, R.D. New combinations and taxa in the revision of Characeae. *Taxon* **1962**, *11*, 7–25. [CrossRef]

25. Proctor, V.W.; Griffin, D.G.; Hotchkiss, A.T. A Synopsis of the Genus *Chara*, Series Gymnobasalia (Subsection Willdenowia RDW). *Am. J. Bot.* **1971**, *58*, 894. [CrossRef]

26. Corillion, R.; Guerlesquin, M. Notes phytogéographiques sur les Charophycées d'Egypte [Flora geographical notes on the Charophyceae of Egypt]. *Rev. Algol.* **1971**, *10*, 177–191.

27. Guerlesquin, M. Recherches sur *Chara zeylanica* Klein ex Willd. (Charophycées) d'Afrique occidentale [Research on *Chara zeylanica* Klein ex Willd. (*Charophyceae*) from West Africa]. *Rev. Algol.* **1971**, *10*, 231–247.

28. van Raam, J.C. *Annotated Bibliography of the Characeae, 2nd edition. CD.*; Chara Boek: Hilversum, The Netherlands, 2010.

29. van Raam, J.C. A matrix key for the determination of Characeae. *Rostock. Meeresbiol. Beitr.* **2009**, *22*, 53–55.

30. Becker, R. The Characeae (Charales, Charophyceae) of Sardinia (Italy): Habitats, distribution and conservation. *Webbia* **2019**, *74*, 83–101. [CrossRef]

31. Romanov, R.E.; Napolitano, T.; van de Weyer, K.; Troia, A. New records and observations to the Characean flora of Sicily (Italy). *Webbia* **2019**, *74*, 111–119. [CrossRef]

32. Mouronval, J.-B.; Baudouin, S.; Borel, N.; Soulié-Märsche, I.; Klesczewski, M.; Grillas, P. *Guide des Characées de France Méditerranéenne [Guide of Characeae of the French Mediterranean Region]*; Office National de la Chasse et de la Faune Sauvage: Paris, France, 2015.

33. Christia, C.; Tziortzis, I.; Fyttis, G.; Kashta, L.; Papastergiadou, E. A survey of the benthic aquatic flora in transitional water systems of Greece and Cyprus (Mediterranean Sea). *Bot. Mar.* **2011**, *54*, 169–178. [CrossRef]

34. Bazzichelli, G.; Abdelahad, N. *Alghe D'acqua Dolce d'Italia: Flora Analitica Delle Caroficee [Freshwater Algae of Italy: Analytical Flora of Charophytes]*; Ministero dell'Ambiente e Della Tutela del Territorio e del Mare, Sapienza Università di Roma: Roma, Italy, 2009.

35. Cirujano, S.; Cambra, J.; Sánchez-Castillo, P.M.; Meco, A.; Flor-Arnau, N. *Flora Ibérica. Algas Continentales. Carófitos (Characeae)*; Real Jardin Botánico: Madrid, Spain, 2008.

36. Blaženčić, Z.; Stevanović, V.; Blaženčić, J.; Stevanović, B. Red data list of charophytes in the Balkans. *Biodivers. Conserv.* **2006**, *15*, 3445–3457. [CrossRef]

37. Duarte, T.; Almeida, T.S.; Albertoni, E.F.; Palma-Silva, C. Role of the propagule bank in reestablishing submerged macrophytes after removal of free-floating plants for recovery of a shallow lake in Southern Brazil. *Ecol. Austral* **2020**, *30*, 239–250. [CrossRef]

38. Langangen, A. Some finds of Charophytes from East-Africa (Zambia, Tanzania, Kenya and Somalia). *Acta Musei Natl. Pragae Ser. B Hist. Nat.* **2015**, *71*, 229–248.

39. Casanova, M.T. An overview of *Chara* L. in Australia (Characeae, Charophyta). *Aust. Syst. Bot.* **2005**, *18*, 25–39. [CrossRef]

40. Leghari, S.M.; Langangen, A. Fresh water algae of Sindh, VI. Charales (Charophyta) from fresh and brackish water of Sindh, Pakistan. *J. Biol. Sci.* **2001**, *1*, 461–465. [CrossRef]

41. Guerlesquin, M. Contribution a l'étude des Characées des Petites Antilles [Contribution to the research of Charophytes of the Lesser Antilles]. *Rev. Hydrobiol. Trop.* **1983**, *16*, 213–233.

42. Zaneveld, J.S. The Charophyta of Malaysia and adjacent countries. *Blumea* **1940**, *4*, 1–223.

43. Mjelde, M.; Swe, T.; Langangen, A.; Ballot, A. A contribution to the knowledge of charophytes in Myanmar; morphological and genetic identification and ecology notes. *Bot. Lett.* **2021**, *168*, 102–109. [CrossRef]

44. Romanov, R.E.; Barinova, S.S. The charophytes of Israel: Historical and contemporary species richness, distribution and ecology. *Biodivers. Res. Conserv.* **2012**, *25*, 67–74. [CrossRef]

45. Muller, S.D.; Rhazi, L.; Soulie-Märsche, I.; Benslama, M.; Bottollier-Curtet, M.; Daoud-Bouattour, A.; Belair, G.D.; Ghrabi-Gammar, Z.; Grillas, P.; Paradis, L.; et al. Diversity and distribution of Characeae in the Maghreb (Algeria, Morocco, Tunisia). *Cryptogam. Algol.* **2017**, *38*, 201–252. [CrossRef]

46. Abdelahad, N.; Piccoli, F. Report on Charophytes from rice fields in northern Italy including the alien species *Chara fibrosa* ssp. *benthamii*. *Plant Biosyst.-Int. J. Deal. All Asp. Plant Biol.* **2018**, *152*, 599–603. [CrossRef]

47. Soulié-Märsche, I.; Triboit, F.; Despréaux, M.; Rey-Boissezon, A.; Laffont-Schwob, I.; Thiéry, A. Evidence of *Chara fibrosa* Agardh ex Bruzelius, an alien species in South France. *Acta Bot. Gall. Bot. Lett.* **2013**, *160*, 157–163. [CrossRef]

48. Bergmeier, E.; Abrahamczyk, S. Current and historical diversity and new records of wetland plants in Crete, Greece. *Willdenowia* **2008**, *38*, 433–453. [CrossRef]

49. Trbojević, I.; Marković, A.; Blaženčić, J.; Subakov Simić, G.; Nowak, P.; Ballot, A.; Schneider, S. Genetic and morphological variation in *Chara contraria* and a taxon morphologically resembling *Chara connivens*. *Bot. Lett.* **2020**, *38*, 1–14. [CrossRef]

50. Langangen, A.; Ballot, A.; Nowak, P.; Schneider, S.C. Charophytes in warm springs on Svalbard (Spitsbergen): DNA barcoding identifies *Chara aspera* and *Chara canescens* with unusual morphological traits. *Bot. Lett.* **2019**, *41*, 1–8. [CrossRef]

51. Nowak, P.; Schubert, H.; Schaible, R. Molecular evaluation of the validity of the morphological characters of three Swedish *Chara* sections: Chara, Grovesia, and Desvauxia (Charales, Charophyceae). *Aquat. Bot.* **2016**, *134*, 113–119. [CrossRef]

52. Schneider, S.C.; Nowak, P.; Ammon, U.; von Ballot, A. Species differentiation in the genus *Chara* (Charophyceae): Considerable phenotypic plasticity occurs within homogenous genetic groups. *Eur. J. Phycol.* **2016**, *51*, 282–293. [CrossRef]

53. Casanova, M.T.; Karol, K.G. A revision of *Chara* sect. *Protochara*, comb. *et stat. nov.* (Characeae: Charophyceae). *Aust. Syst. Bot.* **2014**, *27*, 23–27. [CrossRef]
54. Langangen, A.; Leghari, S.M. Some charophytes (Charales) from Pakistan. *Stud. Bot. Hung.* **2001**, *32*, 63–85.
55. Casanova, M.T. Taxonomy of Daly River Macroalgae. In *Recommendations for Nutrient Resource Condition Targets for the Daly River*; Schult, J., Townsend, S., Douglas, M., Webster, I., Skinner, S., Casanova, M.T., Eds.; Charles Darwin University: Darwin, UK, 2008.
56. Lee, E.-Y.; Kwang Chul, C.; Sang-Rae, L.; Kim, Y.H.; Lee, J.E.; Jee-Hwan, K. Re-examination of the morphological characteristics and *rbcL* gene sequences of *Chara zeylanica* Willdenow (Charales, Charophyceae) in Korea. *Aquat. Bot.* **2016**, *134*, 97–102. [CrossRef]
57. Borges, F.R.; Necchi, O., Jr. Taxonomy and phylogeny of *Chara* (Charophyceae, Characeae) from Brazil with emphasis on the midwest and southeast regions. *Phytotaxa* **2017**, *302*, 101. [CrossRef]
58. Casanova, M.T.; Becker, R. *Lamprothamnium sardensis* sp. nov.: A new species of *Lamprothamnium* for Europe (Characeae, Streptophyta). *Phytotaxa* **2021**. submitted.
59. Nowak, P.; van de Weyer, K.; Becker, R. The occurrence of sexual *Chara canescens* (Charales, Charophyceae) in Sardinia. *Webbia* **2019**, *74*, 103–109. [CrossRef]
60. Pullmann, G.D.; Crawford, G. A Decade of Starry Stonewort in Michigan. *Lakline Rep. Summer* **2010**, *2010*, 36–42.
61. Langangen, A. *Chara fibrosa* Agardh ex Bruzelius, a charophyte new to the European flora. *Allionia* **2000**, *37*, 249–252.
62. Langangen, A. Charophytes (Charales) from Crete (Greece) collected in 2010. *Fl. Medit.* **2012**, *22*, 25–32. [CrossRef]
63. Wasserchemische Gesellschaft in der GDCH, DIN (Ed.) *DIN EN ISO 7980. Wasserbeschaffenheit [Water Quality]*; Beuth-Verlag: Berlin, Germany, 2000.
64. Wasserchemische Gesellschaft in der GDCH, DIN (Ed.) *DIN EN ISO 38406. Deutsche Einheitsverfahren zur Wasser-, Abwasser- und Schlammuntersuchung, Kationen (Gruppe E) [German Standard Methods for the Examination of Water, Waste Water and Sludge—Cations (Group E)]*; Beuth-Verlag: Berlin, Germany, 1993.
65. Wasserchemische Gesellschaft in der GDCH, DIN (Ed.) *DIN EN ISO 38405. Deutsche Einheitsverfahren zur Wasser-, Abwasser- und Schlammuntersuchung, Anionen (Gruppe D) [German Standard Methods for the Examination of Water, Waste Water and Sludge—Anions (Group D)]*; Beuth-Verlag: Berlin, Germany, 1993.
66. Hall, T. BioEdit: A user-friendly biological sequence alignment editor and analysis program for Windows 95/98/NT. *Nucleic Acids Symp. Ser.* **1999**, *41*, 95–98.
67. Altschul, S. Basic local alignment search tool (BLAST). *J. Mol. Biol.* **1990**, *215*, 403–410. [CrossRef]
68. Kumar, S.; Stecher, G.; Li, M.; Knyaz, C.; Tamura, K. MEGA X: Molecular evolutionary genetics analysis across computing platforms. *Mol. Biol. Evol.* **2018**, *35*, 1547–1549. [CrossRef]
69. Ronquist, F.; Huelsenbeck, J.P. MrBayes 3: Bayesian phylogenetic inference under mixed models. *Bioinformatics* **2003**, *19*, 1572–1574. [CrossRef] [PubMed]
70. Rambaut, A.; Drummond, A.J.; Xie, D.; Baele, G.; Suchard, M.A. Posterior summarisation in Bayesian phylogenetics using Tracer 1.7. *Syst. Biol.* **2018**, *67*, 901. [CrossRef] [PubMed]
71. Sakayama, H.; Kasai, F.; Nozaki, H.; Watanabe, M.M.; Kawachi, M.; Shigyo, M.; Nishihiro, J.; Washitani, I.; Krienitz, L.; Ito, M. Taxonomic reexamination of *Chara globularis* (Charales, Charophyceae) from Japan based on oospore morphology and *rbcL* gene sequences, and the description of *C. leptospora* sp. nov. *J. Phycol.* **2009**, *45*, 917–927. [CrossRef] [PubMed]
72. Karol, K.G.; Alix, M.S.; Scribailo, R.W.; Skawinski, P.M.; Sleith, R.S.; Sardina, J.A.; Hall, J.D. New records of the rare North American endemic *Chara brittonii* (Characeae), with comments on its distribution. *Brittonia* **2018**, *70*, 277–288. [CrossRef]
73. Kato, S.; Sakayama, H.; Sano, S.; Kasai, F.; Watanabe, M.M.; Tanaka, J.; Nozaki, H. Morphological variation and intraspecific phylogeny of the ubiquitous species *Chara braunii* (Charales, Charophyceae) in Japan. *Phycologia* **2008**, *47*, 191–202. [CrossRef]
74. Hall, J.D.; Fučíková, K.; Lo, C.; Lewis, L.A. An assessment of proposed DNA barcodes in freshwater green algae. *Cryptogam. Algol.* **2010**, *31*, 529–555.
75. McCourt, R.M.; Karol, K.G.; Casanova, M.T.; Feist, M. Monophyly of genera and species of Characeae based on *rbcL* sequences, with special reference to Australian and European *Lychnothamnus barbatus* (Characeae: Charophyceae). *Aust. J. Bot.* **1999**, *47*, 361–369. [CrossRef]
76. Pérez, W.; Casanova, M.T.; Hall, J.D.; McCourt, R.M.; Karol, K.G. Phylogenetic congruence of ribosomal operon and plastid gene sequences for the Characeae with an emphasis on *Tolypella* (Characeae, Charophyceae). *Phycologia* **2016**, *56*, 230–237. [CrossRef]
77. Sanders, E.R.; Karol, K.G.; McCourt, R.M. Occurrence of *matK* in a *trnK* groupII intron in charophyte green algae and phylogeny of the Characeae. *Am. J. Bot.* **2003**, *90*, 628–633. [CrossRef] [PubMed]

 *plants*

Article

Genetic Diversity of *Stratiotes aloides* L. (Hydrocharitaceae) Stands across Europe

Barbara Turner [1,*], Steffen Hameister [1], Andreas Hudler [2] and Karl-Georg Bernhardt [1]

[1] Department of Integrative Biology and Biodiversity Research, University of Natural Resources and Life Sciences, Gregor-Mendel-Straße 33, 1180 Vienna, Austria; sthameister@hotmail.com (S.H.); karl-georg.bernhardt@boku.ac.at (K.-G.B.)
[2] Tiroler Umweltanwaltschaft, Meranerstraße 5/III, 6020 Innsbruck, Austria; a.hudler@tiroler-umweltanwaltschaft.gv.at
* Correspondence: barbara.turner@boku.ac.at

Abstract: Intense land use and river regulations have led to the destruction of wetland habitats in the past 150 years. One plant that is affected by the reduction in appropriate habitats is the macrophyte *Stratiotes aloides* which has become rare in several areas. The preservation of genetic diversity within a species is a prerequisite for survival under changing environmental conditions. To evaluate the level of genetic diversity within and among populations of *Stratiotes aloides*, we investigated samples from waterbodies across Europe using AFLP. Low genetic diversity among samples from the same population was found, proving that stands consist of few clones which propagate clonally. Nevertheless, most populations showed differences compared to other populations indicating that there is genetic diversity within the species. The analyzed samples formed two groups in STRUCTURE analyses. The two groups can be further subdivided and mainly follow the major river systems. For conserving the genetic diversity of *Stratiotes aloides*, it would thus be preferable to focus on conserving individuals from many different populations rather than conserving selected populations with a higher number of individuals per population. For reintroductions, samples from the same river system could serve as founder individuals.

Keywords: AFLP; conservation; genetic diversity; river systems; *Stratiotes aloides*; wetland habitats

Citation: Turner, B.; Hameister, S.; Hudler, A.; Bernhardt, K.-G. Genetic Diversity of *Stratiotes aloides* L. (Hydrocharitaceae) Stands across Europe. *Plants* **2021**, *10*, 863. https://doi.org/10.3390/plants 10050863

Academic Editor: Angelo Troia

Received: 30 March 2021
Accepted: 21 April 2021
Published: 25 April 2021

Publisher's Note: MDPI stays neutral with regard to jurisdictional claims in published maps and institutional affiliations.

1. Introduction

The monotypic genus *Stratiotes* includes the sole living species *S. aloides* L. (water soldier) and is a member of the Hydrocharitaceae which belong to the order Alismatales within monocots. During the Tertiary and Quaternary periods, there were up to twenty different species of the genus *Stratiotes* in Europe and Asia [1] (and references therein). The free-floating aquatic macrophyte is perennial with leaves up to 40 cm long and 4 cm wide which are arranged in rosettes. Depending on the season, the plants are emerged or submerged [2]. During the vegetative and reproductive period of a year, the plants are mostly emergent with the rhizoids free in the water or loosely attached to the soil. In autumn, the plant submerges in order to overwinter at the bottom of the water until the following spring [2]. Besides sexual reproduction, the dioecious plants also propagate via vegetative organs (turions and offshoots). Since vegetative reproduction is much more common in *S. aloides*, stands in one waterbody are often formed by clonal individuals of the same sex [2,3]. As long as individuals from different sexes are not transferred from one waterbody to another by floods and high waters, sexual reproduction is very rare. As a typical flood plain species, it inhabits slow-moving or stagnant waters such as ponds, canals, ditches and oxbow waters where it often dominates macrophyte communities [1]. Stands of water soldiers are frequently inhabited by macroarthropod fauna of which several species are of conservation concern [4–6]. *Stratiotes aloides* is distributed from northern Middle Europe in the West to Siberia in the East.

Wetlands are among the most endangered habitats in Central Europe and at the same time, among the hotspots of European biodiversity [7,8]. Back waters are part of natural flood plains. They are independent habitat types with a special flora and fauna. Natural backwaters are caused by the dynamics of the watercourses, which cause seasonal fluctuations of the water level and thus, a temporary regional flood. Today, those dynamics no longer exist in the low- and high-water areas of our modern cultural landscape. Human settlement in floodplains, river straightening, power plant construction and other land uses have led to the systematic destruction of these habitats since the end of the 19th century [9]. Due to anthropogenic influences, there has been an increasing decline, since wetlands have been drained and replaced by grassland [10]. Natural back-waters are endangered by sinking groundwater tables and a lack of flow dynamics [9]. Not only have the habitats themselves been destroyed, but water quality has also decreased, especially due to the increase in nitrogen and nitrates, and has led to a further decrease in the biodiversity of wetland habitats [11,12]. Additionally, wrong management such as clearings of fish-ponds and ditches [3,13] leads to a decrease in water soldier populations. Due to the reduction in appropriate habitat, *S. aloides* has started to decline and is extinct at its southern and western distribution range [14] (and references therein). Apart from the already mentioned threats for wetland habitats and the biodiversity within them, introduced alien species also have to be mentioned as a severe threat to biodiversity in wetland habitats [15,16].

When *Stratiotes* waters are regularly flooded, the drifting away of parts of the population results in a transfer of plants to other areas and thus, to a genetic transfer and exchange between populations. Due to river regulations, flooding in riparian landscapes has decreased significantly. Only through extreme floods might it still be possible for *Stratiotes* to colonize new habitats via water ways [3]. Besides flowing water, vectors such as water birds play an important role in the dispersal of macrophytes (e.g., [17]). Although no study to our knowledge has directly investigated the dispersal of *Stratiotes* by birds, several authors mention the possibility of birds as dispersal vectors for *Stratiotes* [1,2,18]. Especially in regions such as central and eastern Europe, western Europe and secondary ranges in North America, where *Stratiotes* is mainly found in lakes and ponds with no water ways connecting these waterbodies, dispersal by birds seems to be likely. While vegetative parts seem to be too large to be transported by birds, seeds, if present, could possibly be dispersed endo- as well as exo-zoochorically, by birds [18]. However, independent genetic exchange through the transfer of individuals is unlikely in the regulated river areas of Europe. For example, the Austrian water soldier stocks are up to 55 km apart. Due to this geographical isolation, sexual reproduction between populations is no longer possible, because *Stratiotes* needs a pollination distance of less than one kilometer [19]. The maintenance of an evolutionary reproductive community, given through sexual reproduction or through the penetration of daughter individuals into other areas, and thus, the preservation of genetic diversity within and between the water soldier populations, is a prerequisite for the survival of the *Stratiotes* populations in changing environmental conditions [11,20]. Since *Stratiotes* reproduces clonally for the most part and the possibilities of gene flow through sexual reproduction and transfer of individuals are limited, it is assumed that there is a reduction in the genetic diversity of the species [21]. Knowledge about genetic diversity within a species and between populations of a species is necessary for in situ and ex situ conservation and following conservation concepts [22].

Here, in this study, we aim to investigate the genetic diversity of *Stratiotes aloides* populations across Europe to obtain insights into the circumference of the genepool of the species. These results could be helpful to find answers to conservation issues such as status of a population in a particular locality or possible source populations for recolonizations in habitats where *Stratiotes aloides* has already become extinct.

2. Results

After excluding 102 replicates, the final matrix used for analyses contained 345 individuals and 1320 fragments.

Results based on uncorrected p-distances and Hamming distances gave the same clustering patterns in neighbor-joining (NJ) dendrograms and principal coordinate analysis (PCoA). The same was found for the pair of Dice distances and Jaccard distances. Therefore, we used only results based on uncorrected p-distances and Dice distances for further analyses.

2.1. Neighbor Joining

NJ dendrograms based on uncorrected p and Dice distances both showed a star-like shape with a backbone of relative short branches lacking bootstrap support greater than 75% (Figure 1). They differed slightly in clustering patterns, but all of the differing branching patterns did not receive high bootstrap support in either of the two analyses. The groups found in STRUCTURE analyses and in PCoA are partly found in the NJ dendrograms. The two groups "Baltic + Hungary" (BH) and "Central European Highlands and plains 1 + Romania" (CER) based on STRUCTURE analyses (K = 5) are supported with high bootstrap values in the NJ dendrograms (BH: 99.7% in Dice, 89.8% in uncorrected p; CER: 99.9% in Dice and uncorrected p). Here, we present only unrooted trees due to the low resolution of their backbone.

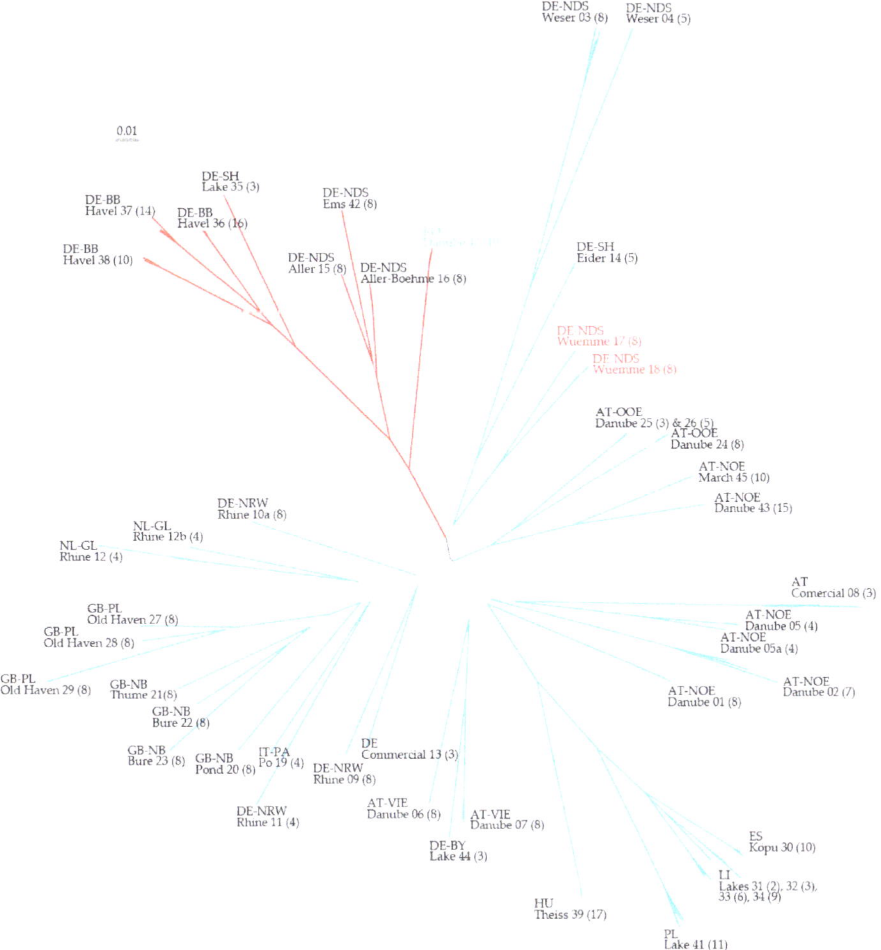

Figure 1. Unrooted NJ dendrogram based on Dice distances. Colors according to STRUCTURE results (K = 2); red—central highlands and plains; blue—rest of the sample regions.

2.2. STRUCTURE

STRUCTURE analysis gave the highest value of ΔK for K = 3 plus a few other suboptimal K values (Figure S1a) in the analysis of the reduced dataset (one or two representative individuals per population). However, the latter contained clusters with negligible membership ("empty" clusters). Visualization of K = 45 based on the STRUCTURE results (reduced dataset) showed six clusters which are subsets of the clusters in K = 3 (Figure S1b). STRUCTURE analysis of the whole dataset gave the highest value of ΔK for K = 2 and a suboptimum for K = 5 (Figure S2). The fastSTRUCTURE results gave a model complexity that maximizes marginal likelihood of 33 (this corresponds to the highest value of ΔK in STRUCTURE). These 33 potential clusters circumscribe mainly the sampling localities/populations with some populations being fused together (Figure S3). However, both NJ and PCoA analyses based on different distance measures are in correlation with clustering based on STRUCTURE rather than those based on fastSTRUCTURE. The main grouping found in STRUCTURE analyses (K = 2) and PCoA separates the samples into two groups and some admixed individuals (Figure 2). Group 1 includes only samples, but not all, from waterbodies within the catchment of the central European highlands and plains (populations 15; 16; 35–38; 42). The populations from Romania (Danube) and rivers Wümme and Eider (central European highlands and plains) appeared admixed. The rest of the samples (British rivers, Rhine, Danube, Baltic and eastern–central, two populations from the central European highlands and plains) forms the second group. A deeper look at the clustering patterns in PCoA and STRUCTURE analyses shows that both main groups can be further subdivided. Within the group of the central European highlands and plains (CE), samples from the Havel lowering in Brandenburg form a cluster together with the individuals from Lake Tolk in Schleswig-Holstein, which forms the core CE group. Samples from rivers Aller and Ems in Lower Saxony (CE-AE) appear to be admixed between the core CE group, Danube and Weser. Individuals from rivers Wümme and Eider appear to be related to populations from the Danube region and Weser. The two populations from the river Weser form an individual group with around 1/3 the impact of the British and Rhine populations. Within the second, much bigger group, samples from British rivers form a group as well as the samples from the Baltic and eastern central rivers together with the population from the Theiss river in Hungary (BH). Samples from Danube waters in Austria form a group with more or less impact from the Rhine, Weser and BH. Individuals from waterbodies along the Rhine river are a mixture between the British and the Danube genepools. The same was found for the population from the Botanical Garden of the University in Padua, which should originally be from the Po river. The only population that cannot be assigned to any of the groups is the population from the Danube estuary in Romania because this population shows impacts from Weser, Baltic, CE and Danube genepools.

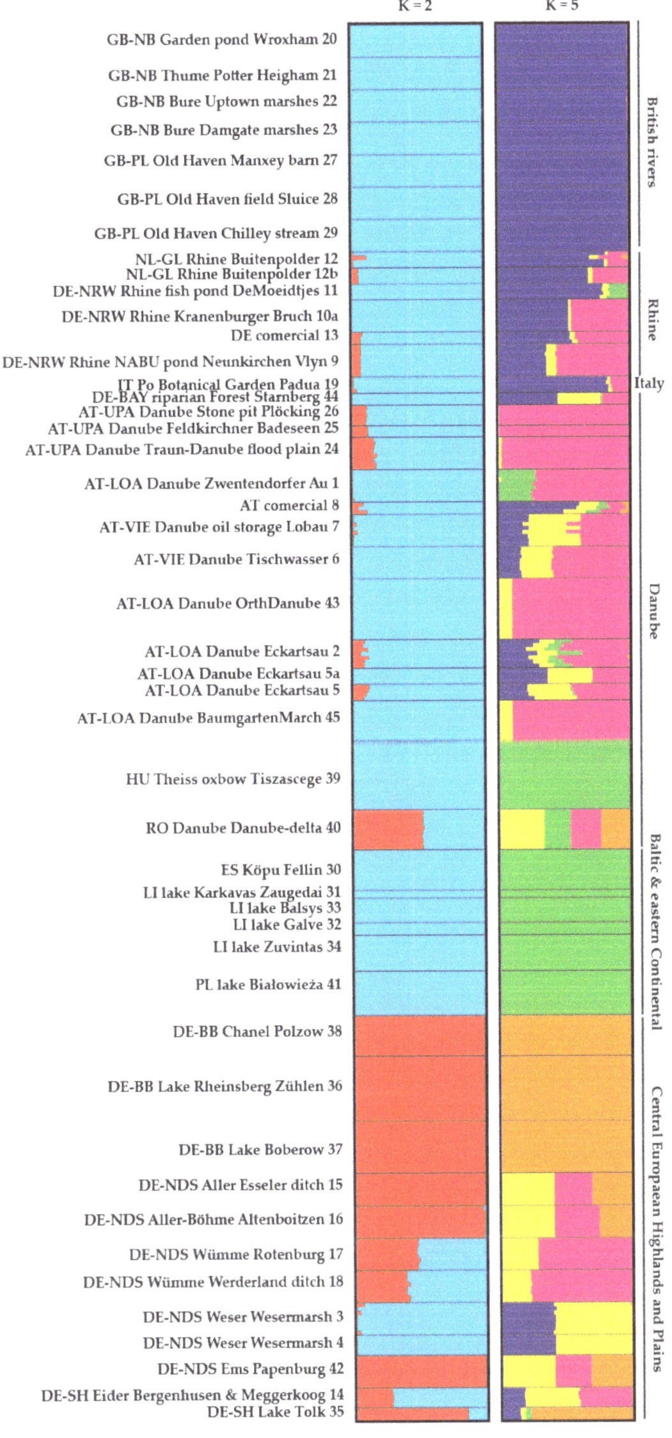

Figure 2. STRUCTURE results (K = 2 and K = 5) of the whole dataset.

2.3. Principal Coordinate Analyses

Principal coordinate analyses based on the two distance methods (uncorrected p and Dice) gave very similar clustering patterns, with uncorrected p-distances showing a higher total sum of coordinates (Table S1). The first coordinate (uncorrected p: 57%; Dice: 49%) separates the two main groups found in STRUCTURE analyses from each other with the admixed samples in between the two groups. The second coordinate (uncorrected p: 22%; Dice: 25%) separates the two main groups into two subgroups each. (Figure 3). The CE group is separated into the core CE group and the Aller-Ems group (CE–AE). The second group is separated into the BH group and a continuum of samples from British waterbodies, Rhine, Danube, Po and Weser. Among the PCoA based on pairwise F_{ST} distances from hierarchical AMOVAs, those based on groupings according to the STRUCTURE results gave the highest values, and also gave the highest values over all PCoA.

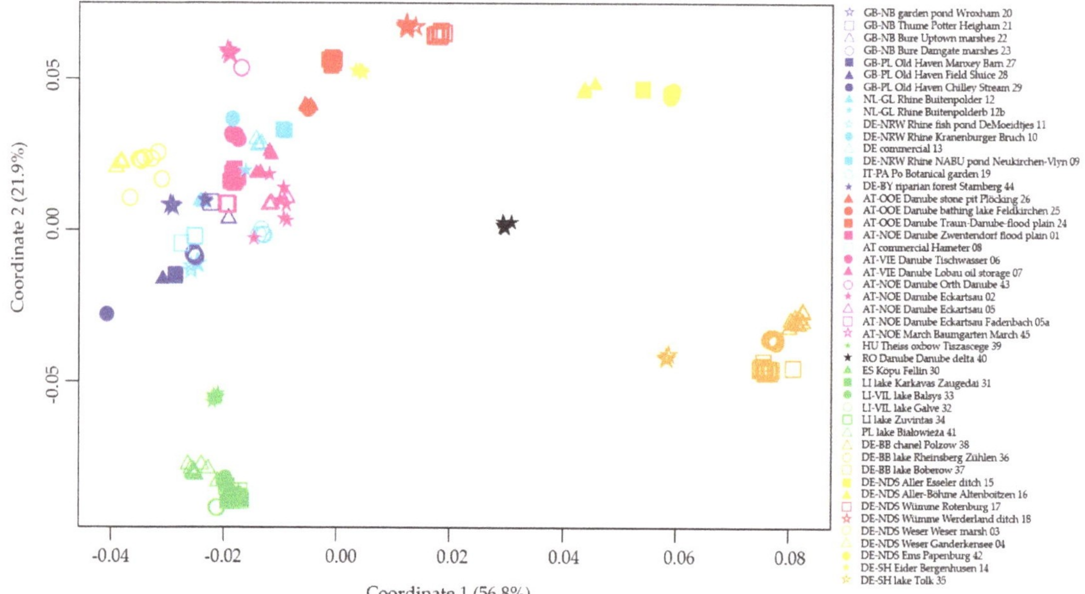

Figure 3. PCoA based on uncorrected p-distances.

2.4. AMOVA and Population Statistics

In order to quantify the amount of genetic variation between populations, we have performed AMOVAs. When keeping all sampling sites as separate populations, the analysis showed 97% of the molecular variance occurred between the populations and F_{ST} value of 0.97 (Table S1). If populations are assigned according to STRUCTURE results (K = 2), the amount of molecular variance between populations drops to 33% and the F_{ST} value to 0.33. To investigate alternative groupings apart from the one based on STRUC-TURE results, we also conducted AMOVAs for groupings based on fastSTRUCTURE results and river systems. Both of these groupings gave higher F_{ST} values than the grouping based on STRUCTURE results (Table S1). Average gene diversity over loci in non-hierarchical AMOVA was 0.214; in hierarchical AMOVA based on populations, average gene diversity varied between 0.047 within the commercial samples from Stauden Hameter (population 8) and 0.000 within samples from Potter Heigham (population 21) and Chilley Stream (population 29).

Nei's H-value (unbiased expected heterozygosity) was estimated with uHe = 0.213 in the overall analysis of all samples together. Analysis of separate populations gave the highest Nei's H-value of H = 0.031 for the commercial samples from Austria (Stauden Hameter population 8) and the lowest value (H = 0.000) for populations from the UK (Potter Heigham, population 21; and Chilley Stream, population 29). Shannon's index was estimated to be I = 0.344 in the overall analysis of all samples together. In the analysis of separate populations, the highest and lowest values were found in the same populations for Nei's H (for details, see Table S2).

2.5. Mantel Tests

Mantel tests based on different distance matrices showed between 2.3 and 100% correlation (R^2) among the tested pairs of matrices (Table S3). The highest correlations were found between matrices based on uncorrected p-values, Dice distances and binary distances calculated with GenAlEx (r = 0.95–1). The lowest correlations were observed between the matrices based on the genetic data and the matrix containing the geographical data (r = 0.15–0.31), indicating that there is no or only little correlation between the geographic distance and genetic distance of the samples in our dataset. The only pair of datasets where a correlation (r = 0.64) between geographic distance and genetic-based AMOVA distances was observed is the pair of geographic distances and AMOVA distances based on grouping according to river systems.

3. Discussion

Here, in this study, we examined the genetic diversity of *Stratiotes aloides* populations from different water systems across Europe. As this species propagates mainly vegetatively, genetic diversity within populations is expected to be low. Due to missing connections (river regulations and lack of flooding) between the water systems, genetic diversity between populations is expected to be high.

Indeed, we did find a high F_{ST} value (0.97) when analyzing the populations separately which shows a high level of genetic differences between the populations and a very low level of genetic differences within the populations, indicating that the populations consist mainly of clones of few genetically different individuals. Considering that the populations propagate vegetatively, and that today, there is no gene flow between the populations via transfer of individuals from one population to another, the relations between the populations could show historical connections between populations. This explains why no, or if only a medium, correlation between geographic distance and genetic distance of the populations is found. The observed correlation between geographic distances and genetic distances based on AMOVA grouped by water bodies has to be viewed with some precaution, as the grouping based on water bodies is, of course, a geography-related grouping and will, therefore, already have a slight bias towards a stronger correlation. Nevertheless, we can still see that there is some correlation between geographic distance and genetic distance when we look at it at the level of waterbodies. The populations from waterbodies of the central European highlands and plains in particular have a genepool which is different from the genepool shared by populations from other regions in Europe. However, it looks like that geneflow between populations has occurred. The fact that water soldier populations from waterbodies along the Rhine seem to be a mixture of genepools from British and Danube genepools might be explained by the historic watercourse of the river Rhine with pervious headwaters of the Danube being directed to the Rhine and a common delta of the Rhine and Thames [23]. A second hypothesis for the connections between populations from British rivers, the Rhine and Italian rivers is long distance dispersal of seeds by migrating water birds (for an example of migration routes of ducks across Europe, see [24]). The connections between the populations from Poland, Baltic countries and the river Theiss in Hungary might also be explained by transfer of plant material by birds [25]. There are not much data available about the dispersal of water soldier fruits by animals, but Efremov et al. [1] and Orsenigo et al. [14] give an overview of the current knowledge of dispersal of *Stratiotes* and Cook and Urmi-König [2] as well as Forbes [18] mention birds as possible dispersals vectors. Summed up, animals can disperse *Stratiotes aloides* fruits exo- and endo-zoochorically and if they are migrating over longer distances, seeds and thus genetic information can be transferred between localities. A further point that has to be kept in mind, when investigating relationships among European water soldier populations, is the fact that *Stratiotes aloides* has a long history as an ornamental plant [1,2]. Unexpected and probably by natural means, difficult to explain relationships between populations could be the result of human-mediated transfer of plant material. As the earliest known fossils of *Stratiotes aloides* date back around 45 million years [26], the observed groups could be the result of repeated glaciation and deglaciation events in Europe [1]. Summing up, we found the investigated populations of *Stratiotes aloides* across Europe to form two main groups which can be further subdivided. Roughly, the two groups can be referred to as a central northern Europe-group (CE) and a western–southern–eastern Europe group.

Previous studies of *Stratiotes aloides* across its distributional range showed a much higher level of genetic diversity within the examined populations [14]. As the sampling regions of the study of Orsenigo et al. [14] are not the same as in our study, the main clustering structures of European populations cannot be fully compared. However, clustering of samples from the Rhine in The Netherlands and the Po in Italy, together with some similarities to populations from the Danube in Bavaria, was observed in both studies. Comparable genetic differences within and between populations of dioecious Hydrocharitaceae were found in *Ottelia acuminata* where high levels of genetic differences between the investigated populations were found, but little diversity within the populations [27].

All still present-day populations of *Stratiotes aloides* found in Europe are remnants of much larger and connected populations. For example, in the Danube flood plains around Vienna, *Stratiotes aloides* was still very common by the mid-19th century, around 100 years later, this species was already mentioned to be rare [28] (and references therein). This example shows that previous large and vital populations became rare and fragmented within the last 150 years.

One possible hypothesis for explaining the differentiation of the samples into two groups could be differences in ploidy level. Orsenigo et al. [14] mention that different ploidy levels (diploid and tetraploid) were observed in *Stratiotes aloides*. Unfortunately, the material available for our study was not appropriate for chromosome counts and genome size measurements.

Summing up the results and viewing them in light of conservation issues, we can conclude that for conserving the genetic diversity of *Stratiotes aloides*, it would be preferable to focus on conserving individuals from many different populations all over its distributional range, rather than conserving selected populations with a higher number of individuals per population. For reintroductions, samples from closely located populations, or at least from populations from the same river system, could serve as founder individuals. As sexual reproduction is rare in natural populations, ex situ collections of samples of both sexes might be established to facilitate sexual reproduction and thus, maintain or even slightly increase genetic diversity in *Stratiotes aloides*. Apart from protecting and conserving *Stratiotes aloides* as a species, protection of the species as a habitat for fauna species such as the dragonfly *Aeshna viridis* [29] and the black tern *Chlidonias niger* [30], which fully or at least mainly depend on *Stratiotes* [14], is important.

4. Materials and Methods

4.1. Material

Material was continuously collected between 2012 and 2018 all over Europe wherever populations of *Stratiotes* were found. Depending on the size of the populations and on the accessibility of the individuals, between 5 and 20 individuals per population were sampled. Wherever possible, individuals from the whole waterbody were collected (e.g., North and South shore, etc.). Short (approx. 5–7 cm) pieces of leaves were collected and immediately dried in silica gel. From several populations, herbarium specimens were collected and deposited in the herbarium of the University of Natural Resources and Life Sciences, Vienna

(WHB). Herbarium accession numbers are indicated in the table of accessions (Table 1). In total, we included 345 individuals from 46 populations into the final analysis. As previous studies [14] showed that there is no detectable genetic difference between the two sexes, we did not pay attention to the sex of the collected individuals (for some populations, information about sex is available and can be requested from the authors).

Table 1. Table of accessions.

Pop nr	River System [1]	Country	Region	Location	Nr. Indivs.	Year	Collector	HBV Acc. Nr	Coordinates
8	commercial	Austria	commercial	Stauden Hameter	3	2012	(Hameister S)		N 48°17′03.03″ E 16°02′19.84″
1	Danube	Austria	Lower Austria	Floodplain Zwentendorf, Obere Placken	8	2012	Bernhardt K-G	56059 57014	N 48°22′14.00″ E 15°47′47.00″
2	Danube	Austria	Lower Austria	Eckartsau Fadenbach	7	2012	Hermann		N 48°08′03.96″ E 16°45′45.03″
5	Danube	Austria	Lower Austria	Eckartsau Fadenbach	4	2012	Bernhardt K-G		N 48°08′10.50″ E 16°46′52.80″
5a	Danube	Austria	Lower Austria	Eckartsau Fadenbach	4	2012	Hameister S		M 48°08′10.50″ E 16°46′52.80″
6	Danube	Austria	Vienna	Tischwasser	8	2012	Hameister S		N 48°11′34.49″ E 16°28′54.84″
7	Danube	Austria	Vienna	Oilstrage Lobau	8	2012	Hameister S		N 48°10′48.55″ E 16°29′47.30″
24	Danube	Austria	Upper Austria	Traun-Danube-floodplain	8	2013	Hameister S Hudler A		N 48°15′16.90″ E 14°23′18.20″
25	Danube	Austria	Upper Austria	Bathing lake Feldkirchen	3	2013	Hameister S Hudler A		N 48°19′41.20″ E 14°03′45.90″
26	Danube	Austria	Upper Austria	Stone-pit Plöcking	5	2013	Hameister S Hudler A		N 48°26′35.00″ E 14°00′14.20″
43	Danube	Austria	Lower Austria	Orth an der Donau/Steinafurt	15	2015	Lapin K		N 48°08′31.90″ E 16°41′03.70″
45	Danube	Austria	Lower Austria	Baumgarten ad March, Maritz South	10	2018	Gregor L		N 48°18′50.00″ E 16°53′12.00″
13	commercial	Germany	commercial	Holzum	1	2013	(Hameister S)		N 51°46′36.13″ E 06°24′12.77″
13	commercial	Germany	commercial	Stauden Förster	2	2013	(Hameister S)		N 52°25′10.68″ E 13°01′11.81″
9	Rhine	Germany	Nordrhein-Westfalen	NABU pond Neukirchen Vlyn	8	2013	Hameister S		N 51°26′48.60″ E 06°32′35.20″
10	Rhine	Germany	Nordrhein-Westfalen	Kranenburger Bruch	8	2013	Hameister S		N 51°47′14.20″ E 06°01′37.50″
11	Rhine	Germany	Nordrhein-Westfalen	Fishpond "De Moeidtjes"	4	2013	Hameister S		N 51°51′04.00″ E 06°10′14.60″
12	Rhine	Netherlands	Gelderland	Buitenpolder (Rhine back water)	8	2013	Hameister S		N 51°54′03.30″ E 06°03′39.90″
12b	Rhine	Netherlands	Gelderland	Buitenpolder (Rhine back water)	8	2013	Hameister S		N 51°54′03.50″ E 06°03′44.60″
14	CHP	Germany	Schleswig-Holstein	Eider-Bergenhusen	1	2013	Rasran L		N 54°22′06.41″ E 09°20′55.39″
14	CHP	Germany	Schleswig-Holstein	Eider-Bergenhusen NABU	2	2013	Rasran L		N 54°22′27.17″ E 09°19′24.49″
14	CHP	Germany	Schleswig-Holstein	Eider-Meggerkoog	2	2013	Rasran L		N 54°21′55.82″ E 09°22′47.45″
35	CHP	Germany	Schleswig-Holstein	Tolk-lake	3	2014	Rasran L		N 54°34′37.37″ E 09°37′37.37″

Table 1. *Cont.*

Pop nr	River System [1]	Country	Region	Location	Nr. Indivs.	Year	Collector	HBV Acc. Nr	Coordinates
3	CHP	Germany	Niedersachsen	Weser marsh Bremen	8	2012	Bernhardt K-G		N 53°08′38.60″ E 08°39′24.60″
4	CHP	Germany	Niedersachsen	Ganderkensee-Werderland	5	2012	Hanke K		N 53°02′03.24″ E 08°32′33.52″
15	CHP	Germany	Niedersachsen	Aller, Esseler ditch	8	2013	Turner F		N 52°42′12.26″ E 09°37′30.91″
16	CHP	Germany	Niedersachsen	Aller (Böhme), Altenboitzen	8	2013	Turner F		N 52°48′49.25″ E 09°32′14.69″
17	CHP	Germany	Niedersachsen	Wümme, Rotenburg	8	2013	Turner F		N 53°05′49.86″ E 09°21′20.31″
18	CHP	Germany	Niedersachsen	Wümme, Werderland ditch	8	2013	Turner F	57455 57456	N 53°08′49.41″ E 08°38′25.49″
36	CHP	Germany	Brandenburg	Havel, Rheinsberg-Zühlen lake	16	2014	Grimm Oldorff S		N 53°03′55.44″ E 12°48′54.84″
37	CHP	Germany	Brandenburg	Havel, Boberow lake	13	2014	Grimm Oldorff S		N 53°10′57.11″ E 13°01′12.76″
38	CHP	Germany	Brandenburg	Havel, Chanel Polzow	10	2014	Grimm Oldorff S		N 53°07′03.17″ E 13°01′07.05″
42	CHP	Germany	Niedersachsen	Ems, Channel Papenburg	8	2015	Tremetsberger K	64641	N 53°04′27.15″ E 07°27′03.61″
44	Danube	Germany	Bavaria	Isar, Riparian forest Starnberg	3	2016	Bernhardt K-G	67455 67456	N 48°01′37.10″ E 11°23′32.60″
19	Italian	Italy		Po, Botanical Garden Padua	4	2013	Bernhard K-G, Hameister S		N 45°23′55.94″ E 11°52′50.69″
20	British	Great Britain	Norfolk	Garden pond, Norfolk Broads Wroxham	8	2013	Leaney B		N 52°42′21.02″ E 01°24′04.56″
21	British	Great Britain	Norfolk	Thume, Norfolk Broads, Potter Heigham	8	2013	Leaney B		N 52°42′14.34″ E 01°34′31.31″
22	British	Great Britain	Norfolk	Bure, Norfolk Broads, Uptown Marshes	8	2013	Leaney B		N 52°39′46.43″ E 01°32′31.98″
23	British	Great Britain	Norfolk	Bure, Norfolk Broads, Damgate Marshes	8	2013	Leaney B		N 52°37′54.15″ E 01°33′34.24″
27	British	Great Britain	East Sussex	Old haven, Pevensey Level, Manxey Barn	8	2013	Birch J		N 50°49′16.21″ E 00°21′3.45″
28	British	Great Britain	East Sussex	Old haven, Pevensey Level, Field Sluice	8	2013	Birch J		N 50°49′16.21″ E 00°21′03.45″
29	British	Great Britain	East Sussex	Old haven, Pevensey Level, Chilley Stream	8	2013	Birch J		N 50°49′16.21″ E 00°21′03.45″
30	BEC	Estonia	Vijandi	Köpu, Fellin	10	2014	Vellak K		N 58°20′09.00″ E 25°20′08.00″
31	BEC	Lithuania	Utena	Karkavas lake, Zaugedai	2	2014	Bernhardt K-G	62094	N 55°06′22.30″ E 25°40′08.78″
32	BEC	Lithuania	Vilnius	Galve lake	3	2014	Bernhardt K-G		N 54°39′00.20″ E 24°55′49.90″
33	BEC	Lithuania	Vilnius	Balsys lake	6	2014	Bernhardt K-G		N 54°47′01.50″ E 25°20′00.90″

Table 1. *Cont.*

Pop nr	River System [1]	Country	Region	Location	Nr. Indivs.	Year	Collector	HBV Acc. Nr	Coordinates
34	BEC	Lithuania	Alytus	Zuvintas lake	9	2014	Bernhardt K-G	62091 62092 62093	N 54°27′26.40″ E 23°38′18.40″
39	Danube	Hungary	BH	Theiss oxbow, Tiszascege.	17	2014	Hameister S Oschatz		N 47°40′45.20″ E 20°59′01.90″
40	Danube	Romania	Tulcea	Danube-delta; E Tulcea. NE Murighiol.	10	2015	Bernhardt K-G	64193	N 45°08′38.10″ E 29°19′30.70″
41	BEC	Poland	Podlachien	Białowieża; Palace Park	11	2015	Wernisch MM	64088	N 52°42′05.32″ E 23°50′42.42″

[1] Grouping of waterbodies into larger European river systems is based on classifications in Trockner et al. [31]; BEC: Baltic and Eastern Central; CHP: Central Highlands and Plains

4.2. DNA Extraction

DNA was extracted from 20 mg silica gel dried leaf material per individual. The material was ground into a fine powder in 2 mL tubes together with three glass beads in a Tissue-Lyser (Qiagen, Germantown, MD, USA) with $20\ \text{s}^{-1}$ for 5 min. Extraction of DNA was performed via QIAcube (Qiagen) using the DNeasy Plant Mini Kit (Qiagen), mainly according to the manufacturer's protocol. Exceptions were elution of DNA from the columns, which was performed with two steps of 50 µL of elution buffer each. RNA was digested after DNA extraction using 1 µg RNAse A and incubated at 37 °C for 30 min.

Quality control of the DNA extracts was performed photometrically using a NanoDrop 2000 spectrometer. To check RNA digestion, samples were loaded on a 0.8% agarose gel.

4.3. AFLP

All DNA extracts that met the quality criteria were adjusted to 100 ng/µL and used for AFLP fingerprinting. Preparation of AFLP samples mainly followed the original protocol [32] with slight modifications.

Restriction of genomic DNA with two restriction enzymes (*Eco*R I and *Mse* I) and ligation of double-stranded adaptors to the resulting restricted fragments were performed in one step in a thermal cycler (37 °C for 2 h followed by a hold at 10 °C). Reactions comprised 1.1 µL 10× T4 DNA ligase buffer (Promega, Madison, WI, USA), 1.1 µL 0.5 M NaCl, 0.55 µL BSA (1 mg/mL; New England BioLabs, Ipswich, MA, USA), 50 µM *Mse* I adaptors (genXpress, Selangor, Malaysia), 5 µM *Eco*R I adaptors (genXpress), 1 U *Mse* I restriction endonuclease (New England BioLabs, Ipswich, MA, USA), 5 U *Eco*R I restriction endonuclease (New England BioLabs), 67 U T4 DNA ligase (Promega), and 5.5 µL DNA (100 ng/µL) and were made up to a total volume of 11 µL with sterile water. Ligated DNA fragments were diluted 10-fold with TE buffer (0.1%). Preselective amplification reactions contained 1 µL 10× polymerase buffer (ThermoFisher Scientific, MA, Waltham, USA), 0.2 U DreamTaq DNA polymerase (ThermoFisher Scientific, Waltham, MA, USA), 0.1 µL dNTPs (0.25 µM; ThermoFisher Scientific), 0.55 µL preselective primer pairs (*Eco*R I -A and *Mse* I -C, each 5 µM; Sigma, St. Louis, MO, USA), 2 µL diluted restriction ligation product, and were brought to a total volume of 10 µL with sterile water. Amplification was carried out with the following profile: 2 min at 72 °C, 20 cycles of 20 s denaturing at 94 °C, 30 s annealing at 56 °C, 2 min extension at 72 °C, and a final extension step for 30 min at 60 °C. The preselective PCR products were diluted 10-fold with sterile water. Reactions for selective amplification contained 1 µL 10× Polymerase buffer (ThermoFisher Scientific), 0.1 U DreamTaq DNA polymerase (ThermoFisher Scientific), 0.1 µL dNTPs (0.25 µM; ThermoFisher Scientific), 0.55 µL *Mse* I-primer (5 µM; Sigma), 0.55 µL *Eco*R I-primer (1 µM; Sigma), and 2 µL diluted preselective amplification product and were brought to a total volume of 10 µL with sterile water. They were carried out in with the following profile: 2 min at 94 °C, 9 cycles of 10 s at 94 °C, 30 s at 65–57 °C (reducing the temperature at 1 °C per cycle), 2 min at 72 °C, 25 cycles of 10 s at 94 °C, 30 s at 56 °C, 2 min at 72 °C and a final extension for 30 min at 60 °C. All PCR steps and incubations were carried out in an Eppendorf Mastercycler Gradient. The selective PCR products were purified using Sephadex G-50 Superfine (Cytiva Life-Sciences, Marlborough, MA, USA) applied to a MultiScreen-HV 96-Well Plate (Millipore) in three steps of 200 μL each ($5\times g$ sephadex in 60 mL $1\times$ TE-buffer) and settled at $750\times g$ (1, 1 and 5 min, respectively). The same speed was used for centrifugation of the samples (selective PCR products: 3.7 μL of NED, 3.15 μL of FAM and 4.3 μL of VIC), again for 5 min. One microliter of the eluate was combined with 15 μL HiDi and 0.25 μL LIZ 600 (Applied Biosystems, ThermoFisher Scientific, Waltham, MA, USA) and denatured for 3 min at 95 °C before running them on a capillary sequencer (GA3500, Applied Biosystems, ThermoFisher Scientific).

The selective primer pairs (FAM-*Eco*RI-ACT/*Mse*I-CTA, VIC-*Eco*RI-ACG/*Mse*I-CTA and NED-*Eco*RI-ACC/*Mse*I-CTA) were chosen after testing seven different primer combinations in a preliminary test. The selected primer combinations generated clear and not too many bands, thus decreasing the risk of fragments co-migrating by chance, but still with sufficient variability to distinguish the samples.

Reproducibility was checked by repeating ca. 23% of the samples.

4.4. Scoring and Phylogenetic Analysis

Sizing and scoring of the data were performed with GeneMarker v2.4.0 (SoftGenetics, State College, PA, USA). After pre-analysis using default settings, sizing profiles of all samples were checked and where necessary, manually corrected. Most of these corrections concerned the 20 bp peak of the size standard. These peaks were often not correctly recognized by the GeneMarker program. High-quality sizing profiles (score > 90) were obtained for all samples. A panel of scorable fragments was established for each primer combination, and fragments between 30 and 600 bp were scored. The relative fluorescent unit (RFU) threshold was set at 40. Automatic scoring was conducted using Local Southern peak call, peak saturation, baseline subtraction, spike removal, pull up correction, and a stutter peak filter of 5% [33]. The results were exported as a presence/absence matrix. The outcome of the automatic scoring was manually checked and corrected for errors. These errors mostly concerned peaks for which shape was atypical. In total, 447 samples corresponding to 345 individuals were scored. From 78 individuals, replicate samples were performed (between two and four replicates per individual). Peak shifts between different analyses dates of the same individuals were used to correct and align all fragment analyses over the whole timespan of the project. These corrections were performed manually and very carefully to avoid artefacts within the dataset. Most of these corrections were small shifts of the majority of peaks by one or two base pairs. For the final analyses, we ended up with 345 individuals, for which high-quality fragment profiles for all three primer combinations could be obtained.

All three primer combinations were combined in a single matrix and analyzed together. Different distance measures were tested for their power to resolve relationships with our dataset. Distance matrixes were calculated in PAUP* v4.0a167 [34] (Nei–Li distance) and SplitsTree v4.15.1 [35] (uncorrected p, Dice, Jaccard and Hamming). Phylogenetic relationships based on previously mentioned distance matrices were reconstructed using SplitsTree to create unrooted NJ dendrograms. To assess the robustness of branches, NJ-bootstrap (NJ-BS) analyses were performed using SplitsTree.

To visualize the pattern of genetic clustering of individuals and populations, we plotted principal coordinate analysis (PCoA) using the R packages "ecodist" [36] and "scatterplot3d" [37] based on an individual uncorrected p matrix, and, respectively, on AMOVA-derived pairwise F_{ST} distances calculated with Arlequin v3.5.2.2 [38].

To investigate further significant groupings of the included individuals, we used the programs STRUCTURE v.2.3.4 [39–42] and fastSTRUCTURE v 1.0 [43]. STRUCTURE was initially run for K = 1–50 with a subset of one or two individuals per population to keep analysis time in a reasonable frame. Based on those results, a second STRUCUTRE analysis with the full dataset (345 individuals) for K = 1–8 was run. We ran STRUCTURE with 10 replicates each and a model based on admixture and independent allelic frequencies, without considering information regarding sampling localities. Each run had 100,000 iterations with 10% additional burn in. The calculation of delta K (ΔK) [44] and preparation of the input file for Clumpp were performed with Harvester [45]. Production of a combined file from the ten replicates of the best K was performed using Clumpp v1.1.2 [46] with the Greedy search algorithm. The graphical representation of STRUCTURE results was prepared with Distruct v1.1 [47]. FastSTRUCTURE was ran with the full dataset for K = 1–50. The calculation of ΔK and graphical representation of results were performed with the functions "chooseK.py" and "distruct.py", both implemented in the fastSTRUCTURE package.

Both non-hierarchical and hierarchical analyses of molecular variance (AMOVA) and calculations of population statistics were conducted using Arlequin v3.5.2.2 [38]. The Excel plugin GenAlEx v6.503 [48] was also used for calculating population statistics, AMOVAs, PCoA and Mantel tests. For hierarchical AMOVAs, groups have been defined based on different possible clustering according to populations (sampling locality), river systems (grouped according to Trockner et al. [31]) and STRUCTURE results. Mantel tests [49] were performed based on distance matrices calculated with SplitsTree, pairwise F_{ST} values from AMOVAs, binary distances calculated with GenAlEx and geographic distances (calculated with Geographic Distance Matrix Generator v 1.2.3; [50]). Calculations of Nei's heterozygosity [51], Shannon's information index [52] and percentage of polymorphic fragments was performed with GenAlEx.

Supplementary Materials: The following are available online at https://www.mdpi.com/article/10.3390/plants10050863/s1, Figure S1a: Delta K values of the reduced dataset (one or two individuals per population), Figure S1b: Visualization of STRUCUTRE results for K = 3 and K = 45 from the reduced dataset, Figure S2: Delta K values of the complete dataset, Figure S3: Visualization of fastSTRUCTURE results for K = 33, Table S1: Results of PCoA and AMOVA, Table S2: Overview of frequency values from population statistics, Table S3: Overview and comparison of Mantel test results. File S1: Data matrix containing AFLP data.

Author Contributions: K.-G.B. and S.H. conceptualized the project. A.H. acquired funding. S.H. and A.H. coordinated sample collection. S.H. coordinated the project including the lab work. B.T. analyzed the data, wrote the manuscript and prepared figures and tables. All authors have read and agreed to the published version of the manuscript.

Funding: This research was funded by Hochschuljubiläumsfonds of the City of Vienna, grant number H-293872/2014 given to Andreas Hudler.

Data Availability Statement: The data presented in this study are available as supplementary material File S1.

Acknowledgments: Special thanks to all collectors for their support with material collection. The authors would also like to acknowledge Agnes Steyrer who did her master's thesis in the framework of this project and conducted together with Karin Fohringer the lab work.

Conflicts of Interest: The authors declare no conflict of interest. The funders had no role in the design of the study; in the collection, analyses, or interpretation of data; in the writing of the manuscript, or in the decision to publish the results.

References

1. Efremov, A.N.; Sviridenko, B.F.; Toma, C.; Mesterházy, A.; Murashko, Y.A. Ecology of *Stratiotes aloides* L. (Hydrochoritaceae) in Eurasia. *Flora* **2019**, *253*, 116–126. [CrossRef]
2. Cook, C.; Urmi-König, K. A revision of the genus *Stratiotes* (Hydrocharitaceae). *Aquat. Bot.* **1983**, *16*, 213–249. [CrossRef]
3. Küry, D. Krebsschere (*Stratiotes aloides*) in Naturschutzweihern der Schweiz. *Bauhinia* **2009**, *21*, 49–56.
4. Rantala, M.J.; Ilmonen, J.; Koskimäki, J.; Suhonen, J.; Tynkkynen, K. The macrophyte, *Stratiotes aloides*, protects larvae of dragonfly *Aeshna viridis* against fish predation. *Aquat. Ecol.* **2004**, *38*, 77–82. [CrossRef]
5. Suutari, E.; Salmela, J.; Paasivirta, L.; Rantala, M.J.; Tynkkynen, K.; Luojumäki, M.; Suhonen, J. Macroarthropod species richness and conservation priorities in *Stratiotes aloides* (L.) lakes. *J. Insect Conserv.* **2009**, *13*, 413–419. [CrossRef]

6. Kalkman, V.J.; Boudot, J.-P.; Bernard, R.; Conze, K.-J.; De Knijf, G.; Dyatlova, E.; Ferreira, S.; Jović, M.; Ott, J.; Riservato, E.; et al. *European Red List of Dragonflies*; Publications Office of the European Union: Luxembourg, 2010.
7. Sommerwerk, N.; Hein, T.; Schneider-Jakoby, M.; Baumgartner, C.; Ostojić, A.; Siber, R.; Bloesch, J.; Paunović, M.; Tockner, K. The Danube river basin. In *Rivers of Europe*; Tockner, K., Uehlinger, U., Robinson, C.T., Eds.; Elsevier: Amsterdam, The Netherlands, 2009; pp. 59–112.
8. Schmidt-Mumm, U.; Janauer, G.A. Macrophyte assemblages in the aquatic-terrestrial transitional zone of oxbow lakes in the Danube floodplain (Austria). *Folia Geobot.* **2016**, *51*, 251–266. [CrossRef]
9. Gerken, B. *Auen, Verborgene Lebensadern der Natur*, 1st ed.; Rombach GmbH + Co Verlagshaus KG: Freiburg im Breisgau, Germany, 1988; p. 131.
10. Lange, G.; Lecher, K. Kleine Gewässer und landwirtschaftliche Vorfluter. In *Gewässerregelung Gewässerpflege*, 3rd ed.; Lange, G., Lecher, K., Eds.; Vieweg + Teubner Verlag: Wiesbaden, Germany, 1993; pp. 244–252.
11. Smolders, A.J.P.; Lamers, L.P.M.; den Hartog, C.; Roelofs, J.G.M. Mechanisms involved in the decline of *Stratiotes aloides* L. in the Netherlands: Sulphate as a key variable. *Hydrobiologia* **2003**, *506–509*, 603–610. [CrossRef]
12. Abeli, T.; Rossi, G.; Smolders, A.J.P.; Orsenigo, S. Nitrogen pollution negatively affects *Stratiotes aloides* in Central-Eastern Europe. Implications for translocation actions. *Aquat. Conserv.* **2014**, *24*, 724–729. [CrossRef]
13. Kunze, K. *Erpobung von Managementmaßnahmen in Bremen zum Erhalt der Krebsschere*; Abschlusstagung, Hochschule Bremen: Bremen, Germany, 2010.
14. Orsenigo, S.; Gentili, R.; Smolders, A.J.P.; Efremov, A.; Rossi, G.; Ardenghi, N.M.G.; Citterio, S.; Abeli, T. Reintroduction of a dioecious aquatic macrophyte (*Stratiotes aloides* L.) regionally extinct in the wild. Interesting answers from genetics. *Aquat. Conserv.* **2017**, *27*, 10–23. [CrossRef]
15. Zedler, J.B.; Kercher, S. Causes and consequences of invasive plants in Wetlands: Opportunities, opportunists, and outcomes. *Crit. Rev. Plant Sci.* **2004**, *23*, 431–452. [CrossRef]
16. Bolpagni, R. Towards global dominance of invasive alien plants in freshwater ecosystems: The dawn of the Exocene? *Hydrobiologia* **2021**. [CrossRef]
17. Kleyheeg, E.; Fiedler, W.; Safi, K.; Waldenström, J.; Wikelski, M.; Liduine van Toor, M. A Comprehensive model for the quantitative estimation of seed dispersal by migratory Mallards. *Front. Ecol. Evol.* **2019**, *7*, 40. [CrossRef]
18. Forbes, R.S. Assessing the status of *Stratiotes aloides* L. (water-soldier) in Co. Fermanagh, Northern Ireland (v.c. H33). *Watsonia* **2000**, *23*, 179–196.
19. Smolders, A.J.P.; den Hartog, C.; Roelofs, J.G.M. Observations on fruiting and seed-set of *Stratiotes aloides* L. in The Netherlands. *Aquat. Bot.* **1995**, *51*, 259–268. [CrossRef]
20. Silvertown, J. The evolutionary maintenance of sexual reproduction: Evidence from the ecological distribution of asexual reproduction in clonal plants. *Int. J. Plant Sci.* **2008**, *169*, 157–168. [CrossRef]
21. Lambertini, C.; Gustafsson, M.H.G.; Frydenberg, J.; Speranza, M.; Brix, H. Genetic diversity patterns in *Phragmites australis* at the population, regional and continental scales. *Aquat. Bot.* **2008**, *88*, 160–170. [CrossRef]
22. Loeschcke, V.; Tomiuk, J.; Jain, S.K. Introductory remarks: Genetics and conservation biology. In *Conservation Genetics*; Loeschcke, V., Tomiuk, J., Eds.; Birkhäuser Verlag: Basel, Switzerland, 1994; EXS; Volume 68, pp. 3–8.
23. Uehlinger, U.; Arndt, H.; Wantzen, K.M.; Leuven, R.S.E.W. The Rhine river basin. In *Rivers of Europe*; Tockner, K., Uehlinger, U., Robinson, C.T., Eds.; Elsevier: Amsterdam, The Netherlands, 2009; pp. 59–112.
24. Tolf, C.; Bengtsson, D.; Rodrigues, D.; Latorre-Margalef, N.; Wille, M.; Figueiredo, M.E.; Jankowska-Hjortaas, M.; Germundsson, A.; Duby, P.-Y.; Lebarbenchon, C.; et al. Birds and viruses at a crossroads—Surveillance of influenza A virus in Portuguese waterfowl. *PLoS ONE* **2012**, *7*, e49002. [CrossRef] [PubMed]
25. Kranichschutz Deutschland. Available online: https://www.kraniche.de/en/crane-migration.html (accessed on 18 April 2021).
26. Bennike, O.; Hoek, W. Late Glacial and early Holocene records of *Stratiotes aloides* L. from northwestern Europe. *Rev. Palaeobot. Palynol.* **1999**, *107*, 259–263. [CrossRef]
27. Guo, J.-L.; Yu, Y.-H.; Zhang, J.-W.; Li, Z.-M.; Zhang, Y.-H.; Volis, S. Conservation strategy for aquatic plants: Endangered *Ottelia acuminata* (Hydrocharitaceae) as a case study. *Biodivers. Conserv.* **2019**, *28*, 1533–1548. [CrossRef]
28. Schratt-Ehrendorfer, L. Geobotanisch-ökologische Untersuchungen zum Indikatorwert von Wasserpflanzen und ihren Gesellschaften in Donaualtwässern bei Wien. *Stapfia* **1999**, *64*, 23–162.
29. Suhonen, J.; Suutari, E.; Kaunisto, K.M.; Krams, I. Patch area of marophyte *Stratiotes aloides* as a critical resource for declining dragonfly *Aeshna viridis*. *J. Insect Conserv.* **2013**, *17*, 393–398. [CrossRef]
30. Beintema, A.J.; van der Winden, J.; Baarspul, T.; de Krijger, J.P.; van Oers, K.; Keller, M. Black terns *Chlidonias niger* and their dietary problems in Dutch wetlands. *Ardea* **2010**, *98*, 365–372. [CrossRef]
31. Trockner, K.; Tonolla, D.; Uehlinger, U.; Siber, R.; Robinson, C.T.; Peter, F.D. Introduction to European rivers. In *Rivers of Europe*; Tockner, K., Uehlinger, U., Robinson, C.T., Eds.; Elsevier: Amsterdam, The Netherlands, 2009; pp. 59–112.
32. Vos, P.; Hogers, R.; Bleeker, M.; Reijans, M.; Van de Lee, T.; Hornes, M.; Frijters, A.; Pot, J.; Peleman, J.; Kuiper, M.; et al. AFLP: A new technique for DNA fingerprinting. *Nucleic Acids Res.* **1995**, *23*, 4407–4414. [CrossRef] [PubMed]
33. Safer, S.; Tremetsberger, K.; Guo, Y.-P.; Kohl, G.; Samuel, M.R.; Stuessy, T.F.; Stuppner, H. Phylogenetic relationships in the genus *Leontopodium* (Asteraceae: Gnaphalieae) based on AFLP data. *Bot. J. Linn. Soc.* **2011**, *165*, 364–377. [CrossRef] [PubMed]

34. Swofford, D.L. *PAUP*: Phylogenetic Analysis Using Parsimony (and Other Methods) Version 4*; Sinauer Associates: Sunderland, MA, USA, 2003.
35. Huson, D.H.; Bryant, D. Application of phylogenetic networks in evolutionary studies. *Mol. Biol. Evol.* **2006**, *23*, 254–267. [CrossRef] [PubMed]
36. Goslee, S.C.; Urban, D.L. The ecodist package for dissimilarity-based analysis of ecological data. *J. Stat. Softw.* **2007**, *22*, 1–19. [CrossRef]
37. Ligges, U.; Mächler, M. Scatterplot3d—An R package for visualizing multivariate data. *J. Stat. Softw.* **2003**, *8*, 1–20. [CrossRef]
38. Excoffier, L.; Lischer, H.E.L. Arlequin suite ver 3.5: A new series of programs to perform population genetics analyses under Linux and Windows. *Mol. Ecol. Resour.* **2010**, *10*, 564–567. [CrossRef]
39. Pritchard, J.K.; Stephens, M.; Donelly, P. Inference of population structure using multilocus genotype data. *Gentetics* **2000**, *155*, 945–959.
40. Falush, D.; Stephens, M.; Pritchard, J.K. Inference of population structure using multilocus genotype data: Linked loci and correlated allele frequencies. *Genetics* **2003**, *164*, 1567–1587. [PubMed]
41. Falush, D.; Stephens, M.; Pritchard, J.K. Inference of population structure using multilocus genotype data: Dominant markers and null alleles. *Mol. Ecol. Notes* **2007**, *7*, 574–578. [CrossRef] [PubMed]
42. Hubisz, M.J.; Falush, D.; Stephens, M.; Pritchard, J.K. Inferring weak population structure with the assistance of sample group information. *Mol. Ecol. Resour.* **2009**, *9*, 1322–1332. [CrossRef]
43. Raj, A.; Stephens, M.; Pritchard, J.K. fastSTRUCTURE: Variational inference of population structure in large SNP data sets. *Genetics* **2014**, *197*, 573–589. [CrossRef]
44. Evanno, G.; Regnaut, S.; Goudet, J. Detecting the number of clusters of individuals using the software STRUCTURE: A simulation study. *Mol. Ecol.* **2005**, *14*, 2611–2620. [CrossRef]
45. Earl, D.A.; Von Holdt, B.M. STRUCTURE HARVESTER: A website and program for visualizing STRUCTURE output and implementing the Evanno method. *Conserv. Genet. Resour.* **2012**, *4*, 359–361. [CrossRef]
46. Jakobsson, M.; Rosenberg, N.A. CLUMPP: A cluster matching and permutation program for dealing with label switching and multimodality in analysis of population structure. *Bioinformatics* **2007**, *23*, 1801–1806. [CrossRef]
47. Rosenberg, N.A. DISTRUCT: A program for the graphical display of population structure. *Mol. Ecol. Notes* **2004**, *4*, 137–138. [CrossRef]
48. Peakall, R.; Smouse, P.E. GenAlEx 6.5: Genetic analysis in Excel. Population genetic software for teaching and research—An update. *Bioinformatics* **2012**, *28*, 2537–2539. [CrossRef] [PubMed]
49. Mantel, N. The detection of disease clustering and a generalized regression approach. *Cancer Res.* **1967**, *27*, 209–220. [PubMed]
50. Ersts, P.J. *Geographic Distance Matrix Generator (Version 1.2.3)*; American Museum of Natural History, Center for Biodiversity and Conservation: New York, NY, USA; Available online: http://biodiversityinformatics.amnh.org/open_source/gdmg (accessed on 12 January 2021).
51. Nei, M. Analysis of gene diversity in subdivided populations. *Proc. Natl. Acad. Sci. USA* **1973**, *70*, 3321–3323. [CrossRef]
52. Lewontin, R.C. The apportionment of human diversity. In *Evolutionary Biology*; Dobzhansky, T., Hecht, M.K., Steere, W.C., Eds.; Springer: New York, NY, USA, 1972; pp. 381–398.

Article

The Contribution of Historical and Morphological Studies on Herbarium Specimens to a Better Definition of *Chara pelosiana* Avetta (Charales, Charophyceae)

Anna Millozza and Nadia Abdelahad *

Dipartimento di Biologia Ambientale, Sapienza Università di Roma, 00185 Roma, Italy;
anna.millozza@uniroma1.it
* Correspondence: nadia.abdelahad@uniroma1.it

Abstract: The lectotype of Chara pelosiana Avetta 1898 was designated in 2000 by Langangen, who merged the species with *Chara fibrosa* Agardh ex Bruzelius. *Chara pelosiana* belongs to the section *Agardhia* Wood, but the true identity of the species has yet to be confirmed. The purpose of this work is to show some historical and morphological findings regarding this enigmatic species, on the basis of the analysis of herbarium specimens. The original material, which was studied by Avetta, is missing in Italian herbaria, but portions of it have been found in the Herbarium of Jena. Historical research on botanists related with this species resulted in the discovery of several specimens to be considered "original material", and new unpublished localities in Northern Italy. Morphological observations have been made on portions of herbarium specimens as a contribution to unveil the taxonomic identity of this taxon. The specimens are diplostichous with ecorticate branchlets, have stipulodes in a single row, one or two per branchlet, and spine cell up to 1 mm long.

Keywords: charophytes; rice fields; morphology; haplostephanous species

Citation: Millozza, A.; Abdelahad, N. The Contribution of Historical and Morphological Studies on Herbarium Specimens to a Better Definition of *Chara pelosiana* Avetta (Charales, Charophyceae). *Plants* **2021**, *10*, 2488. https://doi.org/10.3390/plants10112488

Academic Editor: Angelo Troia

Received: 30 August 2021
Accepted: 13 November 2021
Published: 17 November 2021

Publisher's Note: MDPI stays neutral with regard to jurisdictional claims in published maps and institutional affiliations.

1. Introduction

Chara pelosiana was published in 1898 by Avetta [1] upon a specimen with a single row of stipulodes and ecorticate branchlets that was collected in 1886 in rice fields in S. Anna (near S. Cesario, Province of Modena, Northern Italy). *C. pelosiana* is one of the rarest haplostephanous species in Europe [1] (p. 229).

The species was part of Enrico Ferrari's "small but interesting collection of Characeae" [1] (p. 230), collected for the University of Rome's Botanical Institute, and was first studied by Alpinolo Pelosi, a young Natural Sciences student who died prematurely in 1887. Following Pelosi's death, the Ferrari collection, as well as the few notes and observations left by Pelosi, were gathered by Carlo Avetta and stored in Parma [1] (p. 230). However, both the Rome and Parma Herbaria have since lost track of Ferrari's collection and Pelosi's documentation.

Pelosi identified the specimen as a variety of *Chara scoparia* Bauer (actually *Chara baueri* A. Braun) [1] (p. 234). When Avetta examined it, he noticed that the cortex was diplostichous rather than triplostichous, as it is in *C. baueri*, and assumed he was dealing with a new species [1] (p. 232). As a result, in honor of Pelosi, he named the species *Chara pelosiana*.

We started looking for Ferrari's collection in 2009. A *C. pelosiana* specimen collected from S. Anna was discovered at the University of Turin Herbarium [2].

C. pelosiana has only been mentioned once in the Italian literature since Avetta's publication [3] (p. 16). For nearly all of the twentieth century, there was no further record of the species in Italy. Plants that looked like *C. pelosiana* were discovered in 1999 in rice fields in the Province of Ferrara (Northern Italy) [4]. They were named *C. fibrosa* Agardh ex Bruzelius ssp. *benthamii* (A. Braun) Zaneveld, following Soulié-Märsche et al. [5].

It was Langangen who merged *C. pelosiana* with *C. fibrosa*, choosing fragments of the species collected in S. Anna and housed in the Herbarium of Oslo to be the lectotype of

C. pelosiana [6]. In support of the merging of the two species, he cited Nordstedt, who identified the fragment kept at Olso as *C. gymnopitys* A. Braun or *C. flaccida* A. Braun, depending on the colour of the oospores. These two species (as well as a third one, *C. benthamii* A. Braun) were fused into *C. fibrosa* by Zaneveld [7] (p. 153).

Van Raam [8] named Avetta's species *C. fibrosa* var. *pelosiana* (*nom. invalid.* according to [9]), and Krause [10], in agreement with Langangen, attributed *C. pelosiana* to *C. fibrosa* in a note at the end of his book. However, Wood [11,12] considered the species to be a form of *C. baueri*.

C. fibrosa is a species complex, which includes, in addition to the three species of Zaneveld mentioned above, other species, varieties, and forms merged by Wood [13–15].

The purpose of this work is to document historical and morphological findings on *C. pelosiana* herbarium specimens.

2. Materials and Methods

The historical search for *C. pelosiana* was based on an examination of the limited available literature [1–3] and, more importantly, a study of the herbarium materials kept in JE, LD, MOD, PAD, PARMA, PAV, RO, and TO (herbarium acronyms according to [16]). All of these Herbaria are related to botanists who have investigated or collected this species (see Appendix A). Further requests were sent without success to the Italian Herbaria BOLO, CAT, FI, NAP, PAL, and PI, which preserve historical collections of algae.

Furthermore, manuscript documents found attached to the specimens, as well as a selection of letters kept in the Archive of Botanical Garden of the University of Padua [17] provided significant additional information.

For the morphological investigation, portions of *C. pelosiana* from the Herbaria of MOD, TO, and PAV, as well as from the specimens kept in PAD Herbarium, were taken and transferred to Rome for examination and photography. The fragments stored in JE were insufficient for portions to be removed for further study. Morphological observations were made using a Zeiss stereomicroscope equipped with a Leica DFC 42 digital camera. The material was photographed either dry or after being rehydrated and decalcified using a 1N hydrochloric acid solution.

3. Results

3.1. Historical Findings

Unfortunately, neither RO nor PARMA, where Carlo Avetta worked from 1893 until his retirement, kept the original collection. Nevertheless, *C. pelosiana* specimens from the original site and other Italian locations have been discovered in several Italian and foreign Herbaria.

Based on the importance of the exsiccata, the *C. pelosiana* specimens were divided into three groups. The data labels for each specimen were faithfully returned and noted.

3.1.1. Herbaria That Keep Original Material of the Name *Chara pelosiana*

Jena Herbarium (JE): Small fragments of the species established by Avetta are kept in an envelope with the stamp "Herbarium Walter Migula Eisenach" in the top right corner. On the envelope, Avetta wrote "<u>Chara</u> <u>Pelosiana</u> Avetta" (Figure 1A,B).

A postcard from Avetta to Migula, dated November 9, 1898, is attached to the sample, and reports, in French, "Mr. le Prof. Migula | Parma 9-11-98 | Je vous envoye un tout petit échantillon d'une <u>Chara</u> italienne que je viens de décrire comme espèce nouvelle (Malpighia dernière livraison) et au sujet de laquelle je voudrais bien connaitre votre opinion, quelconque elle soit (…) Dr. C. Avetta | Jardin botanique—Parma" [(…) I am sending you a very small sample of an Italian *Chara* that I just described as a new species (Malpighia last issue) and would like to hear your opinion on it, whatever it is (…)] (Figure 1C).

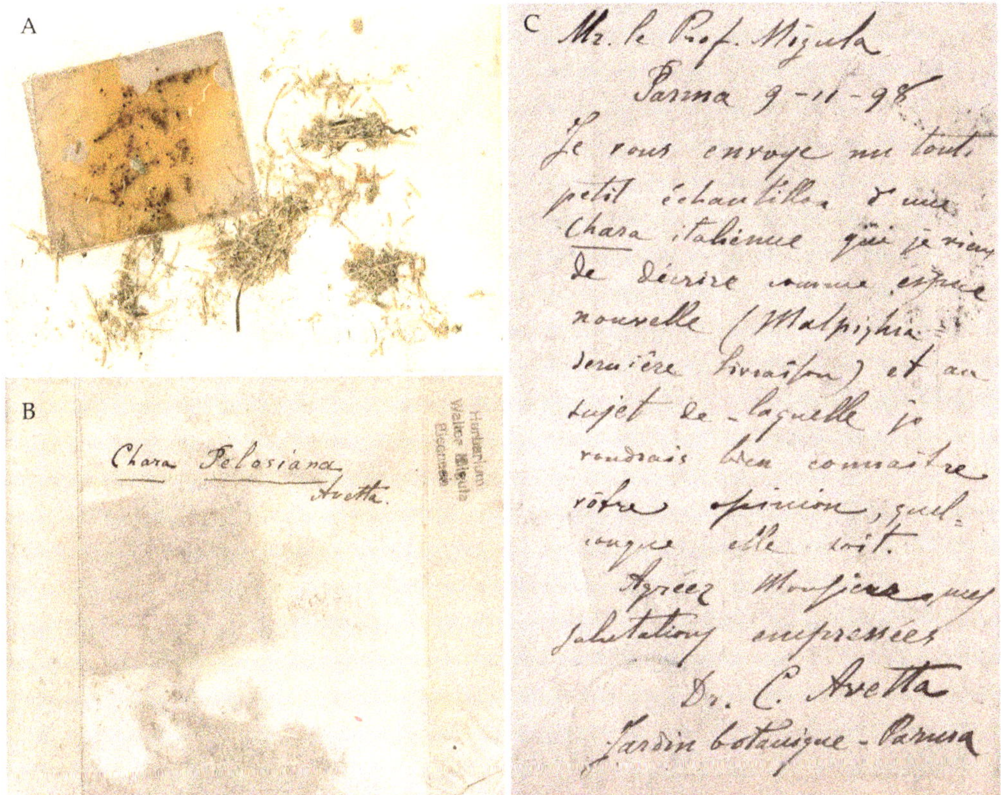

Figure 1. (**A**) JE fragments of *Chara pelosiana* from the missing RO specimen. (**B**) Avetta's handwriting on the specimen envelope. (**C**) Postcard sent by Avetta to Migula on 9 November 1898.

Notes. The fragments are extremely significant because they come from the original material examined by Avetta, although the labels do not mention any locality. According to a footnote in Avetta's paper, they were sent to Migula to obtain his opinion on the validity of the new species [1] (p. 229). In the same footnote, Avetta reports that the new taxon was published without confirmation from Migula, who was away for a few months and could not examine the specimen. As a result, the publication does not provide an illustration of the new species. Avetta goes on to say that the figures will appear in his next note on the Italian Characeae, which, however, was never published.

Modena Herbarium (MOD): The specimen consists of six small fragments pinned to a herbarium sheet (Figure 2A). A preprint label from the Herbarium of Modena with the institution's rubber stamps, "Hortus Reg. Botanicus Mutinensis", provides the following information: "Chara Scoparia Bauer, a Baueri | Valli e risaie di S. Anna presso S. Cesario | 1 8bre [October] 1886" [unknown person *scripsit*] Det. R. Istituto Bot.° [Botanico] di Roma" [a second, unknown person *scripsit*] (Figure 2B).

The preprint revision label was written by Leone Formiggini, who was engaged in revisionary work on the Italian Characeae with Augusto Béguinot (see Appendix A). The label is free within the folder, with the name of the research project, *Characeae Italicae*, at the top and the institution where the project was located, *Patavii, ex R. Instituto botanico*, at the bottom. The revision label bears the information "Chara Pelosiana Avetta *revisit* Dr Leone Formiggini Giugno [June] 1907" [Formiggini *scripsit*]. Béguinot added the comment,

"an potius Lycnothamnus species? et tunc <u>Lycnothamnus</u> <u>Pelosianus</u> Bég. et Form.!"
(Figure 2B).

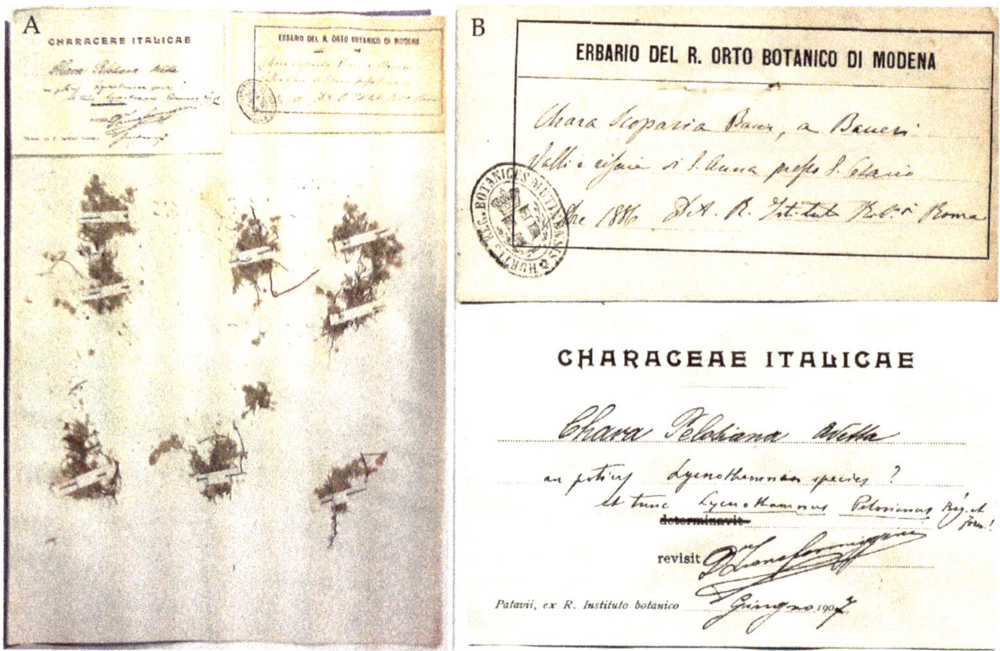

Figure 2. (**A**) MOD specimens of *Chara pelosiana* from the first duplicate of Ferrari's original collection. (**B**) MOD preprint label with unknown handwriting (above) and Formiggini and Béguinot's preprint revision label (below).

Notes. The first duplicate of Ferrari's original set collected in rice fields in the province of Modena is kept in MOD. When the Botanical Institute of Rome requested a collection of Characeae from this area, Ferrari was still working at the University of Modena (see Appendix A). Unfortunately, no Ferrari autograph labels can be seen in MOD.

The Ferrari Characeae collection consists of 26 specimens of different species collected near Modena between 1878 and 1886, 14 of which were collected in 1886 (Table 1). All specimens have Modena Herbarium preprint labels with the stamp "Hortus Reg. Botanicus Mutinensis". The collection is not numbered, there is no indication of *Legit*, and all of the labels were handwritten by two unidentified people. The homogeneity of the compilation becomes apparent when comparing the first handwriting, which included the binomial, locality, and date, as if the labels were filled in all at once by an amanuensis, rather than by a botanist. The second anonymous handwriting only provided information about who made identification, in this case, an Institute, the Botanical Institute of Rome.

Finally, it should be noted that Formiggini and Béguinot disagreed regarding the correct position of *C. pelosiana*, which, according to Béguinot, could be *Lychnothamus pelosianus* (see also below, the letter from Formiggini to Migula kept at JE, and the discussion).

Table 1. Duplicate of Ferrari's Characeae collection from the Modena Herbarium (MOD). The specimens, collected in the Modena area in 1886, are listed in chronological order. For the abbreviations of names and authors, we reproduced the original labels. The *Chara pelosiana* specimen revised by Formiggini is marked in bold.

Date of Collection	Collection Locality	Determination According to the Botanical Institute of Rome	Revision of Leone Formiggini	Formiggini's Revision Date
21 May	Castelvetro	*Chara foetida* A B. *longibracteata* A B *Laxior* A B	*Chara foetida* A. Br. f. *subinermes* β *longibracteata* A. Br.	May 1907
June	Rio di Valle Urbana	*Chara foetida* A B. *subinermis longibracteata*	*Chara foetida* A. Br. f. *subinermis* β *longibracteata* A. Br.	May 1907
21 September	Marshes at Villa S. Faustino	*Chara hispida* L. β *brachyphylla* A. B? (sic)	*Chara hispida* L. f. *macracantha* v. (sic)	May 1907
23 September	Nonantola, in the rice fields Sacerdoti	*Chara foetida* ABr. *subinermis* ABr. *longibracteata* ABr.	*Chara foetida* A. Br. f. *subinermis* ιι *typica* Mig. mixed with some fragments of *C. fragilis* Desv.	May 1907
23 September	Nonantola, in the rice fields Borsari	*Lychnothamnus stelliger* (Bauer) A. Br. var. *major* A. Br.	*Tolypellopsis stelligera* (Bauer) Migula v. *ulvoides* A. Br.	May 1907
23 September	Nonantola, in the rice fields Borsari	*Chara hispida* L *macrantha* A Br. *elongata* A Br.	*Chara hispida* L. f. *macracantha* α *typica*	May 1907
23 September	Nonantola, in the rice fields Sacerdoti along the ditch of the forest	*Lychnothamnus stelliger* (Bauer) A Br.	*Tolypellopsis stelligera* (Bauer) Migula v. *ulvoides* A. Br.	Apr 1907
26 September	Ditch above S. Marino near Carpi	*Chara hispida* L. sterile	*Chara hispida* L.	May 1907
26 September	Ditch above S. Marino near Carpi	*Chara foetida* ABr. *subinermis* ABr. *longibracteata* A.Br. *laxior* ABr.	*Chara foetida* A Br. f. *subinermis* β *longibracteata* ABr	May 1907
28 September	Rice fields Boretti at Villa S. Agnese	*Chara foetida* AB. *subinermis* ABr. *longibracteata* A.B. *laxior* ABr.	*Chara foetida* A Br. f. *subinermis* κ *clausa* A Br.	May 1907
28 September	Rice fields Boretti at Villa S. Agnese	*Chara foetida* AB. *brevibracteata* AB. *expansa* AB	*Chara foetida* A. Br. f. *paragymnophylla* δ *brevibracteata* Mig.	May 1907
1 October	Ditches between Castelfranco Emilia and Valli di St Anna	*Chara foetida* A. B. *subinermis—longibracteata laxior* AB.	*Chara foetida* ABr. f. *subinermis* β *longibracteata* ABr.	May 1907
1 October	Rice fields of S. Anna near S. Cesario	*Chara hispida* L *micrantha* AB. *microphylla* AB.	*Chara hispida* L. f. *micrantha* π *brachyphylla*	May 1907
In the valleys and rice fields of S. Anna near S. Cesario		***Chara Scoparia* Bauer, a *Baueri***	***Chara pelosiana* Avetta ***	**June 1907**

* an potius *Lycnothamnus* species? et tunc <u>*Lycnothamnus Pelosianus*</u> Bég. et Form.! [Beguinot's revision on the same revision label, see Figure 2B].

Turin Herbarium (TO): The specimen, which is nearly entirely fragmented, was found free within a folder with three labels. The first two labels, which are pinned to the herbarium sheet and almost joined by a third pin to form a single label, were handwritten by Ferrari (Figure 3. The first provides information about the specimen: "N° 3 | Chara | Nelle valli e risaie di St Anna presso S. Cesario. | Prov di Modena | 1 Ottobre 1886. Leg: E Ferrari". The second reports the new binomial and references Avetta's publication on a printed label from the Herbarium of Turin: "N° 3 Chara Pelosiana Avetta | Ved. Malp.

anno XII pag: 231. anno 1898". The third label, which was found free within the folder, is Formiggini's preprint revision label: "Chara Pelosiana Avetta, *revisit* D[r] Leone Formiggini, Xmbre [December] 1908".

Figure 3. TO specimen of *Chara pelosiana* (No. 3) from the second duplicate of Ferrari's original collection. Ferrari's two handwritten labels (on the right) and Formiggini's revision label (on the left).

Notes. A second duplicate of the original set collected by Ferrari from rice fields in the province of Modena in 1886 is housed in TO, where Ferrari became the Curator of the Herbarium in November 1887 (see Appendix A). There are 25 Characeae specimens in the collection, 24 of which are numbered (Table 2). TO contains 19 additional Characeae specimens that were collected by Ferrari between 1887 and 1905, mainly from Piedmont and Valle d'Aosta.

Table 2. Duplicate of Ferrari's Characeae collection from the Turin Herbarium (TO). The specimens, collected in the Modena area in 1886, are listed in chronological order. For the abbreviations of names and authors, we reproduced the original labels. The two specimens of *Chara pelosiana* revised by Formiggini are marked in bold.

Date of Collection	Collection Locality	Collecting Number	Determination of Ferrari	Determination/Revision/Confirmation of Formiggini	Formiggini's Revision Date
22 April	In the ditches around Carpi	12	*Nitella capitata* Nees	*Nitella capitata* (N.ab. Es.) Ag.	December 1908
May	Villa Albareto ditches, site called "i Tagliati"	21	*Chara*	*Chara foetida* f. *subinermis* a *normalis* Mig.	January 1909
21 May	"Bosco Bontempelli" in ponds of water (Colli di Castelvetro)	24	*Chara foetida* A. Br sub. var. *longibracteata* A. B β *laxior* A. Br	*Vidit = Confirmavit*	-
June	Sassuolo: along Rio di Valle Urbana	23	*Chara foetida* A. Br a *subinermis. longibracteata* A. Br.	*Chara foetida* A.Br. *f. subinermis* β *longibracteata* A. Br.	June 1910
20 September	Nonantola, in the rice fields Borsari	20	*Chara hispida* L. C. *micrantha* A. Br C. *elongata* A. Br.	-	-
21 September	Marshes at S. Faustino	13	*Chara hispida* a. *brachyphylla* A. Br	*Chara hispida* L. f. *micracantha—brachyphylla* A.Br.	December 1908
21 September	Marshes at Villa S. Faustino	14	*Chara foetida* A. Br β *longibracteata* A. Br	*Vidit = Confirmavit*	-
21 September	Marshes at S. Faustino	15	*Chara foetida* A. Br	*Vidit = Confirmavit*	December 1908
23 September	**Nonantola, in the rice fields Sacerdoti**	-	***Chara***	***Chara pelosiana* Avetta**	**January 1909**
23 September	Nonantola, in the rice fields Sacerdoti	16	*Chara foetida* A. Br a *subinermis* β. *longibracteata* A. Br.	*Vidit = Confirmavit*	-
23 September	Nonantola, in the rice fields Borsari	17	*Lychnothamnus stelliger* A. Br. var. *major* A. Br.	*Tolypellopsis obtusa* (Desv.) Bèg. et Formigg. var. *ulvoides* (Bert.) Bèg. et Formigg.	December 1908
23 September	Nonantola: in the rice fields Sacerdoti along the ditch of the forest	18	*Chara*	*Chara fragilis* Desv. f. *microptila* β *Hedwigii* Ag.	January 1909
23 September	Nonantola, in the rice fields Sacerdoti	19	*Lychnothamnus stellinger* (sic) A. Br. var. *major* A. Br.	*Tolypellopsis obtusa* (Desv.) Bèg. et Formig. var. *ulvoides* (Bert.) Bèg. et Formigg.	December 1908
23 September	Nonantola: in the rice fields Sacerdoti along the ditch of the forest	22	*Chara*	*Chara foetida* A. Br. f. *subinermis* A. Br. *typica* Mig.	January 1909
26 September	Ditches above S. Marino near Carpi	10	*Chara hispida* L.	*Chara hispida* L. f. *microcantha* λ *vulgaris*	December 1908

Table 2. *Cont.*

Date of Collection	Collection Locality	Collecting Number	Determination of Ferrari	Determination/Revision/ Confirmation of Formiggini	Formiggini's Revision Date
26 September	Ditches above S. Marino near Carpi	11	*Chara foetida* A. Br β *longibracteata* A. Br. β *laxa*	*Vidit = Confirmavit*	-
28 September	Rice field fondo Borretta (sic) at Villa S. Agnese	7	*Chara foetida* A. Br β *longibracteata* A. Br β *laxa*	*Vidit = Confirmavit*	-
28 September	Rice fields fondo Borretti at Villa St Agnese	8	*Chara foetida* A. Br β: *longibracteata* A. Br. *expans.* A. Br	*Vidit = Confirmavit*	-
28 September	Rice fields fondo Borretti Villa S. Agnese	9	*Nitella tenuissima* Desv.	*Nitella tenuissima* (Desv.) Coss. et Germ. f. *major* Mig.	December 1908
1 October	Ditches between Castelfranco Emilia and Valli di St Anna	1	*Chara foetida* A. B. | *C. subinermis* A. Br. | *C. longibracteata* A. Br. | *C. laxior* A.Br	*Vidit = Confirmavit*	-
1 October	In the valleys and rice fields of S. Anna at San Cesario	2	*Chara*	*Chara hispida* L. f. *micrantha* π *brachyphylla*	January 1909
1 October	**In the valleys and rice fields of S. Anna near San Cesario**	**3**	***Chara*** | ***Chara pelosiana*** **Avetta Ved. Malp. anno XII. pag: 231. anno 1898.**	***Chara pelosiana*** **Avetta**	**December 1908**
1 October	In the valleys and rice fields of S. Anna near San Cesario	4	*Chara hispida* Thuil var. *microphylla* Schumach	-	-
1 October	In the valleys and rice fields of S. Anna near San Cesario	5	*Chara*	*Chara ceratophylla* Wallr.	January 1909
1 October	In the valleys and rice fields of S. Anna near San Cesario	6	*Chara hispida* L. var *microphylla* Schumach.	-	-

The *C. pelosiana* specimen kept in TO is particularly valuable because the labels were handwritten by Ferrari, the original collector, and refer to both specimen No. 3 and the type locality, S. Anna [1] (p. 234).

TO also preserves two further specimens collected by Ferrari at S. Anna on September 19, 1899 (identified by Formiggini as *C. foetida* A. Br. and *C. fragilis* Desv. f. *subinermis* β *Hedwigii* Ag.), one year after Avetta's publication and thirteen years after the first collection, suggesting that Ferrari tried unsuccessfully to again find *C. pelosiana*.

Oslo Herbarium (O): The lectotype of *C. pelosiana* was designated by Langangen and is kept in the Oslo Herbarium [6]. The specimen, which is kept in an envelope, was identified by Otto Nordstedt. The original label reports, "Chara Pelosiana Avetta. | Valli e risaie di S. Anna presso S. Cesario Prov. di | Modena 18 1/10 86 [October,1 1886] | Leg. E. Ferrari. | Italien" [unknown person *scripsit*].

On the same label, Nordstedt made the following observation: "Si nucleus sporangii niger sit, | = Ch. gymnopitys Al. Braun, | Si nucleus sporangii luteo rufus sit, | = Ch. flaccida Al. Braun | Determ. O. Nordstedt" (lectotype) (Figure 4).

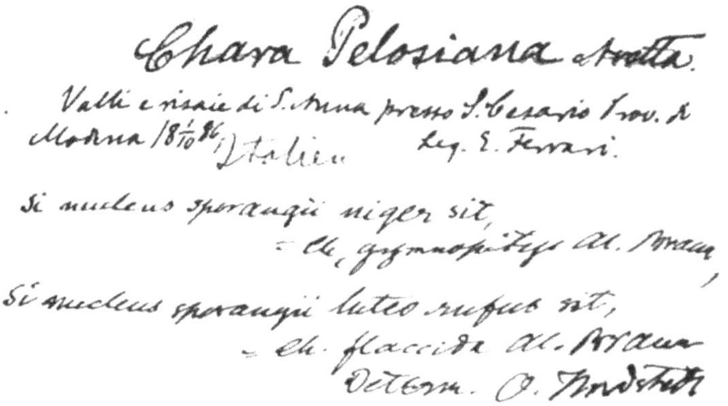

Figure 4. Label of the Oslo lectotype of *Chara pelosiana* with Nordstedt's observation; from [6].

Notes. Langangen did not mention how the *C. pelosiana* specimen reached the Oslo Herbarium. He only recalled that Nordstedt, a charophyte authority at the time, was in close contact with the phycologist N. Wille from Kristiania (Oslo).

We also searched for *C. pelosiana* in the Lund Herbarium (LD), which houses the original Nordstedt herbarium, but found nothing.

3.1.2. Herbaria That Keep Specimens of *C. pelosiana* Collected from New Localities

Pavia Herbarium (PAV): A large amount of material is kept free in a folder with a free label written in pencil: "Risaje Campo maggiore | 16/8/86 | Traverso e Kruch" [unknown person *scripsit*] (Figure 5A). Within the folder, there is also a free preprint revision label reporting, "Chara Pelosiana Avetta, *determinavit* Dr Leone Formiggini, Giugno [June] 1907". [Formiggini *scripsit*] (Figure 5A).

Notes. This is the first unpublished specimen of *C. pelosiana* from a new station, the rice fields of Pavia Province (Campo Maggiore), which is about 150 km from the type locality (S. Anna). Giacomo Traverso and Osvaldo Kruch (see Appendix A) collected the specimen on August 16, 1886, a month and a half before the specimen from S. Anna was collected.

Turin Herbarium (TO): A second specimen of *C. pelosiana* is mounted on an herbarium sheet with a printed label from the Herbarium of Turin pinned to the sheet reporting, "Chara | Nelle risaie di Nonantola | nel fondo Sacerdoti | 23 7bre [September] 1886, E Ferrari" [Ferrari *scripsit*] (Figure 5B). Formiggini's preprint label is pinned to the sheet as well: "Chara Pelosiana Avetta, *determinavit* Dr Leone Formiggini, Gennaio [January] 1909". [Formiggini *scripsit*] (Figure 5B).

Notes. This is the second unpublished specimen of *C. pelosiana*. It was collected by Ferrari from the same area in the province of Modena, the Nonantola rice fields, a week before the specimen of S. Anna was collected.

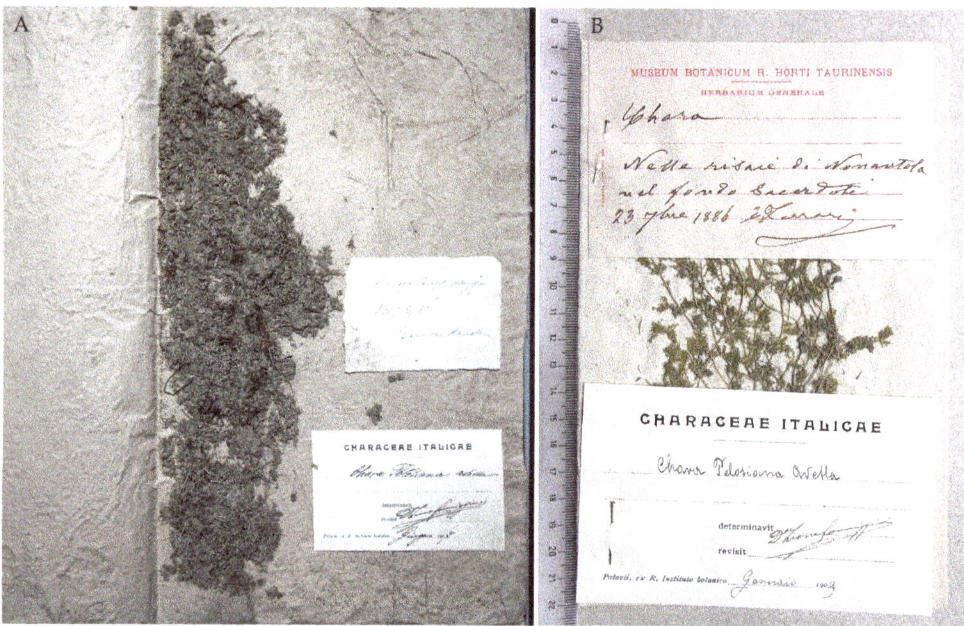

Figure 5. Unpublished specimens of *Chara pelosiana*. (**A**) PAV specimen collected by Traverso and Kruch from Campo Maggiore (Province of Pavia) in August 1886. Below is Formiggini's revision label (**B**) TO specimen collected by Ferrari from Nonantola in September 1886 with Ferrari's label (above) and Formiggini's revision label (below).

3.1.3. Herbaria That Keep Portions of Specimens of *C. pelosiana* Removed from MOD and PAV

Padua Herbarium (PAD): There are only a few MOD fragments (Figure 6A), which are kept in a recycled paper envelope with the indications written directly on it: "Chara Pelosiana Avetta | (sub Ch. Scoparia Braun a Baueri (sic) | Valli e risaie di S. Anna presso S. Cesario | X 1886 | Ex Hb. R. Ortobot.Mut(inensis) [Ex Herbario Regius Hortus Botani-cus Mutinensis]" [Béguinot *scripsit*] (Figure 6B).

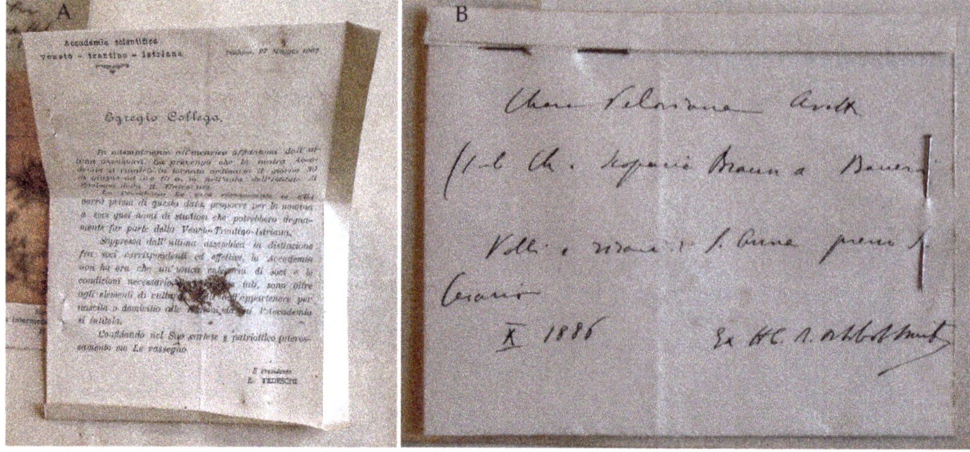

Figure 6. (**A**) PAD portion of *Chara pelosiana* removed from MOD specimen. (**B**) Béguinot's indications on the envelope.

Instead, there are numerous fragments from PAV (Figure 7A). They are kept in an envelope with Formiggini's preprint label pinned to it: "Chara Pelosiana Avetta | ex herbario Ticinensis | *determinavit* Dr Leone Formiggini Giugno [June] 1907" (Figure 7B).

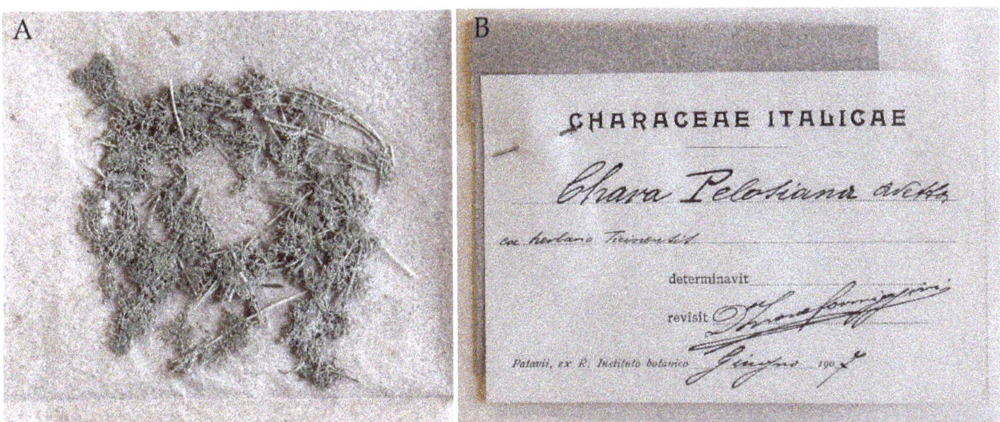

Figure 7. (**A**) PAD portion of *Chara pelosiana* removed from PAV specimen. (**B**) Formiggini's revision label.

Notes. The information on the two envelopes indicates that Béguinot and Formiggini, who both worked at Padua, took samples from MOD and PAV to their Botanical Institute for further investigation.

The portion from PAV specimen was especially valuable in our search for *C. pelosiana* specimens.

Jena Herbarium (JE): In addition to fragments sent by Avetta, the Herbarium of Jena also conserves portions of specimens from MOD and PAV Herbaria.

MOD's portion consists of a few fragments, which are kept in a small envelope with the following indications written in pencil inside: "Chara Pelosiana Avetta | H Mutinensis" [Formiggini *scripsit*]. The revision by Migula is written in pencil on a label glued to the envelope reporting "II Ch. Pelosiana Avetta" with the stamp "Herbarium Walter Migula Eisenach" (Figure 8B).

PAV portion consists of several fragments, which are kept in a bigger envelope with the indications written in pencil inside as well: "ex H. Ticinensis | Ch. Pelosiana Avetta" [Formiggini *scripsit*] (Figure 8A). Migula's revision is also written in pencil outside the envelope. It reports: "Ch. Pelosiana Avetta" with the stamp "Herbarium Walter Migula Eisenach".

A letter from Leone Formiggini to Walter Migula, still in its original envelope, is attached to the specimens (Figure 8C). The letter is written in Italian and dated 9 July 1907:

"Le invio (…) frammenti di due Caracee tratte l'una dall'erbario del R. Istituto Botanico di Modena, l'altra dall'erbario del R. Istituto Botanico di Pavia. La prima corrisponde esattamente oltre a tutto anche per località di raccolta e per data colla specie nuova descritta dal Prof. Avetta sotto il nome di <u>*Chara Pelosiana*</u>, la cui posizione sistematica sarebbe fra la Ch. Coronata e la Ch. Scoparia. La seconda è pure precisa alla precedente pure essendo raccolta in località diversa. A me sembra che questa sia sì nuova, ma vada avvicinata piuttosto al Lychnothamnus e posta in seguito al Lychn. barbatus. Infatti del Lychnothamnus ha tutto l'aspetto, solo appare come un piccolo Lychnothamnus munito di numerose ed assai lunghe spine, oltre che di un completo rivestimento corticale. (…)"

"[I am sending you (…) fragments of two Italian Characeae, one from the Herbarium of the Royal Botanical Institute of Modena and the other from the Herbarium of the Royal Botanical Institute of Pavia. The first corresponds exactly, for locality of collection and date, to the new species described by Prof. Avetta under the name *Chara Pelosiana*, and its systematic position would be between *Ch. Coronata* and *Ch. Scoparia*. The second is very similar to the previous one even though it is collected in a different locality. It seems to me that this is indeed new, but it should be approached rather to *Lychnothamnus* and placed after to the *Lychn. barbatus*. In fact, it has in all the appearance of a *Lychnothamnus*, but it appears as a small Lychnothamnus equipped with numerous and very long spines, as well as a complete cortical covering. (…)].

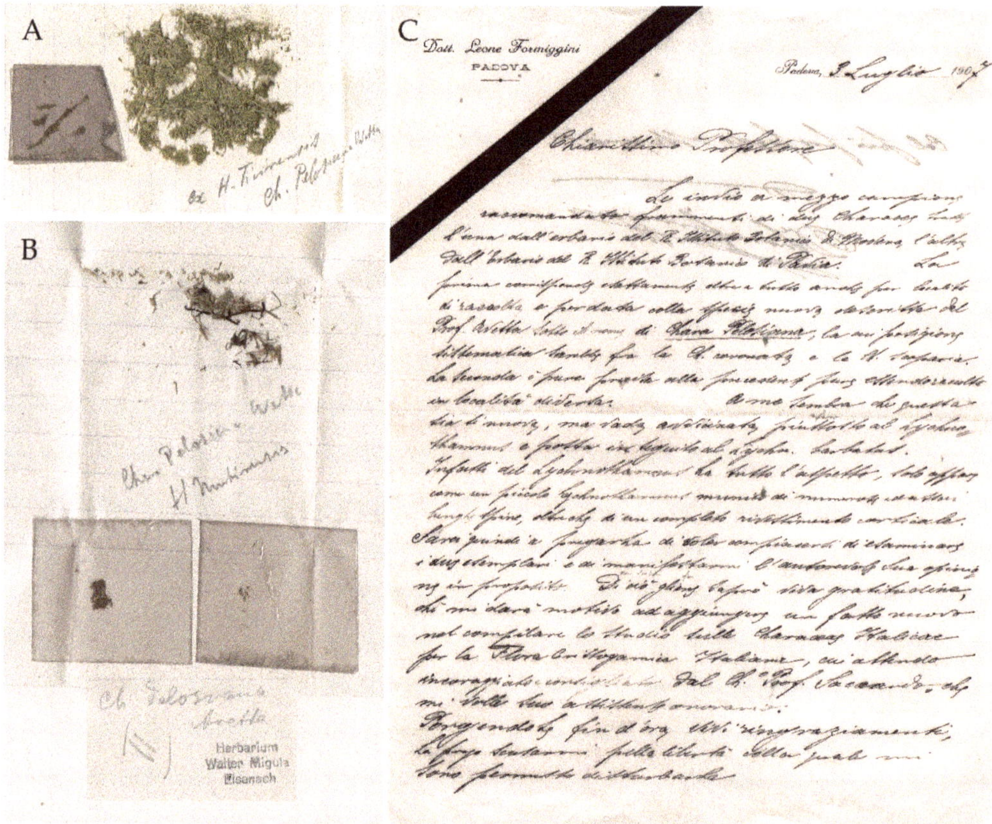

Figure 8. JE portions of *Chara pelosiana* sent by Formiggini to Migula in 1907. (**A**) Portions of PAV specimen with Formiggini's handwritten notes. (**B**) Portions of MOD specimen with Formiggini's handwritten notes (above) and Migula's handwriting (below). (**C**) First page of Formiggini's letter to Migula, dated 9 July 1907.

Notes. According to the documentation, Formiggini submitted fragments of the two *C. pelosiana* specimens kept in PAD, which came from the portions removed from MOD and PAV, to Migula.

3.2. Morphological Findings

Measurements of the axes and lengths of the stipulodes were taken on the dry samples. They are summarized in Tables 3 and 4.

Table 3. Diameters of axes. All values are presented in μm.

Herbarium	Locality	Date of Collection	Collector	Minimum	Maximum	Mean	Number of Axes Measured
MOD	S. Anna	1 October 1886	Probably Ferrari	312	458	383	11
TO	S. Anna	1 October 1886	Ferrari	434	566	486	7
PAV	Campo Maggiore	16 August 1886	Traverso and Kruch	399	578	514	3
TO	Nonantola	23 September 1886	Ferrari	438	566	483	5
PAD	S. Anna	1 October 1886	Ferrari	325	469	401	5

Table 4. Lengths of stipulodes. All values are presented in μm.

Herbarium	Locality	Date of Collection	Collector	Minimum	Maximum	Mean	Number of Stipulodes Measured
MOD	S. Anna	1 October 1886	Probably Ferrari	965	1274	1151	4
TO	S. Anna	1 October 1886	Ferrari	1277	1470	1349	3
PAV	Campo Maggiore	16 August 1886	Traverso and Kruch	962	1614	1255	5
TO	Nonantola	23 September 1886	Ferrari	1265	1337	1301	2
PAD	S. Anna	1 October 1886	Ferrari	1194	1312	1253	2

These tables show that there are no consistent differences between the herbarium samples. Their axes have similar minimum and maximum diameters. The stipulodes' lengths follow the same pattern. Additionally, preliminary phylogenetic analyses of partial chloroplast gene sequence data from the Nonantola and S. Anna collections stored in TO supports their con-specificity (Kenneth G. Karol, pers. comm.).

A stereomicroscope examination of these samples also revealed that they are morphologically comparable (Figure 9A–G and Figure 10A–E). Therefore, a single description can be extended to all the material removed.

The specimens, which are more or less heavily calcified, are almost all fragmented. Despite this, they do not appear to be longer than 7–8 cm (Figure 10B), as reported by Avetta [1] (p. 231), who observed them closer to the time of collection than we did. The axis diameter has mean values from 383 to 514 μm (Table 3). All the axes are corticated, diplostichous, isostichous or slightly tylacanthous (Figures 9C and 10D,E), and bear spine cells (Table 5) generally longer than the axis, sometimes in a whorl (Figure 10E) and sometimes curved towards the axis (Figure 9A). Stipulodes are in a single row, perpendicular to the axis, one or two per branchlet, long to 1350 μm (Figures 9D,G and 10A,D). Branchlets are (6)8-9(10) per whorl, totally ecorticate (Figure 10C), wide approximately half of the axes. They are composed of 3-4(6) segments bearing at each node, including the apical nodes, a crown of long bract cells (Figures 9G and 10C). The two bracteoles are longer than the oogonia (Figure 9G). The mean length of the basal branchlet cells is 1.7–2.7 mm in the apical parts of MOD, TO, PAD, and PAV dry samples, and 3.5 mm in the fourth whorl of branchlets of the removed sample collected from S. Anna (TO).

Figure 9. Morphologies of the portions removed from the specimens collected by Ferrari from S. Anna. (1 October 1886). (**A–D**) From the MOD specimen; (**C**) Details of the cortex; (**D**) Details of the stipulodes; (**E,F**) From the TO specimen (N. 3); (**G**) From the PAD portion.

Figure 10. Morphologies of the portions removed from unpublished specimens. (**A,B**) From the TO specimen collected by Ferrari from Nonantola (23 September 1886); (**C–E**) From the PAV specimen collected by Traverso and Kruck from Campo Maggiore (16 August 1886); (**C**) Decalcified portion.

Table 5. Lengths of spine cells. All values are presented in μm.

Herbarium	Localiy	Date of Collection	Collector	Minimum	Maximum	Mean	Number of Spine Cells Measured
MOD	S. Anna	1 October 1886	Probably Ferrari	301	638	456	6
TO	S. Anna	1 October 1886	Ferrari	397	807	557	4
PAD	S. Anna	1 October 1886	Ferrari	463	856	685	5
TO	Nonantola	23 September 1886	Ferrari	157	1.060	590	6
PAV	Campo Maggiore	16 August 1886	Traverso and Kruch	336	879	645	12

The plants are monoecious with conjoined gametangia. The oogonia are 425–450 μm long (excluding the coronula) and 312–340 μm wide. The oospores are golden-brown, and have, in the dry material collected from S. Anna (TO), *c.* 350 μm in length, and 235 μm in width (Figure 9F). The spiral turns are 7–8. The antheridia in the portions removed from the PAV specimen are 250–270 μm in diameter.

In light of the information presented in Tables 1 and 2, and considering the nomenclatural changes, the *C. pelosiana* species found in rice fields in the Modena area in the week of 23 September to 1 October 1886 were accompanied by: *C. vulgaris* L., *C. hispida* sensu auct. nonnull., *C. globularis* Thuiller, *C. tomentosa* L., and *Nitellopsis obtusa* (Desv.) J. Groves.

4. Discussion

4.1. The Studies of C. pelosiana by Pelosi, Avetta, Formiggini, and Béguinot

Letters from the scientific correspondence received by Pier Andrea Saccardo (1845–1920), a professor of botany and prefect of the Botanical Garden of Padua [17], help to explain the interests of the Rome Botanical Institute in Italian Characeae.

In a letter dated 4 November 1886, Pietro Romualdo Pirotta, the director of the Botanical Institute of Rome, requested a loan of Characeae specimens from the Padua Herbarium. The loan was for Alpinolo Pelosi, "a talented young student" who had been working with Characeae for a year and whom Pirotta encouraged to pursue a monographic study of the Italian species. Between the end of 1886 and the beginning of the next year, a great number of Italian specimens from most of the Italian Herbaria, including Ferrari's small collection from the Modena area, were sent to Rome for this purpose [1] (p. 230) [17] (letters: 15 November 1886, 10 February 1887).

After Pelosi's untimely death in August 1887, Carlo Avetta, who was Pirotta's first assistant, was entrusted with the monograph of the Italian Characeae. Despite the difficulty presented by the large amount of material gathered in Rome and the study of a problematic group, Pirotta considered the work practically complete by the beginning of 1893 [17] (letters: 22 December 1890, 25 November 1891, 17 January 1893). However, Avetta had moved to Parma by the end of 1893, and Pirotta was forced to announce the conclusion of the study at the Botanical Institute of Rome, and he returned the loan of the Characeae of Padua [17] (letter: 12 April 1894).

Avetta was not a specialist of Characeae. The collections of the *General Herbarium* in RO house the only specimen of *Chara* collected by him. This specimen was identified by Formiggini and Béguinot (*Ch. crassicaulis*, Colli Astigiani, September 1886, det. Formiggini and Béguinot, *sine data*). Avetta's revisions and determinations of the genera *Nitella*, *Tolypella*, *Lamprothamnium*, and *Lychnothamnus* can be found in the collections of the *General Herbarium* and *Cesati Herbarium* in RO, although they are nearly always unsigned.

Avetta confirmed [1] (p. 230) that his study of Italian Characeae began after Pelosi's death (1887) while he was gathering the materials and notes left by the unlucky student as well as Ferrari's collection. Both records have disappeared from Rome, but there is a record of a payment made to Ferrari in 1886 for his Characeae collection from the Modena area in the RO Archive [18].

Avetta resumed his study of Characeae in 1898, after a period of interruption [1] (p. 230), with the help of the regional collections kept at RO (*Roman Herbarium*). The Register of loans of RO shows a single loan of 44 specimens of Characeae sent to Avetta in Parma in 1898 [19]. The collection was returned to RO only ten years later, at the beginning of 1908, without any revisions. The sending took place a few months before Avetta's publication on Malpighia, suggesting that Avetta still had Pelosi's documentation and Ferrari's collection with him when he left the University of Rome, perhaps as early as 1893, or that if these materials were forwarded to him later, they were sent privately.

Formiggini and Béguinot's views on *C. pelosiana* can be deduced from specimens kept in TO, MOD, PAV, and JE Herbaria as well as from documentation in RO Archive. While Formiggini appears to have agreed on the validity of the new species (see revised labels from MOD and PAV, June 1907), Béguinot's opinion was quite different (*an potius Lycnothamnus species? et tunc Lycnothamnus Pelosianus Bég. et Form.!*, MOD label), as he was in doubt as to whether *C. pelosiana* should be considered a new *Lychnothamnus* species. Nonetheless, in the letter sent to Migula (JE), Formiggini presented Béguinot's doubts as his own, while both attached samples were sent with the binomial *C. pelosiana* written by Formiggini himself. It is unknown as to what Migula's answer was, but the revisions on the specimens, handwritten by Migula, confirm the specimen was identified as *C. pelosiana*.

At the beginning of 1908, shortly after Avetta's loan was returned, Formiggini and Béguinot examined the complete Characeae collection kept in RO, which consisted of 828 specimens from the three Herbaria: *Roman* (58), *General* (383), and *Cesati* (387). The Register of loans contains a detailed list of all species sent to them [20], revealing the absence of *C. pelosiana*, which was therefore no longer part of the RO collections ten years after Avetta's publication.

Despite this, Formiggini considered still *C. pelosiana* to be valid at the beginning of 1909, as revealed by his subsequent revisions in TO: S. Anna (December 1908) and Nonantola (January 1909). Despite not having seen the original *C. pelosiana* material, he confirmed the validity of the new species by examining specimens in TO and MOD. Furthermore, he also recognized the unpublished specimens kept in TO and PAV as *C. pelosiana*. The one of the two new specimens discovered during our research is therefore the one collected by Ferrari from Nonantola (Modena) almost a week before the S. Anna specimen was collected. The other was collected by Traverso and Kruch near Pavia just over a month before the type specimen was found. This indicates two new Italian stations for this species' distribution area.

4.2. The Double Numbering of C. pelosiana

According to Avetta's paper [1] (pp. 234–235), it seems that two specimens were collected by Ferrari: "*Chara* N.° 3 della raccolta Ferrari" and "*Chara* N.° 101. Raccolta di Ferrari. (Nelle valli e risaie di S. Anna presso S Cesario, prov. di Modena, 1 ott. 86)".

To bolster this impression, each of the two assumed *Chara* species were followed by different observations by Pelosi, which were fully published by Avetta [1] (pp. 234–235).

In the absence of the material seen by Avetta and Pelosi's original notes, the examination of the specimens kept in TO and the information acquired in RO were decisive.

TO preserves Ferrari's only numbered collection, consisting of 25 specimens, 21 of which were collected between 20 September and 1 October 1886. This number appears to be correct, as Avetta, the last person to examine the Rome collection, described it as a "small collection" [1] (p. 230).

On the other hand, in TO, we recognized 14 of Pelosi's revision labels, 12 of which are numbered (104, 107, 109, 111, 112, 113, 116, 118, 120, 122, 123, and 124).

In the collections of the *General Herbarium and Cesati Herbarium* in RO, Pelosi numbered his collections (4 out of 14), revisions (50 out of 55), and determinations (10 out of 15), but the numbering system, which ranges from 4 to 130, is seriously lacking and the numbers are frequently repeated.

Nonetheless, based on a comparison of Ferrari's numbering in TO (up to 24) and Pelosi's numbering in TO and RO (up to 130), it is almost certain that the original material of *C. pelosiana* should be regarded as a single gathering: Ferrari's number 3 and Pelosi's revision, identified by number 101.

4.3. Which Is the Correct Identity of Chara pelosiana?

Langangen [6] merged Avetta's species with *Chara fibrosa*. Other authors have considered *C. pelosiana* to be *C. fibrosa* or a variety or form of this species [4,5,8,10].

C. fibrosa belongs to the section *Agardhia* Wood, which mainly includes exotic taxa (subsection *Agardhia*) and the European species *C. pelosiana* (subsection *Braunia*) [11,12].

C. pelosiana specimens found in herbaria or mentioned in literature in Italy [1,4] are all from rice fields.

In Zaneveld key [7], the essential differences between the three subspecies that this author includes in *C. fibrosa* are the colour of the ripe oospores (golden-brown in *C. fibrosa* ssp. *flaccida* and black in the other two: *C. fibrosa* ssp. *benthamii* and *C. fibrosa* ssp. *gymnopitys*) and the number of stipulodes (as numerous as the branchlets in *C. fibrosa* ssp. *benthamii*, twice as numerous as the branchlets in *C. fibrosa* ssp. *gymnopitys*).

In the examined specimens of *C. pelosiana*, the stipulodes were variable in number, sometimes nearly equal and sometimes more or even twice as numerous as the branchlets (Figure 9D,G and Figure 10A). Avetta reported that the number of stipulodes was equal to the number of the branchlets (10–12) [1] (pp. 232–233), while Langangen (despite having difficulty counting them) stated that there were 1–2 stipulodes per branchlet in the fragments of *C. pelosiana* that he observed in Oslo [6] (p. 250). It appears, therefore, the number of stipulodes in *C. pelosiana* is not constant, as has already observed by several authors in other species [7] (p. 154).

In contrast, the colour of the mature oospores in *C. pelosiana* was consistently found to be yellow-brown (Figure 9A,E).

Only one stipulode per branchlet is mentioned in the description of the type material of *C. fibrosa*, and its oospores are described as being "consistently a light golden-brown" [15]. Furthermore, *C. fibrosa* is endemic to the island of Guam (Micronesia) [15].

This investigation of *Chara fibrosa* led to the separation of two previously merged species in the section *Agardhia*, *C. fibrosa* and *C. wightii* (A. Braun) Casanova [15]. Thus, the correct identity of *C. pelosiana* cannot be determined until the section *Agardhia* will be fully revised. Meanwhile, in this work, we used the valid name *Chara pelosiana* established by Avetta [1].

5. Conclusions

This study of herbarium materials of *Chara pelosiana* revealed unexpected original material. Only the *C. pelosiana* specimen described by Avetta [1] is mentioned in the literature. Although Ferrari's original collection is no longer kept in the Herbarium of Rome, two duplicates of the original set were discovered in the Modena and Turin Herbaria, each including a specimen of *Chara pelosiana* that can be considered "original material".

Furthermore, two new additional localities were discovered in the Pavia and Turin Herbaria (rice fields of Campo Maggiore in the Province of Pavia and rice fields of Nonantola in the Province of Modena), providing new information about the Italian distribution area for this rare species.

This paper gives an example of how, in addition to traditional morphological, taxonomic, and systematic research rules, historical herbarium collections can be used to assist ecology, biogeography, and conservation biology research.

Author Contributions: Conceptualization, A.M. and N.A.; methodology, A.M. and N.A.; formal analysis A.M. and N.A.; investigation A.M. and N.A.; writing—original draft preparation, A.M. and N.A.; writing—review and editing, A.M. and N.A.; visualization, A.M. and N.A. All authors have read and agreed to the published version of the manuscript.

Funding: This research received no external funding. The APC was funded by A.M. and N.A.

Acknowledgments: We would like to express our gratitude to Heiko Korsch, Paola Mario, and Maurizio Preti for their assistance and support during our research. We also thank the directors, curators, and technicians of all of the Herbaria, Archives, and Municipal Registry Offices listed below. Herbaria: Umberto Mossetti (BOLO); Chiara Nepi (FI); Heiko Korsch (JE); Patrik Fródén (LD); Daniele Dallai, Marta Mazzanti (MOD); Annalisa Santangelo (NAP); Rossella Marcucci (PAD); Deborah Beghé, Maria Luigia Borghi, Fabrizia Fossati, Linda Spallanzani, Marcello Tomaselli, Corrado Zanni (PARMA); Maurizio Preti, Vanda Terzo (PAV); Lucia Amadei (PI); Giovanna Abbate (RO); Laura Guglielmone (TO). Archives: Paola Mario (University of Padua, Historical Archive of the Library of the Botanical Garden); Donatella Mazzetto, Marco De Poli, Donka Todorova Ilieva (University of Padua, General Archive); Francesco Piovan, Mariarosa Davi (State Archive of Padua); Alessandra Baretta, Maria Piera Milani (University of Pavia, Historical Archive); Fabio Orecchia (Historical Civic Archive of Pavia). Special thank is given to Riccardo Lodovici and Antonella Bernetti for the processing of historical images. We are grateful to our anonymous reviewers and Angelo Troia, our Academic Editor, for their helpful comments which greatly improved our paper, and to Emma O'Connor for thorough language editing.

Conflicts of Interest: The authors declare no conflict of interest.

Appendix A. Biographical Notes

The use of an exclamation point (!) indicates that personal data were retrieved from the Municipal Registry Offices. For bibliographical details see [21,22].

Avetta, Carlo

Turin, 13 March 1861–Turin, 10 March 1941

Botanist. Avetta's academic career began at the University of Rome, where he worked as the *first assistant* to Pietro Romualdo Pirotta (1853–1936), the director of the Botanical Institute, from 1885 until 1893. His second and final position was as a professor and director of the Botanical Garden at the University of Parma (1893–1935). He was first interested in cryptogamic studies, especially fungi, and then in Scioa's African collections, which he studied and illustrated. Anatomy and cytology were two of his most important subjects of study. In Parma, he was particularly interested in the diffusion of medicinal plant knowledge. From his study on Italian Characeae species, only his paper on *Chara pelosiana* remains. His announced second note on Italian Characeae was never published.

Béguinot, Augusto

Paliano (Frosinone), 17 October 1875–Genua, 3 January 1940

Botanist. Béguinot graduated with a degree in Natural Sciences from the University of Rome in 1898. His academic career covered six universities: Padua (1900–1921), where he worked as an assistant to Pier Andrea Saccardo; Ferrara (1918–1921); Sassari (1921), where he became professor; Messina (1922); Modena (1924–1929); and Genua (1929–1940). He published a large number of scientific works on flora, systematics, phytogeography, and the history of botany. His interest in the Characeae was restricted to the time he worked at the University of Padua with Formiggini (see below), and he published two contributions with him [23,24]. Their final paper on the Characeae of Italy was never published.

Ferrari, Enrico

Modena, 3 November 1845–Turin, 2 November 1921

Gardener at the Botanical Garden of Modena. Ferrari began studying plants in a self-taught manner and quickly advanced to become a skilled collector and an expert florist as well as an appreciated contributor to the *Flora del Modenese e del Reggiano* [25] (p. 5). According to the RO Archive (2), Pirotta was most likely the one who asked Ferrari for a collection of Characeae for Pelosi's study in 1886. From 1887 to 1921, he was the Curator of the Turin Herbarium. Ferrari was the main author of the reorganization of Turin's collections, while also contributing significantly to their increase in assiduous excursions, especially in Piedmont and the Valle d'Aosta.

Formiggini, Leone

Padua, 10 December 1879–Padua, 7 June 1963 (!)

Formiggini graduated from Padua on 5 December 1904 with a thesis titled *"Contribution to the knowledge of the Caracee from Padua"* [26]. From 1906 until 1909, he worked as an *honorary assistant* at the University of Padua's Botanical Institute under the direction of Pier Andrea Saccardo (1845–1920) [27]. Formiggini was the author of the most important contribution, to the knowledge of the Italian Charophytes during this period, through his critical reviews of the Charophytes of Sicily [28], Veneto and Mantovano [29], and Lazio [30]. In collaboration with A. Béguinot, he became interested in some vicarious Characeae of the Italian Flora [23,24]. His collaboration with the Botanical Institute was interrupted in 1910, leaving the announced general work on Italian Charophytes unpublished [3]. However, his extensive and accurate work of revision and identification of Italian Charophytes *exsiccata* is still preserved in the Italy's major public and private Herbaria. In a 1939 information-curriculum [31], Formiggini is still listed as a Padua resident, and his lone occupation appears to be that of covering various positions, some honorific, in Padua's Societies, Academies, and Associations.

Kruch, Osvaldo

Pavia, 1 November 1864–Luino (Varese), 7 October 1942

Botanist. Kruch graduated with a degree in Natural Sciences from the University of Pavia on 29 June 1886 [32]. The following year, he conducted postgraduate studies at the Botanical Institute of the University of Rome under the direction of Pietro Romualdo Pirotta. In May 1893, he was named curator of the Rome Herbarium, and in January 1894 he was promoted to *first assistant*, covering the vacancy left by Carlo Avetta. Finally, between 1896 until 1935, he was a professor at the Agricultural Experiment Institute of Perugia (now Faculty of Agriculture). His studies were mostly focused on histology and plant anatomy.

Migula, Walther

Zyrowa, Upper Silesia (now in southern Poland), 4 November 1863–Eisenach (Germany), 23 June 1938

German botanist. Migula graduated from the University of Breslau (1888). He taught at the Karlsruhe Institute of Technology until 1895 and then at the Forestry School in Eisenach (1895–1915). His works on the cryptogamic flora, bacteriology, and plant physiology are well known. *Die Characeen Deutschlands, Oesterreichs und der Schweiz* in *Rabenhorst's Kriptogamen-Flora* (1889/97) and *Synopsis Characearum europaearum* (1898) are considered fundamental works on the Characeae family. He published the *Characeae exsiccatae* (1892–1901) series with P. Sidow (1851–1925) and L.J. Wahlstedt (1836–1917), which includes 150 specimens in six fascicles.

Pelosi, Alpinolo

S. Pancrazio Parmense (Parma), 2 November 1865 (!)–Anguillara Sabazia (Rome), 1 August 1887 (!), not 1888 [1] (p. 230, note 1).

As a student of Natural Sciences at the University of Rome (1885–1887), he became a favorite pupil of Pietro Romualdo Pirotta, the director of the Botanical Institute. In 1885, he began studying the Italian Charophytes on the collections of the Herbarium of Rome and those of the main Italian Herbaria, under the direction of Pirotta (1). The next summer, Pelosi died in the waters of the Lake of Martignano (Rome), where he was collecting algae and other water plants on behalf of the Botanical Institute [33,34]. He was only 21 years old at the time. Many of his revisions, determinations, and observations on Italian Charophytes remain unsigned in the exsiccata of Rome and the main Italian Herbaria.

Traverso, Giacomo

Pegli (Genoa), 8 May 1849 (!)–?
From 1877 to 1909, Traverso was the head gardener at the Pavia Botanical Garden [35,36]. He was the father of Giovanni Battista Traverso (1878–1955), a mycologist, and Onorato Traverso (1881–1960), the head gardener of the Rome Botanical Garden.

References

1. Avetta, C. Nuova specie di Chara. (Chara Pelosiana mihi). *Malpighia* **1898**, *12*, 229–235.
2. Bazzichelli, G.; Abdelahad, N. *Alghe D'acqua Dolce d'Italia. Flora Analitica Delle Caroficee*; Sapienza Università di Roma—Ministero dell'Ambiente e della Tutela del territorio e del mare: Roma, Italy; Centro Stampa Università: Roma, Italy, 2009; pp. 38–39.
3. Formiggini, L. Cenno storico-bibliografico sulle Caracee della Flora italiana. *Bull. Soc. Bot. Ital.* **1909**, *1*, 14–26.
4. Abdelahad, N.; Piccoli, F. Report on Charophytes from rice fields in northern Italy including the alien species *Chara fibrosa* ssp. *benthamii*. *Plant Biosyst. Int. J. Deal. Asp. Plant Biol.* **2017**, *152*, 599–603. [CrossRef]
5. Soulié-Märsche, I.; Triboit, F.; Despréaux, M.; Rey-Boissezon, A.; Laffont-Schwob, I.; Thiéry, A. Evidence of Chara fibrosa Agardh ex Bruzelius, an alien species in South France. *Acta Bot. Gallica Bot. Lett.* **2013**, *160*, 157–163. [CrossRef]
6. Langangen, A. Chara fibrosa Agardh ex Bruzelius, a charophyte new to the European flora. *Allionia* **2000**, *37*, 249–252.
7. Zaneveld, J.S. The Charophyta of Malaysia and adjacent countries. *Blumea* **1940**, *4*, 1–224.
8. Raam van, J.C. *Wood's "Synopsis of the Characeae" Adapted and Modified*; CD Privately Distributed by van Raam in Anticipation of Publication; 2010.
9. Guiry, M.D.; Guiry, G.M. AlgaeBase. World-Wide Electronic Publication, National University of Ireland, Galway. Available online: http://www.algaebase.org (accessed on 27 October 2020).
10. Krause, W. Charales (Charophyceae). In *Süßwasserflora von Mitteleuropa*; Ettl, H., Gartner, G., Heynig, H., Mollenhauer, D., Eds.; Gustav Fischer Verlag: Stuttgart, Germany, 1997; Volume 18, p. 185.
11. Wood, R.D. New Combinations and Taxa in the Revision of Characeae. *Taxon* **1962**, *11*, 7–25. [CrossRef]
12. Wood, R.D.; Imahori, K. *A Revision of the Characeae. Part I: Monograph of the Characeae*; Verlag Von J. Cramer: Weinheim, Germany, 1965; p. 319.
13. Casanova, M.T. Oospore variation in three species of Chara (Charales, Chlorophyta). *Phycologia* **1997**, *36*, 274–280. [CrossRef]
14. Casanova, M.T. An overview of Chara L. in Australia (Characeae, Charophyta). *Aust. Syst. Bot.* **2005**, *18*, 25–39. [CrossRef]
15. Casanova, M.T. Review of the species concepts *Chara fibrosa* and *C. flaccida* (Characeae, Charophyceae). *Aust. Syst. Bot.* **2013**, *26*, 291–297. [CrossRef]
16. Thiers, D. (Ed.) 2008+ *(Continuously Updated). Index Herbariorum: A Global Directory of Public Herbaria and Associated Staff*; New York Botanical Garden's Virtual Herbarium; Available online: http://sweetgum.nybg.org/ih/ (accessed on 28 February 2020).
17. Pirotta, P.R. *Romualdo Pirotta to Pier Andrea Saccardo, 4 and 15 Nov 1886, 10 Feb 1887, 22 Dec 1890, 25 Nov 1891, 17 Jan 1893. Letters. Scientific Correspondence Received from the Prefect P.A. Saccardo. Ar.B.48.A*; Historical Archive of the Library of the Botanical Garden of the University of Padua.
18. Ferrari, E. *17 October [1886] Ferrari E. for the Characee Collection in the Modena Area, 32.00 Lire. Amministrazione dell'Orto e Gabinetto Botanico di Roma, Anni 1877–1903, c. 75 (Administration of the Garden and Botanical Cabinet of Rome, Years 1877–1903)*; Archive of manuscripts conserved at the Museo Erbario, Sapienza University of Rome.
19. Avetta, C. *Characee Sent to Prof. Avetta in Parma on 24 May, 1898. Loan Returned on 02 January, 1908. Museum Botanicum R. Horti Romani | Prestazioni di Piante Secche degli Erbarii | 1886–1909, cc. 101–103 (Records of Loans of Dried Specimens from RO, 1886–1909)*; Archive of manuscripts conserved at the Museo Erbario, Sapienza University of Rome.
20. Saccardo, P.A. *Characeae of the Herbaria sent to Prof. Saccardo-Padua-Royal Botanical Garden, Year 1908. Loan Returned on 01 August, 1908. Museum Botanicum R. Horti Romani | Prestazioni di Piante Secche degli Erbarii | 1886–1909, cc. 222–246. (Records of Loans of Dried Specimens from RO, 1886–1909)*; Archive of manuscripts conserved at the Museo Erbario, Sapienza University of Rome.
21. Stafleu, F.A.; Cowan, R.S. *Taxonomic Literature: A Selective Guide to Botanical Publications and Collections with Dates, Commentaries and Types*, 2nd ed.; Scheltema & Holkema: Utrecht, The Netherlands, 1976–1988.
22. Stafleu, F.A.; Mennega, E.A. *Taxonomic Literature: A Selective Guide to Botanical Publications and Collections with Dates, Commentaries and Types*; Scheltema & Holkema: Utrecht, The Netherlands; Bonn, Germany, 1992–2009; Volume Suppl. I–VIII.
23. Béguinot, A.; Formiggini, L. Ricerche ed osservazioni sopra alcune entità vicarianti nelle Caracee della flora italiana. *Bull. Soc. Bot. Ital.* **1907**, 7–9, 100–116.
24. Béguinot, A.; Formiggini, L. Ulteriori osservazioni sulle Caracee vicarianti della flora italiana. *Bull. Soc. Bot. Ital.* **1908**, 4–6, 78–81.
25. Gibelli, G.; Pirotta, R. *Flora del Modenese e del Reggiano*; Tipografia di, G.T., Ed.; Vincenzi e Nipoti: Modena, Italy, 1882; p. 5.
26. Formiggini, L. *Università degli Studi di Padova, Facoltà di Scienze, Fascicoli degli Studenti, matr. 36/G Formiggini Leone*; General Archive, University of Padua.
27. Formiggini, L. *Archivio del Novecento, Serie Atti del Rettorato, bb. 28, 34, 40, 41*; General Archive, University of Padua.
28. Formiggini, L. Contributo alla conoscenza delle Caracee della Sicilia. *Bull. Soc. Bot. Ital.* **1908**, 4–6, 81–86.
29. Formiggini, L. Revisione critica delle Caracee della Flora Veneta compreso il Mantovano. *Atti Accad. Sci. Veneto-Trentino-Istriana* **1908**, *1*, 110–143.

30. Formiggini, L. Contributo alla conoscenza delle Caracee del Lazio. *Ann. Bot.* **1909**, *7*, 207–211.
31. Formiggini, L. *Padova, Archivio di Stato, Questura, b. 44, fasc. Formiggini Leone*; State Archive of Padua.
32. Kruch, O. *Carriera scolastica, reg. 3210, Osvaldo Kruch*; Facoltà di Scienze Matematiche, Fisiche e Naturali. Historical Archive, University of Pavia.
33. Anonymous. *Il Piccolo Pelosi. La Tribuna, Daily Newspaper, Year V, Nr. 217*; 9 August 1887.
34. Pelosi, A. *Administrative Archive, prot. 82, September 16, 1887*; Archive of manuscripts conserved at the Museo Erbario, Sapienza University of Rome.
35. Traverso, G. *Fascicoli Personale Tecnico-Amministrativo, fasc. Giacomo Traverso*; Historical Archive, University of Pavia.
36. Traverso, G. *Traverso Giacomo, Stato di Famiglia Storico*; Historical Civic Archive of Pavia.

Article

Long-Term Changes in Macrophyte Distribution and Abundance in a Lowland River

Andrej Peternel [1,2], Alenka Gaberščik [1], Igor Zelnik [1,*], Matej Holcar [1] and Mateja Germ [1]

[1] Department of Biology, Biotechnical Faculty, University of Ljubljana, Jamnikarjeva 101, 1000 Ljubljana, Slovenia; andrej.peternel@gmail.com (A.P.); alenka.gaberscik@bf.uni-lj.si (A.G.); matej.holcar@bf.uni-lj.si (M.H.); mateja.germ@bf.uni-lj.si (M.G.)

[2] Environmental Agency of the Republic of Slovenia, Vojkova 1b, 1000 Ljubljana, Slovenia

* Correspondence: igor.zelnik@bf.uni-lj.si

Abstract: The aim of this study was to reveal the changes of macrophyte community over time and along the course of the Ižica River. In 1996, 2000, and 2016, we surveyed the distribution and abundance of macrophyte species in the lowland Ižica River, which originates in the town of Ig and then flows through an agricultural landscape. We calculated the River Macrophyte Index (RMI), which reflects the ecological status of the river. In 2016, ecomorphological conditions of the river, using the Riparian, Channel and Environmental inventory, were also assessed. In just 10.5 km of the river, we identified 27 taxa of macrophytes, among which *Potamogeton natans*, *Sagittaria sagittifolia*, and *P. perfoliatus* were the most abundant. Detrended correspondence analysis showed that, in 1996, the surveyed stretches differed more according to macrophyte composition than in the following years. The assessed environmental parameters explained 43% of the variability of the macrophyte species; riverbank stability explained 20%, riverbed structure 10%, while vegetation type of the riparian zone and bottom type explained 7 and 5%, respectively. The species composition of the macrophyte community revealed significant changes over the years of the riverine ecosystem. Comparison of RMIs in 1996 revealed better conditions in the upper and middle part of the river, while in 2016, the situation was the opposite, since the conditions in the upper part deteriorated significantly over time, while the lower part of the river had the best ecological status. These changes may be due to a considerable increase in the population of the settlement Ig, while better status in the lower course of the river may be a consequence of improvements in the infrastructure and the use of sustainable agricultural practices in the catchment due to the establishment of a formal area of protection.

Keywords: macrophytes; lowland river; long-term changes; environmental parameters; ecological status; Slovenia

Citation: Peternel, A.; Gaberščik, A.; Zelnik, I.; Holcar, M.; Germ, M. Long-Term Changes in Macrophyte Distribution and Abundance in a Lowland River. *Plants* **2022**, *11*, 401. https://doi.org/10.3390/plants11030401

Academic Editors: Angelo Troia and Stefano Accoroni

Received: 18 November 2021
Accepted: 31 January 2022
Published: 31 January 2022

Publisher's Note: MDPI stays neutral with regard to jurisdictional claims in published maps and institutional affiliations.

1. Introduction

Rivers are ecosystems that manifest great dynamics in time and space [1]. Aquatic macrophytes are well adapted to seasonal variations of flow rate and flow velocity [2,3]. Macrophytes and riparian vegetation respond to environmental parameters and internal succession mechanisms across the transverse and longitudinal river dimensions and over time [4]. Macrophytes are involved in energy flow, nutrient cycling, and sedimentation processes, and are essential to the structure and functioning of the river ecosystem [5]. They increase habitat heterogeneity and complexity and affect a variety of organisms such as invertebrates, fish, and water birds [6,7], providing food and refuge. They also affect water quality [8] by uptake of nutrients, particularly those containing phosphorus and nitrogen, both from water and sediment [9]. On one hand, macrophytes contribute to river self-purification process as they store nutrients, but on the other hand, they can exert a significant effect on the eutrophication process, as they release these nutrients during decay [10]. They are especially important in lowland streams, where they may occur in high abundance [11].

Macrophytes show differing sensitivities to various natural and human pressures. These differences in sensitivity make them good indicators of the ecological status of a river [8,12,13], as well as indicators of the presence of different toxic substances in the sediment and the water [6,14–16]. The presence and abundance of macrophytes depend directly on water quality, depth, flow, substrate characteristics, and other environmental factors [17]. Their role is especially important in lowland watercourses since they increase the variability of habitats and physical conditions in a river [18]. In watercourses flowing through an agricultural landscape, macrophyte assemblages are well developed since these watercourses usually have poorly developed riparian zones and high input of nutrients [19,20].

The majority of rivers in Europe have been affected to different extents by human activity [21]. Introduction of new standards in river and catchment management, and new legislation, such as the Water Framework Directive (WFD), will reduce these pressures (Directive 2000/60/EC) [22]. In particular, the rivers flowing through agricultural landscapes are often exposed to high influxes of nutrients, as well as morphological alterations, both of which negatively influence the biodiversity of riverine ecosystems [23]. Beside the valuable role of macrophytes as the indicators of the current human pressure, aquatic macrophytes have been used by many researchers to monitor the long-term changes in rivers [24–29], as well as in lakes [30–32].

In 1996, 2000 and in 2016, we completed surveys to estimate potential changes of the presence, abundance, and distribution of macrophytes in the Ižica River (Slovenia) as it originates in the settled area of the town Ig, and then flows through an agricultural landscape. In addition to the WFD implementation in 2008, a part of the catchment of the Ižica River was protected as a Landscape Park within the same timeframe. On the other hand, the population development index of the town of Ig, which spreads in the narrowest part of the catchment area of the Ižica's source was, between 2002 and 2012, the highest within the Ljubljana metropolitan region [33]. Thus, we hypothesized that the composition of macrophyte community in the river would therefore change, as would the ecological status of the river.

2. Material and Methods

2.1. Study Area

The Ižica River is one of the shortest Slovenian rivers, running through the Ljubljana Moor—a 163-km^2 area of former peatland in the central part of Slovenia (Figure 1) that has been subjected to severe melioration measures in the past. The area lies in a tectonic depression between the Alpine and Dinaric regions, built by alluvial and lacustrine sediments, which are up to 200 m thick [34]. The Alpine region stretches further from Slovenia to Austria and North Italy (Central Europe), whereas the Dinarides is a region which continues further southeast to Croatia, Bosnia and Herzegovina, and Montenegro (Southeastern Europe). Until the end of the 18th century, the area of Ljubljana Moor consisted of a combination of bogs and fens. However, extensive melioration processes in the 19th century changed the area into a mosaic of birch groves, fields, meadows and ditches.

River Ižica has a karst spring characterized by relatively high fluctuations of water discharges. The river's source is in the center of the town of Ig, and it then flows north through an agricultural landscape, joining the Ljubljanica River after 10.5 km. The river has a vast catchment area on the karstic Dinaric plateau, south of the Moor, which is hard to delimit due to underground flows. It is a slow flowing river with predominantly fine-grained sediments and a low longitudinal profile and low erosion potential [34,35]. Its floodplain has been very dynamic over the past millennia due to the major transformation of the landscape. In 2008, the area was protected in the frame of the 135 km^2 Ljubljana Moor landscape park. The entire area of this park belongs to the Natura 2000 site, which is protected by European Commission law.

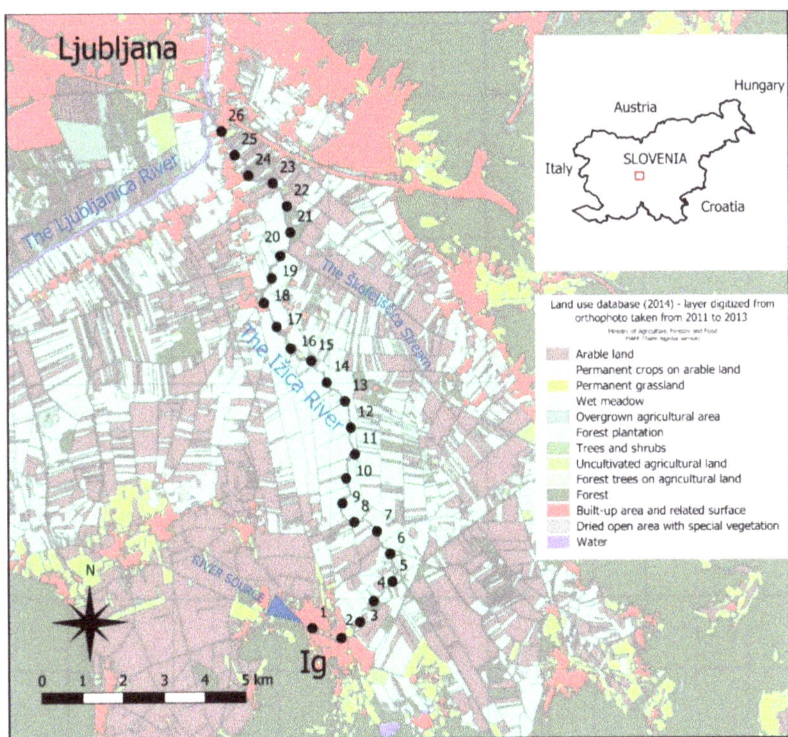

Figure 1. Map of Slovenia with the position of the study area and map of the Ižica River on the Ljubljana Moor. Points represent the starting or ending point of stretches.

2.2. Macrophyte Survey

Surveys were carried out along the whole stream course in 1996, 2000, and 2016. Macrophytes were surveyed at end of July and in August, the peak vegetation period. They were collected from the water with hooks from a small boat. Using GPS, the river was divided into 26 stretches of length 390 ± 10 m. Macrophyte species abundance was estimated as a relative plant biomass using the five-degree scale: 1 = very rare; 2 = rare; 3 = common; 4 = frequent; 5 = abundant or predominant [36]. This approach is widely used in European countries and is a methodology within the WFD [22]. For further data elaboration, these values were transformed to relative plant abundance using a third power function [37]. The classification of the macrophyte species into the functional groups was done according to Janauer et al. [5].

2.3. Assessment of Environmental Conditions

Basic physical and chemical parameters of the water, such as pH, oxygen saturation, oxygen concentration, conductivity and temperature, were measured in the upper (1–8), middle (9–17) and lower (18–26) course of the river with a portable multiprobe (PCD-650, Eutech Singapore). These parameters were investigated simultaneously as macrophyte surveys were performed, and on other dates during the vegetation period. In addition, water samples for analysis of nitrates were collected and analyzed in the laboratory. Water samples were collected from the superficial layers 10–15 cm, which was the same depth as for measurements with the multiprobe. Samples were cooled and filtered through the 0.45-μm glass-fiber filters. The level of NO_3-N was determined spectrophotometrically using HACH Lange tests (see Table 1).

Table 1. Average values and standard deviations of selected abiotic parameters measured in different stretches of the Ižica River.

Year		pH	Conductivity (μS cm^{-1})	NO$_3$-N (mg L^{-1})
1996	average	8.1	466	1.1
	S.D.	0.05	12	0.1
2000	average	7.9	509	1.0
	S.D.	0.2	26	0.2
2016	average	7.8	426	0.9
	S.D.	0.4	17	0.2

In 2016, we also assessed ecomorphological parameters such as riparian vegetation structure, land use, structure of the riverbed. The ecomorphological conditions of the river were assessed in all 26 stretches of the Ižica River using the Riparian, Channel, and Environmental (RCE) inventory proposed by Petersen [38] and modified by Germ et al. [39]. We assessed 12 environmental parameters that define land use beyond the riparian zone, the structure of the riparian zone (width, completeness, and type of vegetation) and stream channel morphology (bank structure, bank undercutting, flow dynamics, the bottom type, the presence of detritus, retention structures, and sediment accumulation). Each parameter is comprised of four quality gradient categories where 1 indicates good, close to a natural condition, and 4 indicates the most highly modified condition. That is not necessarily the case in the stream channel morphology, where the changes may also occur due to landscape characteristics or the longitudinal character of the river. Later, we related these parameters to species composition and the presence and abundance of macrophytes.

2.4. Data Analyses

The relative plant abundance (RPA) was used to calculate the quantitative significance of individual species in the river [36,40]. Based on the presence and abundance of macrophytes, we calculated the River Macrophyte Index (RMI) [41]. It was developed and intercalibrated [42] to assess the ecological status of Slovenian rivers. Macrophyte species were classified into the functional types (see Table 2 for explanation) and their abundances were grouped. Differences in the average proportions of abundances in functional types of aquatic macrophytes over 20 years were tested for significance with Student's *t*-test in MS Excel. The mentioned *t*-tests were also used for testing the differences in RMI values along the course of the river and over time.

The similarity of the macrophyte community composition in the sections and different years was checked with detrended correspondence analysis (DCA) with the program package Canoco 4.5 [43], which was also used to test the influence of environmental factors on the aforementioned composition. These relations were tested by canonical correspondence analysis (CCA), since the unimodal gradients in the matrix of species data were revealed beforehand with DCA, where the eigenvalue for the first axis was 0.52 [44]. We used forward selection, where 499 permutations were performed to rank the relative importance of the explanatory variables.

Table 2. List of aquatic macrophyte taxa recorded in the Ižica River with their codes/abbreviations and functional types (HE—helophytes, AM—amphiphytes, FLH—floating-leaved hydrophytes, FIL—filamentous algae, SM—submerged hydrophytes).

Taxon Name	Code Name	Functional Type
Berula erecta	*Ber ere*	HE
Callitriche spp.	*Cal sp*	SM
Elodea canadensis	*Elo can*	SM
filamentous algae	*Fil alg*	FIL
Fontinalis antipyretica	*Fon ant*	SM
Glyceria fluitans	*Gly flu*	HE
Hippuris vulgaris	*Hip vul*	AM
Iris pseudacorus	*Iri pse*	HE
Lemna minor	*Lem min*	FLH
Mentha aquatica	*Men aqu*	AM
Myosotis scorpioides	*Myo sco*	AM
Myriophyllum spicatum	*Myr spi*	SM
Nasturtium officinale	*Nas off*	HE
Nuphar luteum	*Nup lut*	FLH
Potamogeton crispus	*Pot cri*	SM
Potamogeton lucens	*Pot luc*	SM
Potamogeton natans	*Pot nat*	FLH
Potamogeton nodosus	*Pot nod*	FLH
Potamogeton perfoliatus	*Pot per*	SM
Stuckenia pectinata	*Stu pec*	SM
Ranunculus trichophyllus	*Ran tri*	SM
Ranunculus circinatus	*Ran cir*	SM
Rumex hydrolapathum	*Rum hyd*	HE
Sagittaria sagittifolia	*Sag sag*	AM
Schoenoplectus lacustris	*Sch lac*	HE
Sparganium emersum	*Spa eme*	AM
Veronica anagallis-aquatica	*Ver ana*	AM

3. Results

3.1. Water Quality Parameters

The pH of the river water was around 8 (Table 1). The average values of electrical conductivity along the stream ranged from 457 to 476 µS cm^{-1} in 1996, 495–542 µS cm^{-1} in 2000, and 410–447 µS cm^{-1} in 2016. The concentrations of NO_3-N were largely uniform along the course (Table 1). Average values were 1.1 mg L^{-1} in 1996, while they were 1.0 mg L^{-1} and 0.9 mg L^{-1} in 2000 and 2016, respectively.

3.2. Species Richness and Abundance of Macrophytes

Twenty-seven macrophyte taxa were found in the river (Table 2). However, the recorded number of macrophytes has not changed much over the 20 years, from 24 in 1996, 24 in 2000, to 23 in 2016. On the contrary, the relative plant abundance (RPA) of the most abundant species varies strongly over the sampling years (Figure 2). *Potamogeton natans*, filamentous algae, *P. perfoliatus*, *Sagittaria sagittifolia*, and *P. lucens* reached the highest RPA values. The abundance of *Hippuris vulgaris* was decreasing with time, while the abundance of *S. sagittifolia* was increasing. In 1996 and 2000, *P. natans* was the most abundant species, but in 2016, *S. sagittifolia* dominated. The abundance of *Elodea canadensis* had been slightly decreasing during the studied period. The abundance of the species *Myriophyllum spicatum* decreased between 1996 and 2000, and it could not even be detected in the third survey in 2016. On the other hand, *P. nodosus* was recorded only in 2016. Many species that were newly detected in 2016 exhibit amphibious characteristics (Figure 2) and have significantly increased their proportions (e.g., *S. sagittifolia*, *Veronica anagallis-aquatica*), while the proportion of submerged hydrophytes decreased (Figure 3).

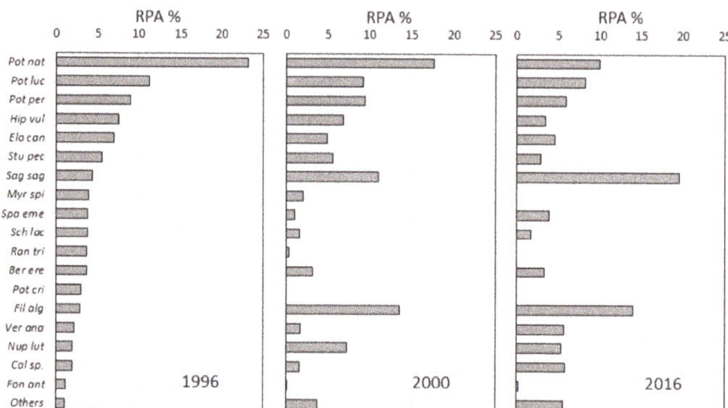

Figure 2. Relative plant abundance (RPA) in 1996, 2000, and 2016. Species with RPA more than 1% abundance are presented. The graphs are based on macrophyte species and abundance in 26 river stretches surveyed in each year. See caption of Table 2 for abbreviations.

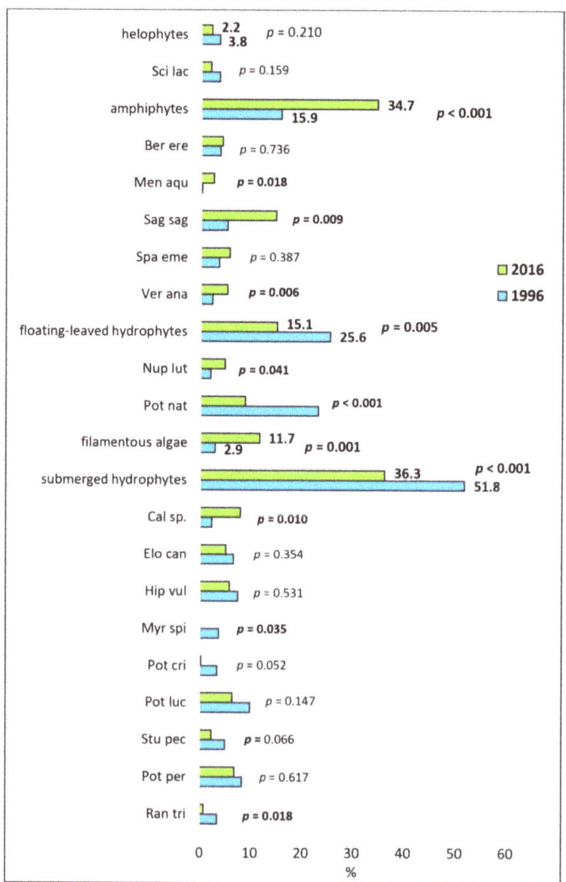

Figure 3. Average proportions of the abundances (in %) of the functional types of aquatic macrophytes, as well as single species with average abundance ≥ 2%. The statistical significance between the years 1996 and 2016 was confirmed with paired *t*-tests. Significant differences over time ($p < 0.05$) are in bold. See caption of Table 2 for abbreviations.

3.3. Changes of Macrophyte Assemblages

The DCA analysis shows the similarity of stretches in terms of the composition of macrophyte assemblages at the peak of the vegetation period in 1996, 2000, and 2016 (Figure 4). The closer the two stretches are on the ordination plot, the more similar the assemblages. In the year 1996, the stretches were more dispersed, but in the subsequent years, the stretches were becoming more uniform.

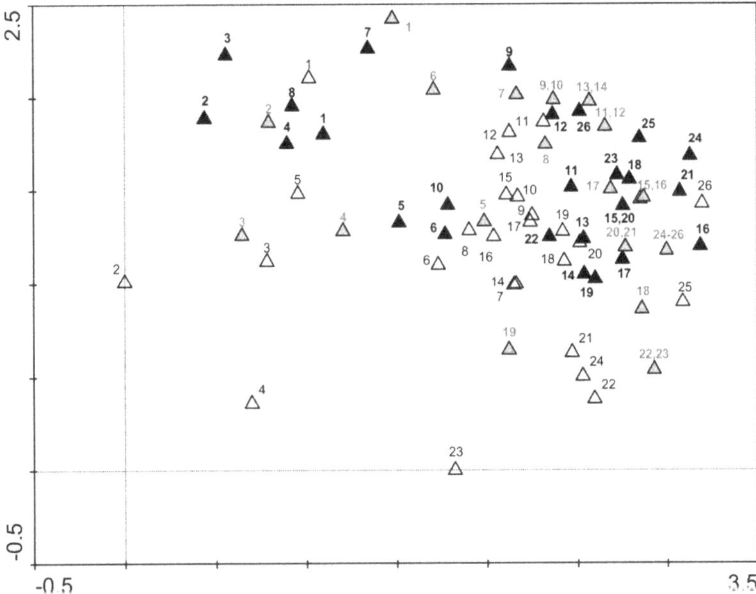

Figure 4. Detrended correspondence analysis ordination diagram showing the similarity among macrophyte assemblages of surveyed stretches in years 1996 (**white triangles**), 2000 (**grey triangles**), and 2016 (**black triangles**). Numbers from 1 to 26 indicate the stretch number (regular, black—1996; bold, grey—2000; bold, black—2016).

The same stretches from the lower half of the Ižica River are close together, and thus we recorded similar macrophyte assemblages in different studied years in this part of the river. Stretches from the upper part of the river are more dispersed, indicating greater differences in macrophyte assemblages during the years. The numbers indicating different stretches on the DCA plot increase from left to right, and indicate the longitudinal effect of the river or the gradual downstream change in the assemblage of macrophyte taxa. Given the presence and abundance of macrophyte taxa, the most different sections in all three years are those in the uppermost flow; the middle and lower course stretches are more grouped.

3.4. Relationships between Species and Environmental Factors

In the year 2016, river stretches were assessed using the RCE inventory of the eco-morphological properties of the river ecosystem. Figure 5 shows the similarity among environmental conditions of surveyed stretches. In general, parameters change along the course of the river, but these changes were not linear as is evident from the distribution of the stretches (Figure 5), which form three clusters and are therefore not evenly distributed. Ecomorphological conditions differed more in the upper course of the river.

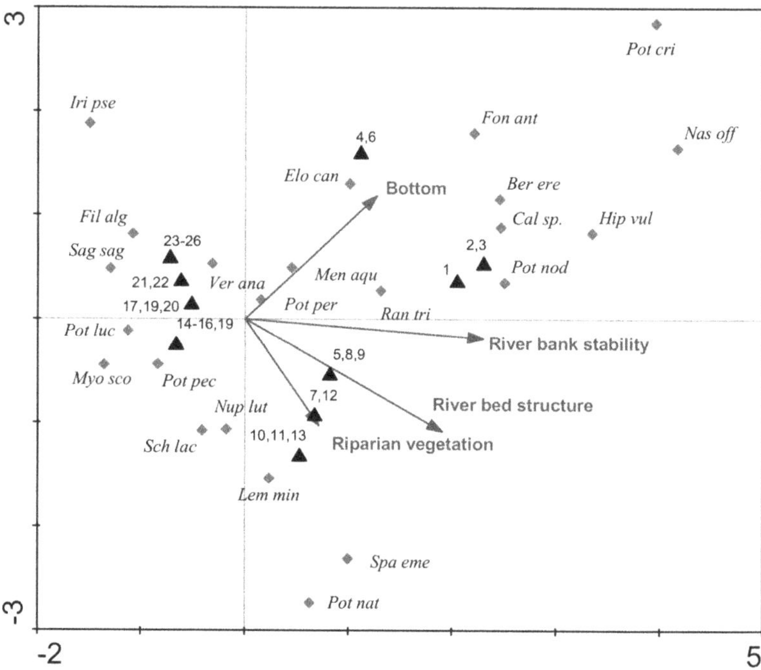

Figure 5. Canonical correspondence analysis (CCA) ordination plot showing the relationship between different locations, macrophytes presence and abundance, and environmental parameters. Abbreviations: *Ber ere*—*Berula erecta*, *Cal* sp.—*Callitriche* spp., *Elo can*—*Elodea canadensis*, *Fil alg*—*filamentous algae*, *Fon ant*—*Fontinalis antipyretica*, *Gly flu*—*Glyceria fluitans*, *Hip vul*—*Hippuris vulgaris*, *Iri pse*—*Iris pseudacorus*, *Lem min*—*Lemna minor*, *Men aqu*—*Mentha aquatica*, *Myo sco*—*Myosotis scorpioides*, *Myr spi*—*Myriophyllum spicatum*, *Nas off*—*Nasturtium officinale*, *Nup lut*—*Nuphar luteum*, *Pot cri*—*Potamogeton crispus*, *Pot luc*—*P. lucens*, *Pot nat*—*P. natans*, *Pot nod*—*P. nodosus*, *Pot per*—*P. perfoliatus*, *Stu pec*—*Stuckenia pectinata*, *Ran tri*—*Ranunculus trichophyllus*, *Sag sag*—*Sagittaria sagittifolia*, *Sch lac*—*Schoenoplectus lacustris*, *Spa eme*—*Sparganium emersum*, *Ver ana*—*Veronica anagallis-aquatica*.

An ordination plot showing the relationship among the composition of macrophyte community and river morphology parameters revealed that four out of twelve parameters were significant, and together explain 43% of species variability. The most influential parameter was riverbank stability, which explained 20% (p = 0.001) of the variability of macrophyte species composition, the riverbed structure explained 10% (p = 0.001), vegetation of riparian zone 7% (p = 0.001), and type of the bottom explained an additional 5% (p = 0.013). The riverbank stability and the riverbed structure are the parameters which correlate most with the first axis and species are clearly distributed along these gradients. Vectors representing the type of the riparian vegetation are most related to the second axis. The stretches of the upstream half of the river show a gradient along the first axis, while the lower part shows the distribution along the second axis and thus the relationship with the vector vegetation of the riparian zone (Figure 5).

3.5. Ecological Status

The values of RMI calculated based on macrophyte species showed changes in the ecological status of different sections of the river (Table 3) and in the ecological status of specific stretches along the course, as well as over the years (Figure 6, Table 4). In 1996, more than half of the river stretches (14) showed very good status, while the rest (12) showed good status, with the better status in the upper half of the flow. The situation had already changed in 2000, as the condition of the source changed to moderate, and the nearby stretches changed to good ecological status (Figure 6, Table 4). In 2000, we classified nine stretches as having very good ecological status, sixteen stretches as good, and one stretch had a moderate ecological status. In 2016, we found only seven stretches that were classified to very good ecological status, eighteen to good ecological status and one to moderate ecological status. The location of the stretches with better ecological status was the opposite as in the year 1996, since in 2016, stretches with very good ecological status were concentrated in the lowest part of the river, which had significantly better status than other parts (Table 3).

Table 3. Average values of RMI in the upper, middle and lower course in different years of the survey, and results of testing for significance (*t*-tests) of these changes along the course of the Ižica River. Significant differences ($p < 0.05$) are in bold.

Year	Average RMI, Upper Section:	*p*	Average RMI, Middle Section:	*p*	Average RMI, Lower Section:	Changes of RMI along the Course:
1996	0.836	0.725 n.s.	0.825	**0.030**	0.742	worst status in lower course
2000	0.74	0.653 n.s.	0.75	0.894 n.s.	0.76	no significant changes
2016	0.71	0.070 n.s.	0.65	**0.0003**	0.78	best status in lower course
three years	0.762	0.492 n.s.	0.745	0.439 n.s.	0.762	no significant changes

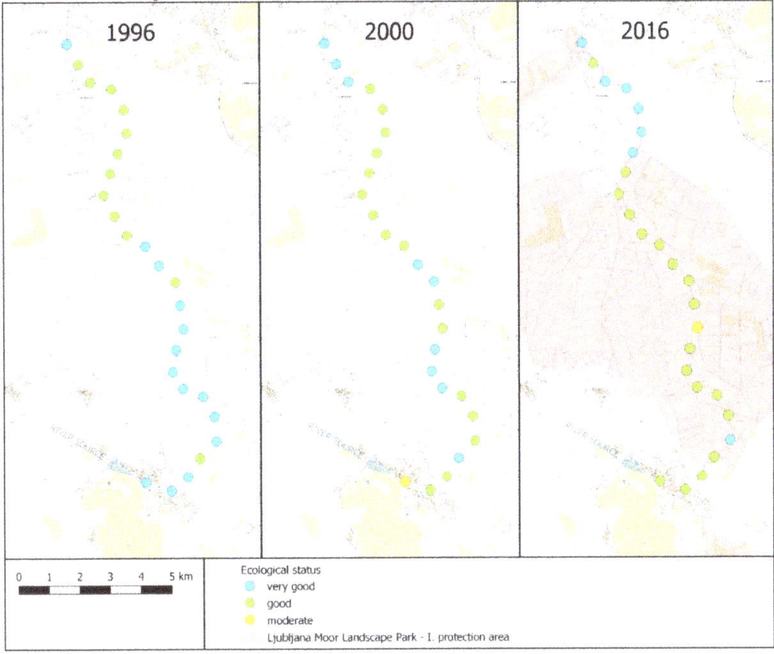

Figure 6. Map of the Ižica River from different years displaying spatial distribution of stretches with different ecological status.

Table 4. Average values of RMI for the entire course of the Ižica River, and for its upper, middle, or lower course in different years of the survey, and results of testing for significance (paired *t*-tests) of these changes over time. Significant differences ($p < 0.05$) are in bold.

year 1996		year 2000		year 2016
average RMI: entire course = 0.80	**p = 0.018**	average RMI: entire course = 0.75 **p = 0.004**	p = 0.151	average RMI: entire course = 0.72
upper = 0.836 middle = 0.825 lower = 0.742		**p = 0.001** **p = 0.003** p = 0.127		upper = 0.71 middle = 0.65 lower = 0.78

4. Discussion

Luxuriant macrophyte growth and species diversity in all three studied years were supported by favorable conditions, which included sufficient light, type of substratum, and non-torrential water regime of the river [45]. Low water velocity and fine sediment favored the growth of aquatic vegetation, especially in the river's lower part, as was also shown in other studies [46,47]. The number of macrophyte taxa in the Ižica River was 24 in 1996, 25 in 2000, and 23 in 2016. High RPA values reached taxa such as *P. natans*, *S. sagittifolia* and *P. lucens*, which were most abundant in all the studied years in the lower part of the Ižica River. Both sediment and plant assemblages were more heterogeneous in the upper section.

We found taxa including *R. trichophyllus*, *Berula erecta*, and *Callitriche* spp., indicating low nutrient levels, as well as species indicating high nutrient levels such as *P. natans* [48], which is ecologically the most tolerant species of all pondweeds [49] with respect to eutrophic conditions or turbid water, and is a typical representative in low current velocity waterbodies. Preston [49] reports that *P. natans* thrives in a variety of ecological conditions, from oligotrophic to eutrophic water with different types of substrates. Due to its floating leaves, *P. natans* reduces light penetration into the water column [50,51] and outcompetes other species. *P. natans* often grows in the company of the species *P. lucens* and *P. nodosus* [52], which were also found in the Ižica River in all three study years. *P. nodosus* and *P. natans* usually thrive in similar ecological conditions, but *P. nodosus* prefers a stony substrate [49], thus it was commonly present in the middle flow of the river. On the other hand, *P. lucens* is often the dominant macrophyte in slow flowing rivers with a fine silty substrate. It is often found in diverse assemblages cohabitating with the species *P. natans*, *N. luteum*, *Sparganium erectum*, *Stuckenia pectinata*, and *P. crispus* [53]. Except the species *P. crispus*, these species were found on a muddy substrate in the lower part of the river Ižica. *P. crispus* can also thrive in parts of the river with a faster flow [54], and it is presumed that this is the main reason that it is found in the upper part, where the flow velocity is higher. *B. erecta* was common along the Ižica River, especially in the upper part. It often occurs in alkaline waters in oligotrophic and mesotrophic states [55]. It could be found as an emergent form in shallow water, with leaves partly floating on the surface in slow streams, and as a submerged form [56]. *S. sagittifolia* and filamentous algae prevailed in the lower part of the river in 2016, in the stretches, where in 1996 and 2000, *P. natans* was dominant.

The invasive alien species *Elodea canadensis* has not increased its abundance in the 20-year period (Figure 3). In fact, its relative abundance even decreased, as this species was the fifth most common species in the year 1996, but in only the ninth place 20 years later (Figure 2). This was previously shown by Kuhar et al. [33]. As reasons for low invasiveness of alien aquatic species, Troia et al. [57] report the low nutrient concentration in the water and a diverse macrophyte community, with efficient competitors. In general, the macrophyte assemblage showed a decrease in abundance of species with a wider ecological range, such as *S. pectinata* ($p = 0.066$) and *M. spicatum* ($p = 0.035$), and an increase in the abundance of taxa with narrower ecological range, such as *Callitriche* sp. ($p = 0.01$) and *S. sagittifolia* ($p = 0.009$), the latter as species with amphibious character (Figure 3). Genus

Callitriche occurred with a high abundance in the upper part of the lowland Ljubljanica River, into which the Ižica River flows [18]. A relatively dry summer and the consequently lower water levels and higher insolation in 2016 [58] may be the reasons for the higher abundance of the amphibious plants, such as *S. sagittifolia*, *Veronica anagallis-aquatica* and *M. aquatica*, compared to other growth forms (Figure 3).

A DCA ordination plot (Figure 4) shows that macrophyte assemblages are changing with time. A longitudinal effect of the downstream change of the macrophyte community is evident. In contrast to stretches of the upper Ižica River, stretches in the lower half of the river were grouped together, showing that, in all years, the macrophyte assemblages of these stretches were similar. This means that the macrophyte community is more homogeneous in the lower part.

Ecomorphological factors with the greatest influence on the species composition of macrophyte community in the river Ižica in 2016 were the vegetation type of the riparian zone, riverbed structure, and the bottom type, that together explained 43% of the species variability. A small number of taxa located in the middle of the ordination plot (Figure 5) correspond to mean values of significant environmental parameters. Hrivnak et al. [47] report that macrophyte community composition in Slovak streams is affected by sediment type, riparian vegetation due to shade of woody vegetation, water depth, NO_2 level and pH. Halabowski and Lewin [20] showed that conductivity, altitude, land use adjacent to the rivers, and the proportion of sand were the most important factors that affected the distribution of macrophytes in rivers in southern Poland. However, Lewin and Szoszkiewicz [13] showed that non-nutrient parameters play an important role in determining macrophyte presence even in rivers with a relatively high input of nutrients.

The river ecological status over three years was estimated using RMI, which is based on the composition of the macrophyte community and considered a list of taxa indicating different ecological status [41]. The comparison of RMI values along the flow in 1996, 2000, and 2016 showed pronounced differences (Figure 6, Table 3). These values were significantly higher ($p = 0.004$) in 1996 (average = 0.80) than in 2016 (average = 0.72) (Table 4). In 1996, the entire river showed a good to very good ecological status, with the worst status in its lowest course (Tables 3 and 4). In 2016, the conditions were significantly worse than in 1996 in the upper course ($p = 0.001$), in the middle course ($p = 0.003$), as well as in the entire river ($p = 0.004$) (Table 4). The stretches with very good ecological status were found mainly in the lower course of the river (Figure 6). This was possibly a consequence of the increasing population of Ig, where the river originates. The population development index in the town of Ig between 2002 and 2012 had been above 125, which is the highest within the Ljubljana metropolitan region [59]. The population of the Municipality of Ig in 1991 was 4498, and in 2015 was 7135, which is a 59% increase [60]. The growth of the settlement was not supported with an adequate municipal sewage collecting system, which has affected the structure of the macrophyte community and the ecological status. Better ecological status in 2016 in the lower part of the river ($p = 0.01$) was possibly the result of the improved infrastructure and introduction of more sustainable agricultural practices in the nearby area, as well as along the Škofeljščica stream (Figure 1), which inflows exactly where the Ižica River leaves the most rigorous area of protection within the Landscape Park and Natura 2000 site. The inflow of the cleaner water improved the ecological status of the Ižica River downstream (Figure 6). The settlements on the eastern bank of the Ižica River were connected to the central wastewater treatment plant (WWTP) of Ljubljana in 2014, so there was no such negative impact on the Ižica River system from the area eastward of the river in 2016.

5. Conclusions

Favorable ecomorphological conditions and moderate concentrations of nutrients in the river support high plant diversity, and we found a high number of macrophyte taxa in the short river Ižica. The ecological status of the river deteriorated significantly ($p = 0.004$) from 1996 to 2016, particularly in the upper part of the river. The most probable reasons for those changes are the karst character of the river source and its catchment area, respectively, and population growth of the town of Ig around the source of the Ižica River, which was not supported with adequate infrastructure for wastewater treatment. Such long-term studies indicate changes in the environment and human attitude to it, and thus present a basis for future management in the watershed of this and similar watercourses.

Author Contributions: Conceptualization, A.G. and M.G.; methodology, A.G. and M.G.; validation, M.G., A.G. and A.P.; formal analysis, A.G., M.H., I.Z. and M.G.; investigation, M.G., A.G. and A.P.; data curation, A.G., M.G., A.P., M.H. and I.Z.; writing—original draft preparation, M.G., A.G. and A.P.; writing—review and editing, M.G., A.G. and I.Z.; visualization, A.G., M.H. and I.Z.; supervision, M.G. and A.G.; funding acquisition, A.G. and I.Z. All authors have read and agreed to the published version of the manuscript.

Funding: This research was funded by the Slovenian Research Agency, within the core research funding Nr. P1-0212, 'Biology of Plants'.

Data Availability Statement: Data could be available on reasonable request and consent of the first and the last author.

Acknowledgments: Authors thank G.W.A. Milne, who checked the language, and to Mateja Dolinšek for her help with fieldwork.

Conflicts of Interest: The authors declare no conflict of interest.

References

1. Boulton, A.J.; Brock, M.A. *Australian Freshwater Ecology: Processes and Management*; Wiley-Blackwell: Gleneagle, Australia, 1999; p. 300.
2. Bornette, G.; Amoros, C.; Chessel, D. Effect of allogenic processes on successional rates in former river channels. *J. Veg. Sci.* **1994**, *5*, 237–246. [CrossRef]
3. Bornette, G.; Puijalon, S. Response of aquatic plants to abiotic factors: A review. *Aquat. Sci.* **2011**, *73*, 1–14. [CrossRef]
4. Lozanovska, I.; Rivaes, R.; Vieira, C.; Ferreira, M.T.; Aguiar, F.C. Streamflow regulation effects in the Mediterranean rivers: How far and to what extent are aquatic and riparian communities affected? *Sci. Total Environ.* **2020**, *749*, 141616. [CrossRef] [PubMed]
5. Janauer, G.; Germ, M.; Kuhar, U.; Gaberščik, A. Ecology of aquatic macrophytes and their role in running waters. In *Macrophytes of the River Danube Basin*; Janauer, G., Gaberščik, A., Květ, J., Germ, M., Exler, N., Eds.; Academia: Prague, Czech Republic, 2018; pp. 13–33.
6. Thomaz, S.M.; da Cunha, E.R. The role of macrophytes in habitat structuring in aquatic ecosystems: Methods of measurement, causes and consequences on animal assemblages' composition and biodiversity. *Acta Limnol. Bras.* **2010**, *22*, 218–236. [CrossRef]
7. Paice, R.L.; Chambers, J.M.; Robson, B.J. Potential of submerged macrophytes to support food webs in lowland agricultural streams. *Mar. Freshw. Res.* **2017**, *68*, 549–562. [CrossRef]
8. Urbanič, G.; Debeljak, B.; Kuhar, U.; Germ, M.; Gaberščik, A. Responses of freshwater diatoms and macrophytes rely on the stressor gradient length across the river systems. *Water* **2021**, *13*, 1814. [CrossRef]
9. Levi, P.S.; Riis, T.; Alnøe, A.B.; Peipoch, M.; Maetzke, K.; Bruus, C.; Baattrup-Pedersen, A. Macrophyte complexity controls nutrient uptake in lowland streams. *Ecosystems* **2015**, *18*, 914–931. [CrossRef]
10. Quilliam, R.S.; van Niekerk, M.A.; Chadwick, D.R.; Cross, P.; Hanley, N.; Jones, D.L.; Vinten, A.J.A.; Willby, N.; Oliver, D.M. Can macrophyte harvesting from eutrophic water close the loop on nutrient loss from agricultural land? *J. Environ. Manag.* **2015**, *152*, 210–217. [CrossRef]
11. Alnoee, A.B.; Levi, P.S.; Baattrup-Pedersen, A.; Riis, T. Macrophytes enhance reach-scale metabolism on a daily, seasonal and annual basis in agricultural lowland streams. *Aquat. Sci.* **2021**, *83*, 11. [CrossRef]
12. Li, L.; Zheng, B.; Liu, L. Biomonitoring and bioindicators used for river ecosystems: Definitions, approaches and trends. *Procedia Environ. Sci.* **2010**, *2*, 1510–1524. [CrossRef]
13. Lewin, I.; Szoszkiewicz, K. Drivers of macrophyte development in rivers in an agricultural area: Indicative species reactions. *Open Life Sci.* **2012**, *7*, 731–740. [CrossRef]
14. Ladislas, S.; El-Mufleh, A.; Gérente, C.; Chazarenc, F.; Andrès, Y.; Béchet, B. Potential of aquatic macrophytes as bioindicators of heavy metal pollution in urban stormwater runoff. *Water Air Soil Pollut.* **2012**, *223*, 877–888. [CrossRef]

15. Mechora, Š.; Germ, M.; Stibilj, V. Selenium and its species in the aquatic moss *Fontinalis antipyretica*. *Sci. Total Environ.* **2012**, *438*, 122–126. [CrossRef] [PubMed]

16. Mechora, Š.; Germ, M.; Stibilj, V. Monitoring of selenium in macrophytes—The case of Slovenia. *Chemosphere* **2014**, *111*, 464–470. [CrossRef]

17. Schulz, M.; Kozerski, H.-P.; Pluntke, T.; Rinke, K. The influence of macrophytes on sedimentation and nutrient retention in the lower river Spree (Germany). *Water Res.* **2003**, *37*, 569–578. [CrossRef]

18. Germ, M.; Janež, V.; Gaberščik, A.; Zelnik, I. Diversity of macrophytes and environmental assessment of the Ljubljanica river (Slovenia). *Diversity* **2021**, *13*, 278. [CrossRef]

19. Germ, M.; Kuhar, U.; Gaberščik, A. Abundance and diversity of Taxa Within the genus *Potamogeton* in Slovenian watercourses. In *Natural and Constructed Wetlands*; Springer International Publishing: Berlin/Heidelberg, Germany, 2016; pp. 283–291.

20. Halabowski, D.; Lewin, I. Impact of anthropogenic transformations on the vegetation of selected abiotic types of rivers in two ecoregions (Southern Poland). *Knowl. Manag. Aquat. Ecosyst.* **2020**, *2020*, 35. [CrossRef]

21. Schinegger, R.; Trautwein, C.; Melcher, A.; Schmutz, S. Multiple human pressures and their spatial patterns in European running waters. *Water Environ. J.* **2012**, *26*, 261–273. [CrossRef]

22. European Commission. E.C. Directive 2000/60/EC of the European Parliament and of the Council of 23 October 2000 Establishing a Framework for Community Action in the Field of Water Policy. Available online: https://eur-lex.europa.eu/eli/dir/2000/60/oj (accessed on 19 October 2021).

23. Vörösmarty, C.J.; McIntyre, P.B.; Gessner, M.O.; Dudgeon, D.; Prusevich, A.; Green, P.; Glidden, S.; Bunn, S.E.; Sullivan, C.A.; Liermann, C.R.; et al. Global threats to human water security and river biodiversity. *Nature* **2010**, *467*, 555–561. [CrossRef]

24. Veit, U.; Kohler, A. Long-term study of the macrophytic vegetation in the running waters of the Friedberger Au (near Augsburg, Germany). *River Syst.* **2003**, *14*, 65–86. [CrossRef]

25. Oťaheľová, H.; Valachovič, M.; Hrivnák, R. The impact of environmental factors on the distribution pattern of aquatic plants along the Danube River corridor (Slovakia). *Limnologica* **2007**, *37*, 290–302. [CrossRef]

26. Hrivnák, R.; Oťaheľová, H.; Valachovič, M. Macrophyte distribution and ecological status of the Turiec River (Slovakia): Changes after seven years. *Arch. Biol. Sci.* **2009**, *61*, 297–306. [CrossRef]

27. Kõrs, A.; Vilbaste, S.; Käiro, K.; Pall, P.; Piirsoo, K.; Truu, J.; Viik, M. Temporal changes in the composition of macrophyte communities and environmental factors governing the distribution of aquatic plants in an unregulated lowland river (Emajõgi, Estonia). *Boreal Environ. Res.* **2012**, *17*, 460–472.

28. Kennedy, M.P.; Lang, P.; Grimaldo, J.T.; Martins, S.V.; Bruce, A.; Hastie, A.; Lowe, S.; Ali, M.M.; Sichingabula, H.; Dallas, H.; et al. Environmental drivers of aquatic macrophyte communities in southern tropical African rivers: Zambia as a case study. *Aquat. Bot.* **2015**, *124*, 19–28. [CrossRef]

29. Schmidt, B.; Janauer, G.A.; Barta, V.; Schmidt-Mumm, U. Breg and Brigach, headstreams of the River Danube: Biodiversity and historical comparison. In *Macrophytes of the River Danube Basin*; Janauer, G., Gaberščik, A., Květ, J., Germ, M., Exler, N., Eds.; Academia: Prague, Czech Republic, 2018; pp. 58–80.

30. Sender, J. The dynamics of macrophytes in a lake in an agricultural landscape. *Limnol. Rev.* **2012**, *12*, 93–100. [CrossRef]

31. Lindholm, M.; Alahuhta, J.; Heino, J.; Hjort, J.; Toivonen, H. Changes in the functional features of macrophyte communities and driving factors across a 70-year period. *Hydrobiologia* **2020**, *847*, 3811–3827. [CrossRef]

32. Lindholm, M.; Alahuhta, J.; Heino, J.; Toivonen, H. Temporal beta diversity of lake plants is determined by concomitant changes in environmental factors across decades. *J. Ecol.* **2021**, *109*, 819–832. [CrossRef]

33. Kuhar, U.; Germ, M.; Gaberščik, A. Habitat characteristics of an alien species Elodea canadensis in Slovenian watercourses. *Hydrobiologia* **2010**, *656*, 205–212. [CrossRef]

34. Budja, M.; Mlekuž, D. Lake or floodplain? Mid-Holocene settlement patterns and the landscape dynamic of the Ižica floodplain (Ljubljana Marshes, Slovenia). *Holocene* **2010**, *20*, 1269–1275. [CrossRef]

35. Urbanc-Berčič, O.; Germ, M. Ižica (Iščica). In *Opredelitev območij evropsko pomembnih negozdnih habitatnih tipov s pomočjo razširjenosti značilnih rastlinskih vrst: Končno poročilo*; Jogan, N., Kotarac, M., Lešnik, A., Eds.; Center za Kartografijo Favne in Flore: Ljubljana, Slovenia, 2004; p. 961.

36. Kohler, A.; Janauer, G.A. Zur Methodik der Untersuchung von aquatischen Makrophyten in Fließgewässern. In *Handbuch Angewandte Limnologie*; Steinberg, C., Bernhardt, H., Klapper, H., Eds.; Ecomed Verlag: Landsberg am Lech, Germany, 1995; Volume VIII-1.1.3, pp. 3–22.

37. Schneider, S.; Melzer, A. The Trophic Index of Macrophytes (TIM)—A new tool for indicating the trophic state of running waters. *Int. Rev. Hydrobiol.* **2003**, *88*, 49–67. [CrossRef]

38. Petersen, R.C. The RCE: A Riparian, Channel, and Environmental Inventory for small streams in the agricultural landscape. *Freshw. Biol.* **1992**, *27*, 295–306. [CrossRef]

39. Germ, M.; Gaberščik, A.; Urbanc-Berčič, O. The wider environmental assessment of river ecosystems. Širša okoljska ocena rečnega ekosistema. *Acta Biol. Slov.* **2000**, *43*, 13–19.

40. Pall, K.; Janauer, G.A. Die Makrophytenvegetation von Flußstauen am Beispiel der Donau zwischen Fluß-km 2552.0 und 2511.8 in der Bundesrepublik Deutschland. *River Syst.* **1995**, *9*, 91–109. [CrossRef]

41. Kuhar, U.; Germ, M.; Gaberščik, A.; Urbanič, G. Development of a River Macrophyte Index (RMI) for assessing river ecological status. *Limnologica* **2011**, *41*, 235–243. [CrossRef]

42. Aguiar, F.C.; Segurado, P.; Urbanič, G.; Cambra, J.; Chauvin, C.; Ciadamidaro, S.; Dörflinger, G.; Ferreira, J.; Germ, M.; Manolaki, P.; et al. Comparability of river quality assessment using macrophytes: A multi-step procedure to overcome biogeographical differences. *Sci. Total Environ.* **2014**, *476–477*, 757–767. [CrossRef] [PubMed]
43. Ter Braak, C.J.F.; Smilauer, P. *CANOCO Reference Manual and CanoDraw for Windows User's Guide: Software for Canonical Community Ordination (Version 4.5)*; Microcomputer Power: Ithaca, NY, USA, 2002.
44. Ter Braak, C.J.F.; Verdonschot, P.F.M. Canonical correspondence analysis and related multivariate methods in aquatic ecology. *Aquat. Sci.* **1995**, *57*, 255–289. [CrossRef]
45. Germ, M.; Gaberščik, A. Macrophytes of the River Ižica—Comparison of species composition and abundance in the years 1996 and 2000. *River Syst.* **2003**, *14*, 181–193. [CrossRef]
46. Vukov, D.; Anačkov, G.; Igić, R.; Janauer, G.A. The aquatic macrophytes of "Mali Derdap" (Danube, rkm 1039-999). *Limnol. Rep.* **2004**, *35*, 421–426.
47. Hrivnák, R.; Ot'ahel'ová, H.; Valachovič, M.; Pal'ove-Balang, P.; Kubinská, A. Effect of environmental variables on the aquatic macrophyte composition pattern in streams: A case study from Slovakia. *Fundam. Appl. Limnol.* **2010**, *177*, 115–124. [CrossRef]
48. Haslam, S.M. *River Plants of Western Europe: The Macrophytic Vegetation of Watercourses of the European Economic Community*; Cambridge University Press: Cambridge, UK, 1987; p. 512.
49. Preston, C.D. Pondweeds of Great Britain and Ireland. In *BSBI Handbook No. 8*; Botanical Society of the British Isles: London, UK, 1995.
50. Chambers, P.A.; Lacoul, P.; Murphy, K.J.; Thomaz, S.M. Global diversity of aquatic macrophytes in freshwater. *Hydrobiologia* **2008**, *595*, 9–26. [CrossRef]
51. Van Vierssen, W. The ecology of communities dominated by *Zannichellia* taxa in western Europe. II. Distribution, synecology and productivity aspects in relation to environmental factors. *Aquat. Bot.* **1982**, *13*, 385–483. [CrossRef]
52. Hollingsworth, P.M.; Preston, C.D.; Gornall, R.J. Isozyme evidence for hybridization between *Potamogeton natans* and *P. nodosus* (Potamogetonaceae) in Britain. *Bot. J. Linn. Soc.* **1995**, *117*, 59–69. [CrossRef]
53. Butcher, R.W. Studies on the ecology of rivers: I. On the distribution of macrophytic vegetation in the rivers of Britain. *J. Ecol.* **1933**, *21*, 58–91. [CrossRef]
54. Bilby, R. Effects of a spate on the macrophyte vegetation of a stream pool. *Hydrobiologia* **1977**, *56*, 109–112. [CrossRef]
55. Robach, F.; Hajnsek, I.; Eglin, I.; Trémolières, M. Phosphorus sources for aquatic macrophytes in running waters: Water or sediment? *Acta Bot. Gall.* **1995**, *142*, 719–731. [CrossRef]
56. Zelnik, I.; Kuhar, U.; Holcar, M.; Germ, M.; Gaberščik, A. Distribution of vascular plant communities in Slovenian watercourses. *Water* **2021**, *13*, 1071. [CrossRef]
57. Troia, A.; Ilardi, V.; Oddo, E. Monitoring of alien aquatic plants in the inland waters of Sicily (Italy). *Webbia* **2020**, *75*, 77–83. [CrossRef]
58. ARSO. ARSO Meteo. Available online: https://www.meteo.si/met/sl/app/webmet/ (accessed on 8 November 2021).
59. Rebernik, D. Population and spatial development of settlements in Ljubljana Urban Region after 2002. *Dela* **2014**, *42*, 75–93. [CrossRef]
60. Statistical Office of the Republic of Slovenia. Available online: https://pxweb.stat.si/SiStatData/pxweb/en/Data/ (accessed on 8 November 2021).

Article

Aquatic Macrophytes Occurrence in Mediterranean Farm Ponds: Preliminary Investigations in North-Western Sicily (Italy)

Patrizia Panzeca [1], Angelo Troia [1,*] and Paolo Madonia [2]

1 Dipartimento di Scienze e Tecnologie Biologiche, Chimiche e Farmaceutiche, Università Degli Studi di Palermo, 90123 Palermo, Italy; patriziapanzeca@gmail.com
2 Istituto Nazionale di Geofisica e Vulcanologia, Sezione di Roma 2, 00143 Roma, Italy; paolo.madonia@ingv.it
* Correspondence: angelo.troia@unipa.it

Abstract: Mediterranean wetlands are severely affected by habitat degradation and related loss of biodiversity. In this scenario, the wide number of artificial farm ponds can play a significant role in the biodiversity conservation of aquatic flora. In the present contribution we show the preliminary results of a study on Mediterranean farm ponds of north-western Sicily (Italy), aimed to investigating the environmental factors linked to the occurrence of submerged macrophytes (vascular plants and charophytes). We studied the aquatic flora of 30 ponds and determined the chemical and isotopic composition of their water bodies on a subset of the most representative 10 sites. Results show that (1) farm ponds host few but interesting species, such as *Potamogeton pusillus* considered threatened at regional level; (2) *Chara vulgaris, C. globularis* and *P. pusillus* behave as disturbance-tolerant species, occurring both in nitrates-poor and nitrates-rich waters, whereas *Stuckenia pectinata* and *Zannichellia palustris* occur only in nitrates-poor waters. Although farm ponds are artificial and relatively poor habitats, these environments seem to be important for the aquatic flora and for the conservation of the local biodiversity, and can give useful information for the use of macrophytes as bioindicators in the Mediterranean area.

Keywords: Mediterranean flora; aquatic flora; carbon isotopes; hydrophytes; conservation; bioindicators; inland waters; water geochemistry

Citation: Panzeca, P.; Troia, A.; Madonia, P. Aquatic Macrophytes Occurrence in Mediterranean Farm Ponds: Preliminary Investigations in North-Western Sicily (Italy). *Plants* **2021**, *10*, 1292. https://doi.org/10.3390/plants10071292

Academic Editor: Pablo García Murillo

Received: 29 May 2021
Accepted: 21 June 2021
Published: 25 June 2021

Publisher's Note: MDPI stays neutral with regard to jurisdictional claims in published maps and institutional affiliations.

1. Introduction

Mediterranean wetlands are severely affected by habitat degradation and related loss of biodiversity [1,2], and for this reason they are protected by various international (EU) and national laws. In spite of this, the knowledge about flora of these wetlands is poor [2], making difficult their proper management.

Sicily is a hotspot of plant biodiversity [3], hosting a plenty of habitats including wetlands (such as saltpans, temporary ponds, coastal ponds, small lakes, etc.), even if a lot of them have been destroyed in recent times, and especially in the last two centuries. The Sicilian landscape is characterized by the occurrence of a wide number of farm ponds, which are supposed to play a significant role for the life of aquatic species, and this is the reason why we focused our attention on their flora, presently poorly investigated.

Farm ponds are usually considered artificial poor habitats, although it has been shown that sometimes they host a significant biodiversity, contributing to maintain rich trophic webs, from producers to consumers and decomposers [4–7]. Although biodiversity of these habitats is usually poorer than that of natural ones [8,9], it has been shown that the two groups are rather comparable as far as hydrophytes and red-listed species is concerned [9].

In the present contribution, we show the first results of a study aimed at individuating the environmental variables linked to the occurrence of aquatic macrophytes (vascular plants and charophytes) in Mediterranean farm ponds of north-western Sicily, with particular attention to the chemical and isotopic characteristics of their waters. Numerous studies have shown the importance of aquatic macrophytes as a key component of lake ecosystems, as they provide refuge and food for various organisms, influence the nutrient availability in water, and enhance the stability of lake shores ([10] and literature cited therein). In detail, some recent studies focused on the possibility of using aquatic macrophytes as bioindicators [11,12], or of testing the effects of high concentrations of nitrates or phosphates in the water (e.g., [13]), but we made special reference to the work of Gallego et al. [14] on the macrophytes in Mediterranean farm ponds.

Our results will be discussed not only with reference to biodiversity and conservation issues, but also to the EU Water Framework Directive 2000/60/EC [15] and to the need of assessing the quality of inland waters through biotic indicators.

2. Study Area Settings

The studied area is located in the western sector of the coastal northern mountainous chain of Sicily (Italy), close to the town of Caccamo, about 35 km SE of the regional capital city of Palermo (Figure 1). It is comprised in the altitudinal belt 200–900 m a.s.l., characterized by a clayey hill landscape, dominated by the SW-NE oriented carbonate massif of Mt. San Calogero (1326 m a.s.l.) and punctuated by sparse outcrops of Mesozoic gypsum (Gessoso-Solfifera Formation). To NW the area is delimited by the Rosamarina basin, created by blocking with a dam the San Leonardo River; its SE limit is the Torto River.

The climate is typical of the Mediterranean area, with mild-humid winters alternating to hot-dry summers. Figure 2 reports the average (1965–1994) monthly air temperatures and rainfall amounts measured at Ciminna, the closest available meteorological station whose orographic location (elevation and distance from the coastline) is similar to that of the investigated area [16]. Minima of temperature (8 °C) are recorded in January, while the hottest months are July and August, with average values over 24 °C. The dry season spans from May to September; the driest month is July, with few millimetres of rain on average.

Land use is characterized by traditional semi-intensive agriculture with a mosaic of cultivated and uncultivated land.

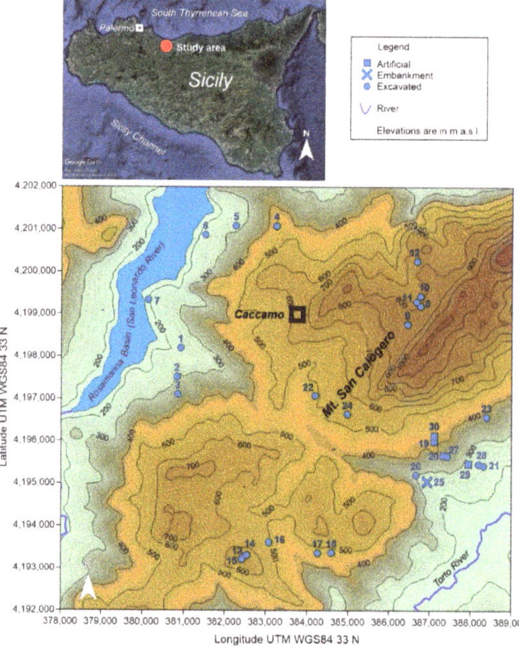

Figure 1. Location of the studied area (at the upper left) and map showing position and constructive characteristics of the ponds.

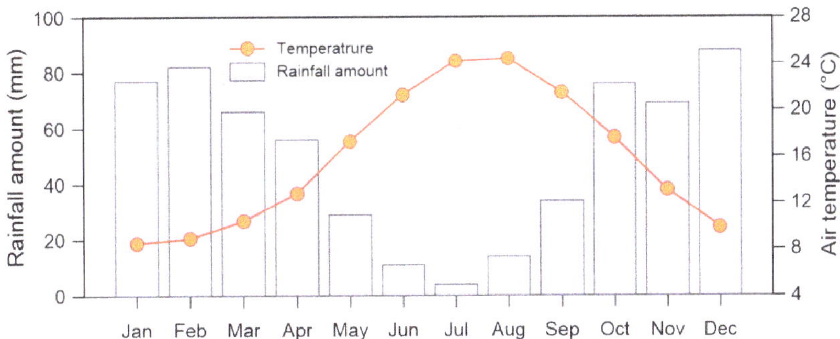

Figure 2. Average (1965–1994) monthly air temperature and rainfall amount; data are from the Ciminna station, the closest available to the study area showing comparable orographic conditions [16].

3. Materials and Methods

Pond selection followed criteria of geo-lithologic representativeness, as well as pond construction-management typologies [14]. Since their construction and management are closely related in our study area, ponds were classified according to these criteria. Artificial ponds (A) are small-sized ponds made by concrete or masonry, which are (usually) intensively managed, showing high water renewal rates and scarce littoral vegetation cover (helophytes and/or riparian vegetation). Excavated ponds (E) exhibit a natural substrate, high water renewal rates, high coverage of littoral vegetation, and are moderately managed. Embankment ponds (D) are obtained by blocking small streams: they have a natural substrate, with a continuous renewal ensured by the flowing water, and rarely managed.

A total of 30 ponds were selected for this study, whose location and characteristics are reported in Figure 1 and Table 1. Their surface areas range from a minimum of 2 m² to a maximum of 2395 m², and their altitudes from 193 to 838 m a.s.l. Only mountain ponds (8–12) fall within a protected area, the Monte San Calogero Regional Natural Reserve, which is also a Natura2000 site (code ITA020033) according to the EU 92/43/CEE "Habitats" Directive. Chemical and isotopic analyses on the stored water bodies were carried out on a subset of 10 sites, representative of the different managing and construction criteria, geo-climatic conditions and water origins. Sites pertaining to this subset are highlighted in bold in Table 1.

Table 1. Sampled sites (in bold those selected for geochemical analyses); geographic coordinates are in decimal degrees WGS84, elevations in m a.s.l., areas in m². "Water" indicates the feeding source: surface run-off (S), direct rain (R), groundwater (G). "Type" is the construction method: excavation (E), concrete or masonry (A), stream barrage (D). "Use" is the utilization: irrigation (IRR), cattle watering (CWA), drinking trough (DTR) and not used (NUS). "Depth": we separated two groups, "S" (shallow ponds, depth 0.5–1.5 m) and "D" (deep ponds, depth 2–8 m). Visibility: we separated two groups, "C" (clear waters) and "T" (turbid waters). The field "SM" reports the number of submerged macrophytes found.

Id	Longitude	Latitude	Elevation	Area	Water	Type	Use	Depth	Visibility	SM
1	13.6451	37.9235	224	638	R-S	E	IRR	D	C	2
2	13.6442	37.9174	309	93	R	E	CWA	D	C	2
3	13.6445	37.9137	349	24	R	E	CWA	S	C	1
4	13.6710	37.9498	438	379	R	E	NUS	D	T	0
5	13.6598	37.9497	280	408	R	E	NUS	D	T	2
6	13.6513	37.9477	220	224	R-S	E	CWA	D	C	1
7	13.6360	37.9337	193	623	R	E	CWA	D	C	1
8	**13.7114**	**37.9332**	**808**	**343**	**G**	**E**	**CWA**	**S**	**C**	**2**
9	**13.7078**	**37.9293**	**838**	**114**	**G**	**E**	**CWA**	**S**	**C**	**1**
10	13.7113	37.9354	786	66	G-R	E	CWA	S	C	2
11	13.7102	37.9341	785	48	G-R	E	CWA	S	C	2
12	13.7103	37.9428	767	213	R	E	CWA	D	T	0
13	13.6638	37.8794	641	42	R	E	CWA	D	T	1

Table 1. *Cont.*

Id	Longitude	Latitude	Elevation	Area	Water	Type	Use	Depth	Visibility	SM
14	13.6644	37.8798	628	113	R	E	NUS	D	T	0
15	13.6630	37.8788	649	43	R	E	CWA	D	T	1
16	13.6704	37.8825	500	129	R	E	IRR	D	C	2
17	13.6914	37.9100	472	273	R	E	NUS	D	T	0
18	13.6877	37.8805	469	422	R	E	NUS	D	T	0
19	13.7159	37.9045	409	64	G	A	IRR	D	C	1
20	13.7186	37.9015	371	16	G	A	NUS	S	C	1
21	13.7297	37.8995	249	527	G-S	E	IRR	D	C	2
22	13.6825	37.9139	486	1034	S	E	IRR	D	C	1
23	13.7303	37.9099	387	401	R	E	IRR	D	T	2
24	13.6914	37.9100	564	2395	S	E	IRR	D	C	2
25	13.7140	37.8959	255	54	S	D	NUS	S	C	1
26	13.7108	37.8973	296	208	S	E	IRR	S	C	1
27	13.7195	37.9014	359	233	G	E	IRR	D	T	0
28	13.7283	37.8998	260	891	G	E	IRR	D	T	0
29	13.7254	37.8998	287	2	G	A	DRT	S	C	1
30	13.7159	37.9057	423	4	G	A	DRT	S	C	0

A single sampling session was carried out on 20 November 2019. Electric conductivity, pH and Eh of water were measured in the field, using electrochemical sensors. Water clarity was measured in situ with a Secchi disc. Samples for the determination of dissolved major and trace elements (Table 1), taken close to the free water surface, were first filtered using 0.45 μm Millipore MF filter and then collected in LD-PE bottles for major element analyses, acidifying with HNO_3 to ca. pH 2 the aliquot destined to cation determination. Untreated aliquots were collected for isotopic and alkalinity determinations, made via titration with HCl (0.1 N).

Water samples were analysed at the lab facilities of Istituto Nazionale di Geofisica e Vulcanologia (INGV), Sezione di Palermo. Major ions were determined by ionic chromatography, using Dionex columns AS14 and CS12 for anions and cations respectively. The determination of $\delta^{18}O$ [17] of water was performed by CO_2-water equilibration technique using a Thermo Delta V Plus instrument, equipped with a Gas Bench II. The results were reported in δ‰ versus the V-SMOW standard, with a precision better than ±0.1‰.

We focused our study to the strictly aquatic macrophytes, corresponding to the hydrophytes of the Raunkiaer's system [18], including submersed, floating-leaved, and free-floating species, and not including the emergent ones ("helophytes", as in [10]). The floristic survey was performed between June and November 2019: all the 30 ponds were sampled a first time between June and July, and a second time between October and November. Macrophytes were collected from the shores, using a grapnel or a rake. The abundance of macrophytes was evaluated, on sight from the shore, using a 5-rank qualitative estimation scale: 1 = very rare; 2 = infrequent; 3 = common; 4 = frequent; 5 = abundant/predominant. Every pond was considered as a single sampling unit. Since our ponds were relatively shallow and small, we are confident that our observations on composition and abundance of aquatic macrophytes, made from the shores, represent composition and abundance of the whole pond. After washing the fresh material to remove sediments and organic matter, fresh (charophytes) or dried (angiosperms) were observed using a stereomicroscope (Leica MZ9.5, maximum magnifications of 60×). Characeae were identified and named following Mouronval et al. (2015) [19], vascular plants were identified and named after Pignatti et al. (2017–2019) [20].

Principal component and correlation analyses were performed using PAST software (version 3.26 [21]). Kendall's tau test was used to verify the correlation between measured parameters (including macrophytes occurrence). Kendall's correlation is a nonparametric measure of the strength of the associations between two variables, which can be also used in the analysis of correlations between quantitative and ordinal parameters in case of a small number of observations. Then, statistical significance was tested, and *p*-values were determined.

4. Results

4.1. Physical and Chemical Characterization of the Ponds

Chemical and isotopic data in the selected subset are presented in Table 2. A graphical representation of the chemical composition of water accumulated in the ponds, based on the Langelier–Ludwig diagram [22] is given in Figure 3. The diagram describes the geochemical facies of a natural water considering the concentrations (expressed in meq L^{-1}) of two couples of main cations and two of anions, and reporting to 50 the sums of anions and cations, respectively.

Table 2. Physical, chemical and isotopic data of the water stored in the pond subset selected for analyses, including presence of macrophytes in each pond. Temperature (T) in °C, electric conductivity (EC) in μS cm^{-1}, Eh in mV, dissolved ions in meq L^{-1} (except NO_3^- and PO_4^{3-} in mg L^{-1}), $\delta^{18}O$ in ‰ V-SMOW, b.d.l. is below detection limit (0.01 meq L^{-1}). Abbreviations for macrophytes species (whose abundance was evaluated using a 5-degree estimation scale, see the text): Cha vul (*Chara vulgaris*), Cha glo (*Chara globularis*), Zan pal (*Zannichellia palustris*), Stu pec (*Stuckenia pectinata*), Pot pus (*Potamogeton pusillus*).

Id	8	9	16	19	20	21	24	28	29	30
T	10.9	12.8	13.1	15.4	14.0	16.2	14.3	15.6	19.3	18.5
EC	572	452	834	805	720	1012	1363	1152	1052	832
pH	7.51	8.59	8.40	7.80	7.51	8.02	8.31	8.03	7.00	6.98
Eh	−50	−108	−98	−66	−50	−77	−93	−78	−23	−22
Na^+	5.40	2.72	1.15	4.86	3.47	2.32	1.62	1.73	5.87	5.03
K^+	0.09	0.06	0.16	0.22	0.24	0.11	0.14	0.13	0.16	0.22
Mg^{2+}	2.24	2.14	1.59	0.66	0.71	1.56	3.87	2.07	1.07	0.66
Ca^{2+}	3.14	1.24	2.70	6.97	5.56	5.73	5.54	6.77	8.75	7.15
F^-	0.01	0.02	0.03	b.d.l.	0.03	0.07	0.03	0.09	0.02	b.d.l.
Cl	1.05	1.84	1.73	1.29	1.34	2.93	3.23	3.27	1.78	1.33
Br^-	b.d.l.	0.05	b.d.l.	b.d.l.	b.d.l.	b.d.l.	b.d.l.	b.d.l.	b.d.l.	b.d.l.
NO_3^-	b.d.l.	b.d.l.	b.d.l.	69.4	50.8	19.2	b.d.l.	45.3	68.8	74.4
SO_4^{2-}	0.13	0.37	5.42	2.06	2.39	5.95	11.0	7.28	3.39	2.17
PO_4^{3-}	b.d.l.	b.d.l.	b.d.l.	b.d.l.	b.d.l.	b.d.l.	b.d.l.	b.d.l.	4.75	0.95
CO_3^{2-} + HCO_3^-	5.40	2.72	1.15	4.86	3.47	2.32	1.62	1.73	5.87	5.03
$\delta^{18}O$	−6.15	−1.64	0.88	−6.33	−5.98	−3.38	0.49	1.26	−5.85	−6.13
Cha vul	4	5		4						
Cha glo			4				4	4	1	
Zan pal	4									
Stu pec							4	4		
Pot pus			4		5					

The analysed water samples fall in a triangular area delimited by the compositions typical of three end members: seawater, carbonate and selenitic waters. This distribution reflects the characteristics of the major sources of dissolved solids. Groundwater circulates into aquifers hosted in limestones and dolostones outcropping at Mt. San Calogero (Figure 1), and water-rock interactions are responsible of the carbonate nature of these waters, whose best representation is found in site 8. This pure carbonate character is progressively modified by the chemical interactions with the atmospheric particulate, also composed by sea spray (upper vertex of the triangular area) and gypsum particles (lower left vertex), which are sources of alkali, chlorine and sulphates.

The dissolution of air particulate, both suspended in the atmosphere or deposited on the ground and leached by the flowing water, gives the chemical imprinting to sites 16, 21, 24 and 28, fed by surface runoff and direct rain, and less dependent on groundwater contributions.

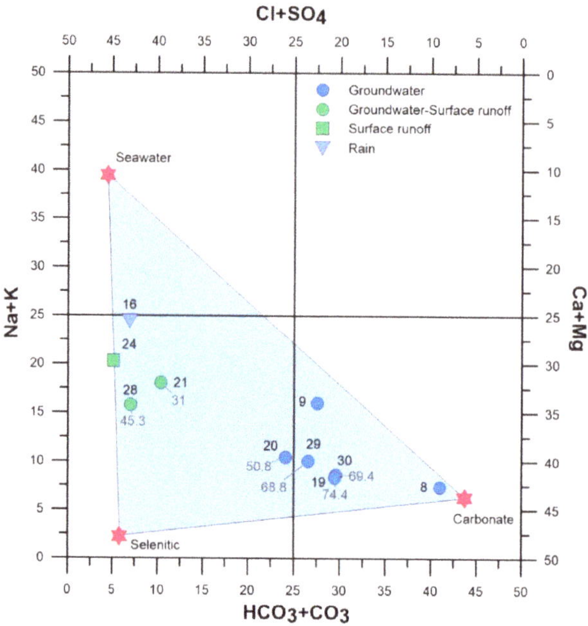

Figure 3. Langelier–Ludwig diagram showing the chemical facies of the analysed waters (numbers of the sites according to Table 1). Grey labels associated to sample points indicate the concentration of nitrates, expressed in mg L^{-1}.

These different feeding sources are also reflected by the concentrations of nitrates (Figure 3, grey labels), which in sites 19, 29 and 30, and in 20 at a lesser extent, are over the important threshold of 50 mg L^{-1} (on this topic see [23]). The higher concentrations are found in ponds fed by groundwater, indicating that nitrates, coming from fertilizers dispersed on the ground, are leached by infiltrating rainwaters and successively delivered to groundwater bodies.

Other information, useful for identifying the main sources feeding the ponds, are given by the plot illustrating the variation with altitude of the oxygen isotopic compositions of the pond water (Figure 4).

Points representative of ponds fed by groundwater (ponds number 8, 19, 20, 29 and 30) show small vertical variations of $\delta^{18}O$: only few decimals of per mil around an average value of $-6‰$, which is congruent with local reference values for groundwater bodies [24]. The sole exception is pond 9: although it is fed by groundwater, the volume of water accumulated inside is so small to be strongly affected by evaporation, thus explaining the shift towards positive values. Values close or over 0‰, typical of surface evaporated water at these latitudes, are shown by sites 9, 16, 24 and 28, confirming that the provenience of water is more linked to surface runoff and/or direct rain; site 21 falls in an intermediate position between the above-described groups. These different positive isotopic shifts remark that evaporation have occurred at different extents, controlled by different factors. Ponds 9, 24 and 28 have low depths and large surface areas, which are conditions fostering direct evaporation, and consequently a positive isotopic shift. An additional influencing factor is the inter-time between two consecutive water extractions from the ponds (renewal time): the longer it is, the most the water will be affected by evaporation.

A principal component analysis has been performed on the data shown in Table 2 (excluding temperatures and macrophytes occurrence), using a correlation matrix since variables are measured in different units (Figure 5). PCA axis 1 explained 53.395% of the total variance, with d$\delta^{18}O$, pH, SO$_4$ and Cl showing positive loadings and Na, T.A., Eh and

NO_3 showing negative loadings. PCA axis 2 explained 26.217% of the total variance, with EC showing positive loading and Br showing negative loading. According to the analysis above, we found sites 19, 20, 29, 30 on the left side of the diagram, representing ponds fed by groundwater and rich in NO_3; sites 16, 21, 24, 28 on the right part of the diagram, representing ponds fed by surface runoff and/or direct rain, and rich in SO_4; ponds 8 and 9 fall in the lower part of the diagram, including (mountain) ponds with low values of EC.

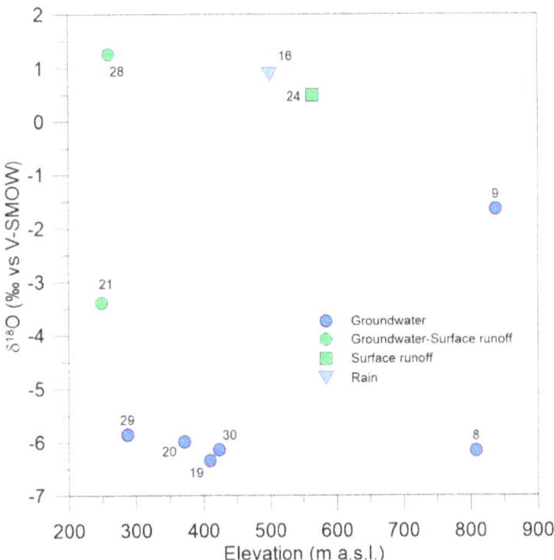

Figure 4. Vertical oxygen isotopic gradient of water accumulated into the studied ponds (numbers of the sites according to Table 1).

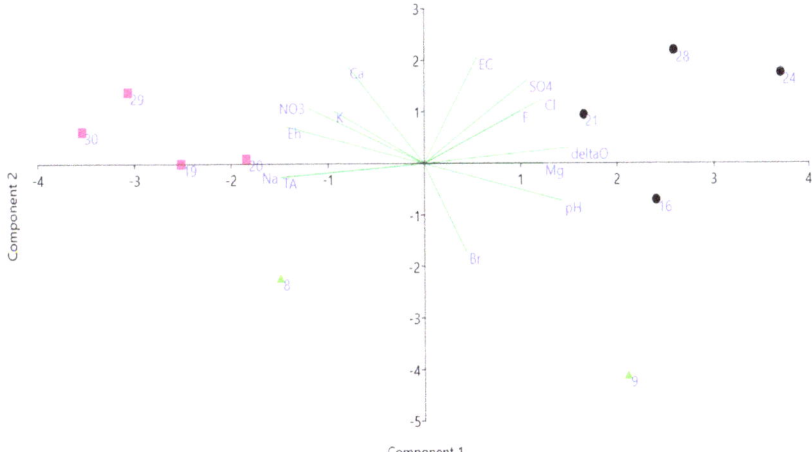

Figure 5. PCA of data shown in Table 2, excluding temperature and macrophytes occurrence. Abbreviations as in Table 2. Pink squares—ponds rich in NO_3, black circles—ponds rich in SO_4, green triangles—ponds with low values of EC (numbers according to Table 1).

4.2. Flora of the Ponds

In 22 out of 30 ponds one aquatic macrophyte at least was retrieved; minimum occurrence was zero, maximum two, with a median of one species. A total of five species was found in the ponds: two charophytes (*Chara vulgaris* L. and *C. globularis* Thuill.) and three angiosperms (*Potamogeton pusillus* L., *Stuckenia pectinata* (L.) Börner, *Zannichellia palustris* L.). All of them are rooted submerged plants; no rooted-floating or free-floating species were found. If one species occurs in a pond, it is a charophyte or an angiosperm, but if two species are present, one is a charophyte and the other one is an angiosperm (we never found two different angiosperms or two different charophytes in the same pond). *Potamogeton pusillus* is a species rare in Sicily, according to the regional red-list [25] where it is listed as vulnerable. No alien aquatic macrophytes were found.

The results of the correlation analysis (using Kendall's tau test) are shown in Figure 6. Regarding the correlations between different macrophytes, there is a significant positive correlation between *Stuckenia pectinata* and *Chara globularis*. Both charophytes show a significant correlation with EC and SO_4, which is positive for *Chara globularis* and negative for *C. vulgaris*. *Stuckenia pectinata* shows a significant positive correlation with SO_4.

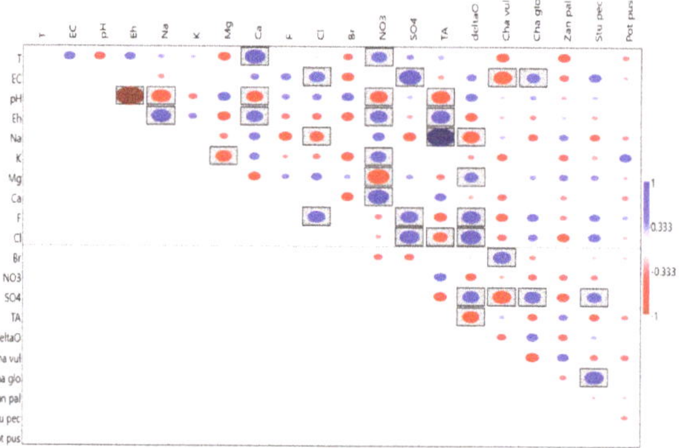

Figure 6. Correlation analysis using Kendall's tau test; $p < 0.05$ are boxed. Abbreviations as in Table 1.

Correlation between *S. pectinata* and *C. globularis* is also evidenced by PCA made on macrophytes' occurrences (Table 2 and Figure 7); we used a var–covar matrix because variables are measured in the same units. PCA axis 1 explained 48.053% of the total variance, axis 2 28.321%; *C. vulgaris* and *Z. palustris* fall in the same quarter, but without a significant correlation, as shown in Figure 6. The studied ponds occupy different areas of the diagram illustrated in Figure 7, partially reproducing the groups based on chemical and physical parameters evidenced in Figure 5: ponds 8 and 9 are on the left side, not far from pond 19, ponds 16, 21 and 24 are on the right side, pond 28 is coincident with pond 30, while ponds rich in NO_3 are scattered.

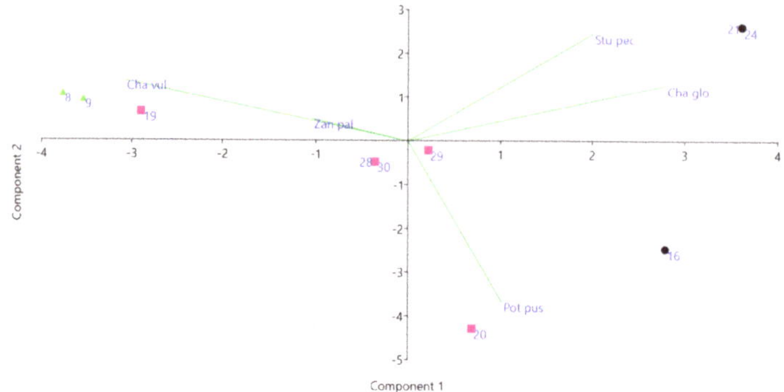

Figure 7. PCA of macrophytes' occurrences shown in Table 2. Pink squares—ponds rich in NO3, black circles—ponds rich in SO4, green triangles—ponds with low values of EC (numbers according to Table 1).

5. Discussion

5.1. Role of Man-Made Farm Ponds for the Biodiversity of the Agricultural Landscape

In Europe, traditional agriculture has created landscapes of considerable conservation value both from ecological and cultural points of view [26,27]. In this context, farm ponds have received considerable attention because of their critical role for the conservation of biodiversity in agricultural landscapes [28–33].

Our preliminary analysis has shown that farm ponds can contribute to increase the overall biodiversity of the agricultural landscape, hosting species not present otherwise. Ponds also act as a refuge station for the rare *Potamogeton pusillus*, included in the Red List of the vascular flora of Sicily. Submerged hydrophyte meadows represent a feeding source and a shelter for the wild fauna, and play a role in water purification [10].

In the framework of the European "Habitats" Directive, although only mountain ponds (sites 8–12) fall within a Natura2000 site (code ITA020033), in the ponds whose bottom is covered totally or partially by charophytes we can recognize a habitat of community interest, i.e., "Hard oligo—mesotrophic waters with benthic vegetation of *Chara* spp." (code 3140).

The number of hydrophytes we found in the pilot area is comparable with other similar contexts (see for example [34]). However, a "rich" pond of our study hosts one charophyte and one angiosperm. Conversely, the Rebuttone gorge near Palermo hosts at least three charophytes and two angiosperms [35], and in the well preserved mountain ponds of the Nebrodi Mountains up to ten vascular hydrophytes (apart from charophytes) can be found in a single pond [36].

Chara globularis is found together with *Stuckenia pectinata* in six ponds, associated to *Potamogeton pusillus* in one site. In both cases, in phytosociological terms, the two co-occurring species do not constitute an association, as—although they are present in the same pond-they constitute two (physically and ecologically) well-separated monospecific communities, to be referred to two different phytosociological classes: *S. pectinata* and *P. pusillus* are to be referred to the class *Potametea pectinati* while *C. globularis* is to be attributed to the class *Charetea intermediae* (for the relationship between the two classes see [35]).

5.2. Use of Aquatic Macrophytes for Monitoring the Quality of Inland Waters in the Mediterranean Area

Data here reported refer to a survey limited both in space and time, not taking into account either seasonal variations or the role of macrophytes on biogeochemical cycle, as those of nitrogen and phosphorus. Nevertheless, their usefulness is remarkable, considering the scarcity of information about the aquatic flora of artificial ponds of the central Mediterranean region, and of Sicily in particular. According to the Water Framework Directive [15], species composition and abundance of macrophytes are biological components to be used in the assessment of the ecological quality of freshwaters. In this scenario, even if the Directive excludes the monitoring of lakes and reservoirs with surfaces <0.2 Km^2 and <0.5 Km^2, respectively, our preliminary data will be useful in testing the application of ecological synthetic indexes, with particular reference to macrophyte indexes suitable for Mediterranean farm ponds.

Phosphates are absent, except in the ponds 29 and 30, where they are present with high concentrations, while nitrate concentrations are high in ponds 29, 30, 19 and 20, with the last two fed by 30. Nitrates remain below 50 mg L^{-1} in ponds 28 and 21, which are fed by 29 entirely or partially, respectively. Nitrates are absent in the other ponds. Based on nitrate concentrations, ponds can be divided in two different groups:

-   Ponds entirely or partially fed by groundwater, rich in nitrates (29, 30, 19, 20) or with values close to the eutrophication threshold (28 and 21); the sole exception is site 8, located in the poorly anthropized area of Monte San Calogero, where nitrates are absent.
-   Ponds fed by surface waters, characterized by low nitrate concentrations (24, 16 and 9).

With the exceptions of pond 19, where *Chara vulgaris* is present, and pond 29, where scattered specimens of *C. globularis* have been found, in both cases associated to more or less abundant filamentous algae, the two charophytes are associated with non-eutrophic waters, confirming that the presence of *C. vulgaris* and *C. globularis* indicates a 'good' ecological status of the waters in shallow lakes [10,37].

Stuckenia pectinata and *Potamogeton pusillus* often grow in co-presence with charophytes that, as just discussed, are indicators of a good ecological status, but in our study area they are absent if nitrates are present in high concentrations. It is well known that *S. pectinata*, commonly associated with eutrophic hard waters, occurs even at low concentrations of nutrients [38]. Our findings then confirm that *S. pectinata* and *P. pusillus* are not exclusive of eutrophic waters, but should be considered disturbing-tolerant and growing in both oligotrophic and eutrophic waters.

Zannichellia palustris occurs in a single pond (associated to *C. vulgaris*), whose water is poor in nitrates: according to the literature it is another disturbing-tolerant species like the previous ones [37].

6. Conclusions

Although preliminary, our study showed that an artificial habitat such as a farm pond can be strategically important for the surviving of aquatic species (plants and charophytes, in this case): here in fact, we found a red-listed angiosperm, and a habitat of community interest. In addition, the presence of those (even few) species can be linked to the quality of the water and in general to the quality of the environment, supplying useful data for the development of ecological indexes that can be applied in artificial and natural ponds of the Mediterranean region.

Author Contributions: Conceptualization, A.T.; methodology, A.T. and P.M.; investigation, P.P., A.T. and P.M.; data curation, P.P.; formal analysis, P.M. and A.T.; writing—original draft preparation, A.T. and P.M.; writing—review and editing, A.T., P.M. and P.P. All authors have read and agreed to the published version of the manuscript.

Funding: The work of A.T. was supported by the Ministero dell'Istruzione, dell'Università e della Ricerca (MIUR) della Repubblica Italiana, fund name 'Incentivi alle attività base di ricerca' (PJ_RIC_FFABR_2017_180267).

Institutional Review Board Statement: Not applicable.

Informed Consent Statement: Not applicable.

Data Availability Statement: All relevant data are presented as tables or can be found in the cited references.

Conflicts of Interest: The authors declare no conflict of interest.

References

1. Galewski, T.; Balkiz, O.; Beltrame, C.; Chazée, L.; Elloumi, M.-J.; Grillas, P.; Jalbert, J.; Khairallah, M.; Korichi, N.; Logotheti, A.; et al. *Biodiversity–Status and Trends of Species in Mediterranean Wetlands*; Mediterranean Wetlands Observatory: Tour du Valat, France, 2012.
2. Leberger, R.; Geijzendorffer, I.R.; Gaget, E.; Gwelmami, A.; Galewski, T.; Pereira, H.M.; Guerra, C.A. Mediterranean wetland conservation in the context of climate and land cover change. *Reg. Environ. Chang.* **2020**, *20*, 67. [CrossRef]
3. Médail, F.; Quézel, P. Biodiversity Hotspots in the Mediterranean Basin: Setting global conservation priorities. *Conserv. Biol.* **1999**, *13*, 1510–1513. [CrossRef]
4. Ruggiero, A.; Céréghino, R.; Figuerola, J.; Marty, P.; Angélibert, S. Farm ponds make a contribution to the biodiversity of aquatic insects in a French agricultural landscape. *Comptes Rendus Biol.* **2008**, *331*, 298–308. [CrossRef]
5. Casas, J.; Toja, J.; Bonachela, S.; Fuentes, F.; Gallego, I.; Juan, M.; León, D.; Peñalver, P.; Pérez, C.; Sánchez, P. Artificial ponds in a Mediterranean region (Andalusia, southern Spain): Agricultural and environmental issues. *Water Environ. J.* **2011**, *25*, 308–317. [CrossRef]
6. Fuentes-Rodríguez, F.; Juan, M.; Gallego, I.; Lusi, M.; Fenoy, E.; León, D.; Peñalver, P.; Toja, J.; Casas, J.J. Diversity in Mediterranean farm ponds: Trade-offs and synergies between irrigation modernisation and biodiversity conservation. *Freshw. Biol.* **2013**, *58*, 63–78. [CrossRef]

7.   Briggs, A.; Pryke, J.S.; Samways, M.J.; Conlong, D.E. Macrophytes promote aquatic insect conservation in artificial ponds. *Aquat. Conserv. Mar. Freshw. Ecosyst.* **2019**, 1–12. [CrossRef]

8.   Della Bella, V.; Bazzanti, M.; Dowgiallo, M.G.; Iberite, M. Macrophyte diversity and physico-chemical characteristics of Tyrrhenian coast ponds in central Italy: Implications for conservation. *Hydrobiologia* **2008**, *597*, 85–95. [CrossRef]

9.   Bubíková, K.; Hrivnák, R. Artificial ponds in Central Europe do not fall behind the natural ponds in terms of macrophyte diversity. *Knowl. Manag. Aquat. Ecosyst.* **2018**, *419*, 8. [CrossRef]

10. Stefanidis, K.; Sarika, M.; Papastergiadou, E. Exploring environmental predictors of aquatic macrophytes in water-dependent Natura 2000 sites of high conservation value: Results from a long-term study of macrophytes in Greek lakes. *Aquat. Conserv. Mar. Freshw. Ecosyst.* **2019**, *29*, 1133–1148. [CrossRef]

11. Zervas, D.; Tsiaoussi, V.; Tsiripidis, I. HeLM: A macrophyte-based method for monitoring and assessment of Greek lakes. *Environ. Monit. Assess.* **2018**, *190*, 326. [CrossRef]

12. Schneider, S.C.; Trajanovska, S.; Biberdžić, V.; Marković, A.; Talevska, M.; Imeri, A.; Veljanoska-Sarafiloska, E.; Đurašković, P.; Jovanović, K.; Cara, M. The Balkan macrophyte index (BMI) for assessment of eutrophication in lakes. *Acta Zool. Bulg.* **2020**, *72*, 439–454.

13. Rodrigo, M.A.; Puche, E.; Rojo, C. On the tolerance of charophytes to high-nitrate concentrations. *Chem. Ecol.* **2018**, *34*, 22–42. [CrossRef]

14. Gallego, I.; Pérez-Martínez, C.; Sánchez-Castillo, P.M.; Fuentes-Rodríguez, F.; Juan, M.; Casa, J.J. Physical, chemical, and management-related drivers of submerged macrophyte occurrence in Mediterranean farm ponds. *Hydrobiologia* **2015**, *762*, 209–222. [CrossRef]

15. European Commission. E.C. Directive 2000/60/EC of the European Parliament and of the Council of 23 October 2000 establishing a framework for community action in the field of water policy. *Off. J. Eur. Comm.* **2000**, *327*, 1–71.

16. *Climatologia Della Sicilia 16*; Regione Siciliana, Assessorato Agricoltura e Foreste, Gruppo IV, Unità di Agrometeorologia: Palermo, Italy, 1998; p. 69. Available online: http://www.sias.regione.sicilia.it/pdf/Climatologia_sicilia.pdf (accessed on 12 April 2021). (In Italian)

17. Andreas, W.H.; Hairigh, A. *18O-Equilibration on Water, Fruit Juice and Wine Using Thermo Scientific GasBench II*; Thermo Fisher Scientific: Brema, Germany, 2008.

18. Raunkiaer, C.C. *The Life Forms of Plants and Statistical Plant Geography*; Oxford University Press: Oxford, UK, 1934.

19. Mouronval, J.B.; Baudouin, S.; Borel, N.; Soulié-Märsche, I.; Klesczewski, M.; Grillas, P. *Guide des Characées de France Méditerranéenne*; Office National de la Chasse et Faune Sauvage: Paris, France, 2015.

20. Pignatti, S.; La Rosa, M.; Guarino, R. *Flora d'Italia*, 2nd ed.; Edagricole—New Business Media: Milano-Bologna, Italy, 2017; Volume 2.

21. Hammer, Ø.; Harper, D.A.T.; Ryan, P.D. Past: Paleontological Statistics Software Package for Education and Data Analysis. *Palaeontol. Electron.* **2001**, *4*, 9.

22. Langelier, W.F.; Ludwig, F. Graphical methods for indicating the mineral character of natural waters. *J. Am. Water Works Assoc.* **1942**, *34*, 335–352. [CrossRef]

23. Paladino, O.; Massabò, M.; Gandoglia, E. Assessment of Nitrate Hazards in Umbria Region (Italy) Using Field Datasets: Good Agriculture Practices and Farms Sustainability. *Sustainability* **2020**, *12*, 9497. [CrossRef]

24. Liotta, M.; Grassa, F.; D'Alessandro, W.; Favara, R.; Gagliano Candela, E.; Pisciotta, A.; Scaletta, C. Isotopic composition of precipitation and groundwater in Sicily, Italy. *Appl. Geochem.* **2013**, *34*, 199–206. [CrossRef]

25. Raimondo, F.M.; Bazan, G.; Troia, A. Taxa a rischio nella flora vascolare della Sicilia. *Biogeographia* **2011**, *30*, 229–239. [CrossRef]

26. Schmitt, T.; Rákosy, L. Changes of traditional agrarian landscapes and their conservation implications: A case study of butterflies in Romania. *Divers. Distrib.* **2007**, *13*, 855–862. [CrossRef]

27. Hartel, T.; Schweiger, O.; Ollerer, K.; Cogalniceanu, D.; Arntzen, J.W. Amphibian distribution in a traditionally managed rural landscape of Eastern Europe: Probing the effect of landscape composition. *Biol. Conserv.* **2010**, *143*, 1118–1124. [CrossRef]

28. Williams, P.; Whitfield, M.; Biggs, J.; Bray, S.; Fox, G.; Nicolet, P.; Sear, D. Comparative biodiversity of rivers, streams, ditches and ponds in an agricultural landscape in Southern England. *Biol. Conserv.* **2004**, *115*, 329–341. [CrossRef]

29. Davies, B.; Biggs, J.; Williams, P.; Whitfield, M.; Nicolet, P.; Sear, D.; Bray, S.; Maund, S. Comparative biodiversity of aquatic habitats in the European agricultural landscape. *Agric. Ecosyst. Environ.* **2008**, *125*, 1–8. [CrossRef]

30. Sebastián-González, E.; Sánchez-Zapata, J.A.; Botella, F. Agricultural ponds as alternative habitat for waterbirds: Spatial and temporal patterns of abundance and management strategies. *Eur. J. Wildl. Res.* **2010**, *56*, 11–20. [CrossRef]

31. Casas, J.J.; Toja, J.; Peñalver, P.; Juan, M.; León, D.; Fuentes-Rodríguez, F.; Gallego, I.; Fenoy, E.; Pérez-Martínez, C.; Sánchez, P.; et al. Farm Ponds as Potential Complementary Habitats to Natural Wetlands in a Mediterranean Region. *Wetlands* **2012**, *32*, 161–174. [CrossRef]

32. Bolpagni, R.; Laini, A.; Stanzani, C.; Chiarucci, A. Aquatic Plant Diversity in Italy: Distribution, Drivers and Strategic Conservation Actions. *Front. Plant Sci.* **2018**, *9*, 116. [CrossRef]

33. Marcenò, C.; Minissale, P.; Sciandrello, S.; Cuena-Lombraña, A.; Fois, M.; Bacchetta, G. The MedIsWet project in Sicily and Sardinia. An opportunity for improving wetland knowledge and conservation. Preliminary results. In *Abstract Book of the 28th Meeting Proceedings of the European Vegetation Survey, Vegetation Diversity and Global Change, Madrid, Spain, 2–6 September 2019*; Gavilán, R.G., Gutiérrez-Girón, A., Eds.; University of Cagliari: Cagliari, Italy, 2019; p. 68.

34. Zelnik, I.; Potisek, M.; Gaberščik, A. Environmental conditions and macrophytes of Karst Ponds. *Pol. J. Environ. Stud.* **2012**, *21*, 1911–1920.
35. Guarino, R.; Marcenò, C.; Ilardi, V.; Mannino, A.M.; Troia, A. One *Chara* does not make Charetea in the Mediterranean aquatic vegetation. *Webbia* **2019**, *74*, 139–147. [CrossRef]
36. Brullo, S.; Minissale, P.; Spampinato, G. Studio fitosociologico della vegetazione lacustre dei Monti Nebrodi (Sicilia settentrionale). *Fitosociologia* **1994**, *27*, 5–50.
37. Poikane, S.; Portielje, R.; Denys, L.; Elferts, D.; Kelly, M.; Kolada, A.; Mäemets, H.; Phillips, G.; Søndergaard, M.; Willby, N.; et al. Macrophyte assessment in European lakes: Diverse approaches but convergent views of 'good' ecological status. *Ecol. Indic.* **2018**, *94*, 185–197. [CrossRef] [PubMed]
38. Søndergaard, M.; Johansson, L.S.; Lauridsen, T.L.; Jørgensen, T.B.; Liboriusse, L.; Jeppesen, E. Submerged macrophytes as indicators of the ecological quality of lakes. *Freshw. Biol.* **2010**, *55*, 893–908. [CrossRef]

Article

Wind Exposure Regulates Water Oxygenation in Densely Vegetated Shallow Lakes

Cristina Ribaudo [1,*], Juliette Tison-Rosebery [2,3], Mélissa Eon [2], Gwilherm Jan [2] and Vincent Bertrin [2,3]

[1] EA 4592 Géoressources & Environnement, F-33600 Pessac, France
[2] INRAE, UR EABX, F-33612 Cestas, France; juliette.rosebery@inrae.fr (J.T.-R.); melissa.eon@inrae.fr (M.E.); gwilherm.jan@inrae.fr (G.J.); vincent.bertrin@inrae.fr (V.B.)
[3] Pôle R&D Écosystèmes Lacustres (ECLA), F-13100 Aix-en-Provence, France
* Correspondence: cristina.ribaudo@ensegid.fr

Abstract: The presence of dense macrophyte canopies in shallow lakes locally generates thermal stratification and the buildup of labile organic matter, which in turn stimulate the biological oxygen demand. The occurrence of hypoxic conditions may, however, be buffered by strong wind episodes, which favor water mixing and reoxygenation. The present study aims at explicitly linking the wind action and water oxygenation within dense hydrophytes stands in shallow lakes. For this purpose, seasonal 24 h-cycle campaigns were carried out for dissolved gases and inorganic compounds measurements in vegetated stands of an oligo-mesotrophic shallow lake. Further, seasonal campaigns were carried out in a eutrophic shallow lake, at wind-sheltered and -exposed sites. Overall results showed that dissolved oxygen (DO) daily and seasonal patterns were greatly affected by the degree of wind exposure. The occurrence of frequent wind episodes favored the near-bottom water mixing, and likely facilitated mechanical oxygen supply from the atmosphere or from the pelagic zone, even during the maximum standing crop of plants (i.e., summer and autumn). A simple model linking wind exposure (Keddy Index) and water oxygenation allowed us to produce an output management map, which geographically identified wind-sheltered sites as the most subjected to critical periods of hypoxia.

Keywords: lake management; modelling; carbon dioxide; aquatic weeds; respiration; hypoxia; Keddy Index; submerged aquatic vegetation (SAV)

Citation: Ribaudo, C.; Tison-Rosebery, J.; Eon, M.; Jan, G.; Bertrin, V. Wind Exposure Regulates Water Oxygenation in Densely Vegetated Shallow Lakes. *Plants* 2021, *10*, 1269. https://doi.org/10.3390/plants10071269

Academic Editor: Angelo Troia

Received: 14 May 2021
Accepted: 18 June 2021
Published: 22 June 2021

Publisher's Note: MDPI stays neutral with regard to jurisdictional claims in published maps and institutional affiliations.

1. Introduction

In lentic shallow water bodies, the diel and seasonal oxygen balance is given by the interplay between the photosynthetic activity of primary producers (net production of dissolved oxygen, DO), their respiration (net DO consumption) and heterotrophic respiration of bacteria and animals (net DO consumption). When present, submerged aquatic vegetation (SAV) induces significant diel fluctuations in oxygen levels [1]. During the day, a supersaturation of oxygen (>100%) is observed due to photosynthesis, while at night oxygen is no longer produced, and consumption processes are predominant due to respiration. This type of nycthemeral variation is particularly emphasized in summer, when plant photosynthetic rate and heterotrophic respiration are at their maximum, and whose net effect largely exceeds the contribution of temperature-dependent oxygen solubility [2,3]. Primary production releases high amounts of oxygen in the water column, allowing for the oxidation of methane (CH_4) by methanotrophic epiphytic bacteria [4,5]. In addition, radial oxygen loss in the rhizosphere [6–8] contributes to reduce benthic CH_4 flux through benthic methanotrophy or oxidation of nitrate [9]. The synthesis of large quantities of biomass occurs through the assimilation of nutrients (including N-compounds, phosphate and carbon dioxide, CO_2); SAV is thus able to uptake gases and nutrients coming from the sediment and the atmosphere, and synthesize them in biomass [10]. However, oxygen dynamics can be altered in densely vegetated stands, such as those dominated by invasive macrophytes:

here, due to a fast cycle of growth and decay, fresh organic matter continuously replenishes the organic bulk in the sediment [11]. In those conditions, elevated biological oxygen demand (BOD) and hypoxia are coupled to accumulation and stratification of nutrients, such as carbon dioxide, methane, ammonium and reactive phosphorous, also during the day [12–15].

Dense plant canopies are known to locally generate bottom shading and modify water circulation, that impeding convective cooling, even within small depths [16–20]. This can affect or, inversely, exacerbate the thermally driven exchange flow of nutrients and DO between pelagic and littoral zones of the lake [21]. Nevertheless, it has been assessed that wind episodes may induce surface and internal flows even within stratified waters [22]. Shallow lakes are continuously subjected to mixing and wave-breaking in function of wind speed [21]; in function of the lake size, slope and bathymetry, wind-exposed lakes are also concerned by periodic seiche events which contribute to water mixing [23,24]. Still, only few studies infer that wind action may induce local turbulent mixing and reaeration even within dense submerged canopies [14,18]. Indeed, the explicit interaction between wind exposure and ecosystem functioning in shallow plant-dominated lakes remains almost undescribed [16,25], despite the growing demand for precision by stakeholders (users, managers, politicians) in the domain of biological invasions.

Invasive aquatic plants management is one of the main issue concerning global changes in freshwaters [26]. In the past, manager's decisions on invaded environments were mainly driven by socio-economical questions relating to tourism, boating and swimming [27]. Recently, invasive plants management started to progressively embrace ecological sciences, with the primary goal of understanding if invaded sites are *de facto* degraded or imperiled regarding to their functioning. As a result, aquatic weeds are more and more in question about their overall role on ecosystem metabolism. Coherently, managers are demanding to ecologists to produce effective and readable tools for ameliorating their interventions [28,29]. We herein report a study specifically addressing this issue, which employs DO saturation as a reliable indicator of net ecosystem metabolism, related to autotrophic and heterotrophic processes [30]. Firstly, we hypothesize that the impact on DO levels within invasive macrophyte stands significantly differs in function of the degree of wind exposure. Secondly, we hypothesize that DO levels are inversely correlated to plant densities and sedimentary organic matter content. Thirdly, we propose a quantitative tool to spatially identify sites that are more risky for hypoxic events, and thus need intervention by managers.

2. Results

2.1. Diel Variations in Vegetated Stands

Sampling sites in Lacanau Lake (hereafter, LAC Lake) were homogeneously distributed within the largest invasive macrophyte stands of the lake, which developed in the most sheltered zones of the lake (Figure 1).

Results from seasonal 24 h-cycle campaigns showed that most of the sites were hypoxic (DO saturation <100%), and that DO depletion also occurred during daylight (Figure 2). Concomitantly, CO_2 was mostly supersaturated and pH acid (pH < 7), with some exceptions during daylight in summer and autumn (Figures S1 and S2). CH_4, NH_4^+ and NO_3^- buildup in the water column appeared both during the night and day (Figures S3–S5). Water temperature ranged from 11.1 ± 0.2 to $26.7 \pm 0.3\ ^\circ C$ (in spring and summer, respectively—Figure S6) and DOC averaged 13.1 ± 0.2 mg L^{-1} on an annual basis.

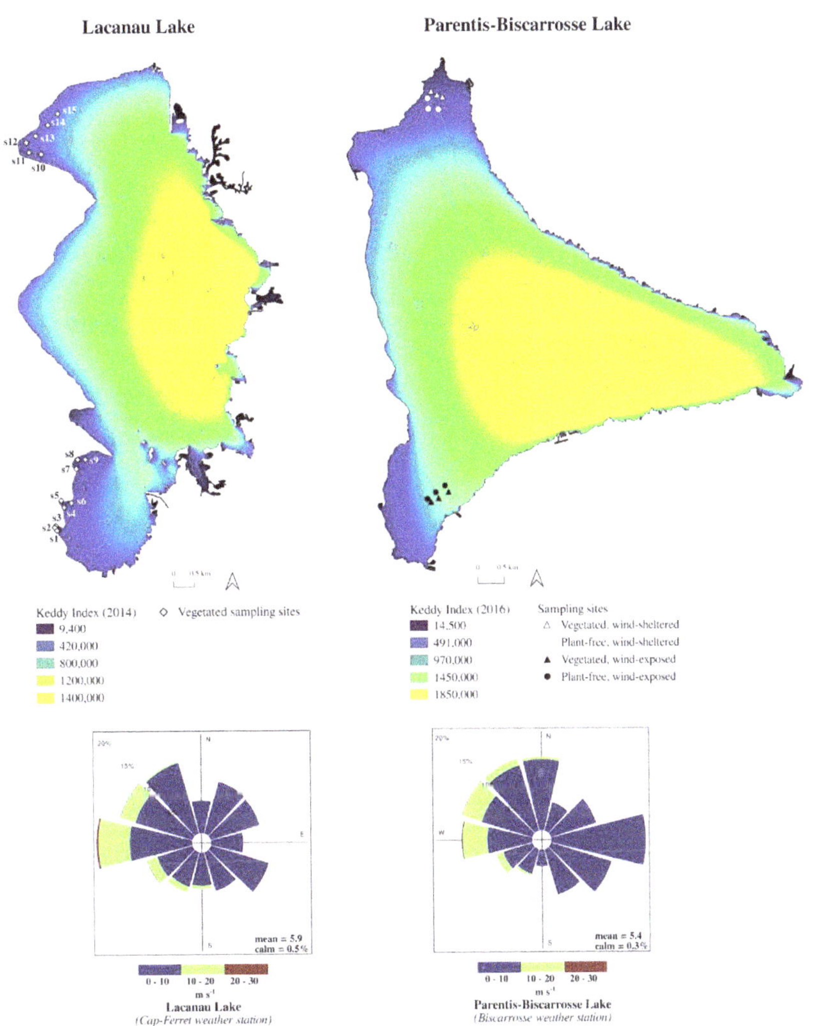

Figure 1. Keddy Index calculated on annual basis for Lacanau Lake (on the **left**) and Parentis-Biscarrosse Lake (on the **right**). The windrose is calculated on wind speed and direction hourly data on an annual basis. Lake bathymetry and sampling sites for seasonal 24-cycle campaigns (LAC Lake), as well as sampling sites for seasonal campaigns at wind-sheltered and -exposed sites (PAR Lake) are reported.

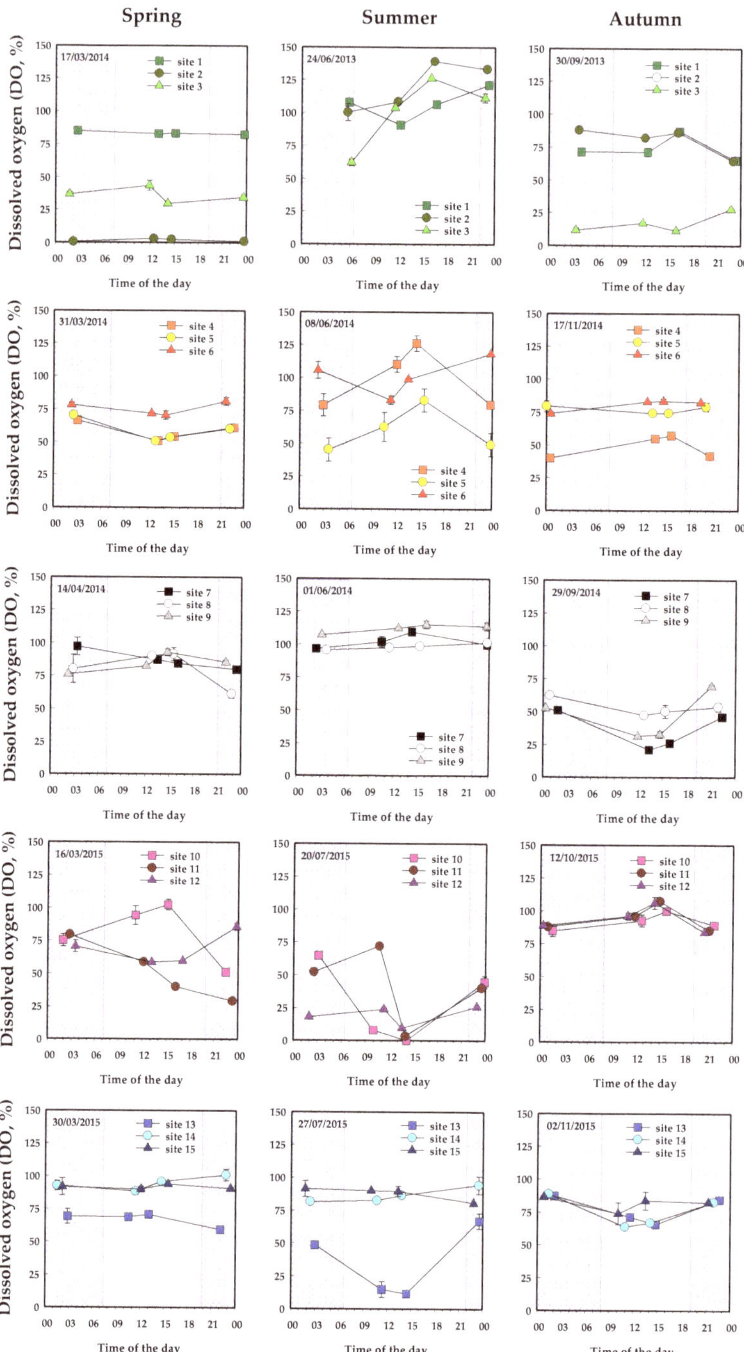

Figure 2. DO results from seasonal 24 h-cycle campaigns in LAC Lake. Measurements were carried out within densely vegetated areas, presenting biomass >100 g_{DW} m^{-2} during samplings. Hatched color indicates night periods.

ANOVA test revealed that, with temperature and ammonium as solely exceptions (day > night), dissolved gases and inorganic compounds measured within plant stands did not differ between day and night; all parameters varied seasonally (Table 1).

Table 1. Summarized results of the two-way ANOVA on physicochemical parameters (diel variation and season as fixed factors; sampling site as random factor). Results refer to seasonal 24 h-cycle campaigns carried out at 15 vegetated sites of LAC Lake.

Source	T (°C)		pH Units		DO (%)	
	df, Residuals	*p*-Value	df, Residuals	*p*-Value	df, Residuals	*p*-Value
Diel variation	1, 204	<0.001	1, 204	n.s.	1, 383	n.s.
Season	2, 204	<0.001	2, 204	<0.001	2, 383	<0.001
Diel × Season	2, 204	n.s.	2, 204	n.s.	2, 383	n.s.

Source	CO$_2$ (%)		CH$_4$ (µM)		NH$_4^+$ (µM)		NO$_3^-$ (µM)	
	df, Residuals	*p*-Value	df, Residuals	*p*-Value	df, Residuals	*p*-Value	df, Residuals	*p*-Value
Diel variation	1, 347	n.s.	1, 415	n.s.	1, 410	<0.05	1, 394	n.s.
Season	2, 347	<0.001	2, 415	<0.001	2, 410	<0.05	2, 394	<0.001
Diel × Season	2, 347	n.s.	2, 415	n.s.	2, 410	n.s.	2, 394	n.s.

2.2. Wind-Sheltered vs. Wind-Exposed Sites

The choice of sampling sites in Parentis-Biscarrosse Lake (hereafter, PAR Lake) was based on two co-occurring conditions: the presence of densely vegetated areas and the difference in wind exposure (Figure 1). Results from seasonal campaigns in PAR Lake showed that dissolved gases and inorganic compounds significantly changed in function of wind exposure (ANOVA, Table 2).

Table 2. Summarized results of the three-way ANOVA on physicochemical parameters (wind exposure, plant presence and season as fixed factors; sampling site as random factor). Results refer to seasonal campaigns carried out at 12 vegetated and plant-free sites of PAR Lake. Please refer to figures for Tukey'HSD test differences between treatments.

Source	T (°C)		pH Units		DO (%)	
	df, Residuals	*p*-Value	df, Residuals	*p*-Value	df, Residuals	*p*-Value
Wind exposure	1, 8	<0.001	1, 8	<0.001	1, 8	<0.001
Plant presence	1, 8	n.s.	1, 8	n.s.	1, 8	n.s.
Season	3, 24	<0.001	3, 24	<0.001	3, 72	<0.05
Wind × Plant	1, 8	n.s.	1, 8	<0.05	1, 8	<0.05
Wind × Seas	3, 24	<0.001	3, 24	<0.001	3, 72	<0.001
Plant × Seas	3, 24	n.s.	3, 24	n.s.	3, 72	<0.05
Wind × Plant × Seas	3, 24	n.s.	3, 24	n.s.	3, 72	<0.001

Source	CO$_2$ (%)		CH$_4$ (µM)		NH$_4^+$ (µM)		NO$_3^-$ (µM)	
	df, Residuals	*p*-Value	df, Residuals	*p*-Value	df, Residuals	*p*-Value	df, Residuals	*p*-Value
Wind exposure	1, 8	<0.001	1, 8	<0.001	1, 8	n.s.	1, 8	<0.05
Plant presence	1, 8	n.s.	1, 8	n.s.	1, 8	n.s.	1, 8	n.s.
Season	3, 72	<0.001	3, 72	<0.001	3, 72	<0.001	3, 72	<0.001
Wind × Plant	1, 8	<0.05	1, 8	n.s.	1, 8	n.s.	1, 8	<0.05
Wind × Seas	3, 72	<0.001	3, 72	<0.001	3, 72	<0.001	3, 72	<0.001
Plant × Seas	3, 72	<0.001	3, 72	<0.05	3, 72	n.s.	3, 72	n.s.
Wind × Plant × Seas	3, 72	<0.001	3, 72	<0.05	3, 72	n.s.	3, 72	n.s.

Differences between sheltered and exposed sites were significant for every physico-chemical parameter, yet only at vegetated sites and in function of the season (in summer and in autumn). Significant differences between vegetated and plant-free sites occurred only at sheltered sites. Tukey's HSD test indicated that water temperature was lower at sheltered sites than at exposed ones (Figure S7). pH values were lower at vegetated and sheltered sites than at exposed ones, only during summer (Figure S8). DO was lower at vegetated and sheltered sites than at exposed ones (Figure 3); CO_2 and CH_4 were higher at vegetated and sheltered sites than at exposed ones (Figures S9 and S10); NH_4^+ and NO_3^- values differed seasonally between sheltered and exposed sites, with no differences between vegetated and plant-free areas (Figures S11 and S12). DOC averaged 6.2 ± 0.2 mg L^{-1} on an annual basis.

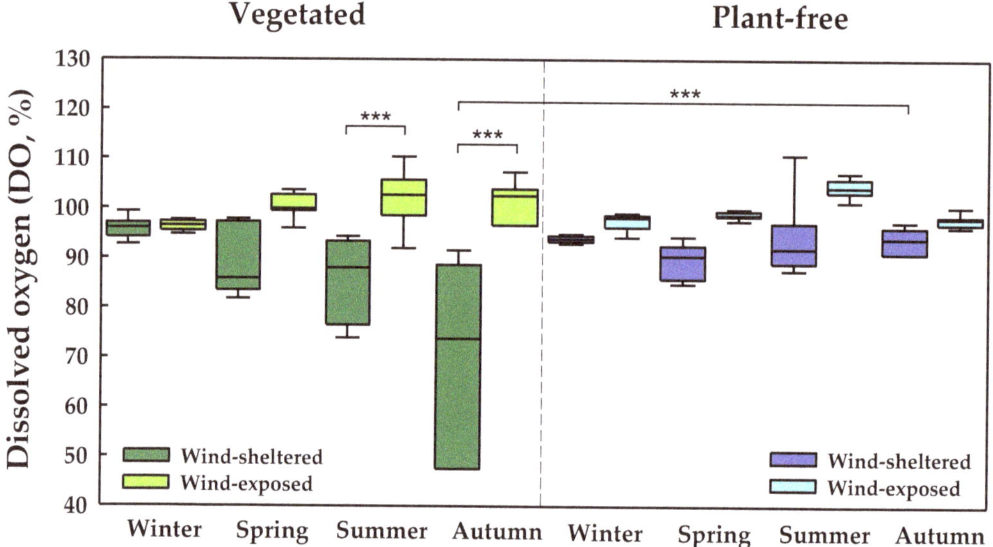

Figure 3. DO results from seasonal campaigns in PAR Lake. Measurements were carried out at wind-sheltered and -exposed sites, in vegetated and plant-free areas. For a better readability, Tukey's HSD results are not reported for the seasonality factor. *** indicates *p*-value < 0.001.

2.3. Dependence of DO Saturation on Plant Biomass and Sedimentary OM

Total biomass varied seasonally at both lakes, with values comprised between 319 ± 245 and 668 ± 414 g_{DW} m^{-2} at LAC Lake (in spring and summer, respectively), and between 1626 ± 132 and 4528 ± 2413 g_{DW} m^{-2} at PAR Lake (in spring at exposed and in autumn at sheltered sites, respectively). OM content in vegetated sediments ranged from 0.7 ± 0.2 to $71 \pm 3\%$ and from 0.7 ± 0.1 to $1.2 \pm 0.1\%$ as LOI, for LAC Lake and PAR Lake, respectively. Linear mixed-effects model, calculated on the two lakes dataset, showed that DO saturation was not dependent on OM sedimentary content on the total plant biomass; only DO values measured in LAC Lake during summer resulted in being negatively correlated to biomass (*p*-value < 0.01).

2.4. Dependence of DO on Wind Exposure and Hypoxia Risk Map Production

The regression of DO saturation against wind exposure, identified with the segmented function in R, showed a structural breakpoint at Keddy Index = 2.9 (Figure 4). We considered this breakpoint as a threshold of hypoxia risk, i.e., low risk above this value and high risk below. This threshold is assumed to be the minimum wind exposure which would be able to decrease the risk of hypoxia in dense submerged plant stands.

Further, in order to produce a hypoxia risk map, Keddy Index was calculated for each 4 h-long period ($n = 2190$) on each pixel cell ($n = 4031$ for LAC Lake and $n = 14{,}438$ for PAR Lake) matching with densely vegetated areas, presenting biomass >50 g_{DW} m^{-2}, mapped at the lake scale (1.19 km^2 and 4.17 km^2 in LAC and PAR Lakes, respectively, from 31) (Figure 5). Hypoxia risk was above 50% in 70 ha of plants stands (corresponding to 60% of the total vegetated surface) in LAC Lake and in 50 ha in PAR Lake (12% of the total vegetated surface). This risk was above 75% in 11 ha of plants stands (9% of the total vegetated surface) in LAC Lake and in 11ha in PAR Lake (3% of the total vegetated surface).

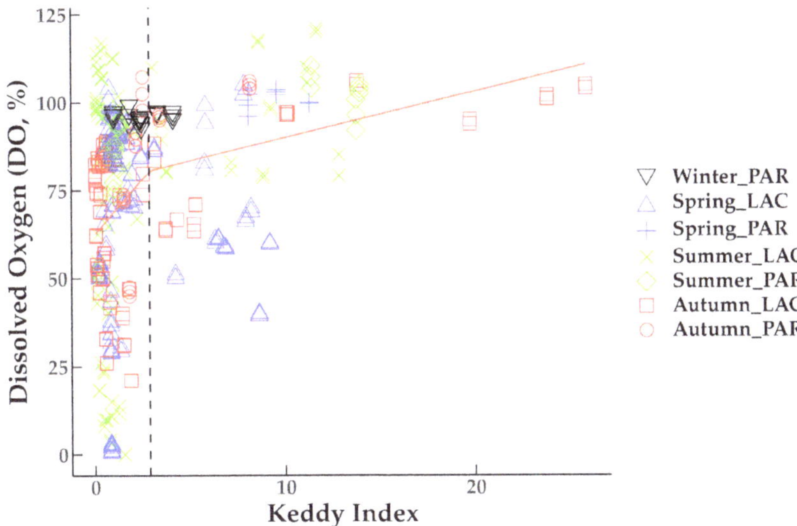

Figure 4. Dependence of DO saturation on wind exposure (Keddy Index). Dashed line is the breakpoint indicating the threshold between low-risk and high-risk of hypoxia/anoxia events in densely vegetated areas.

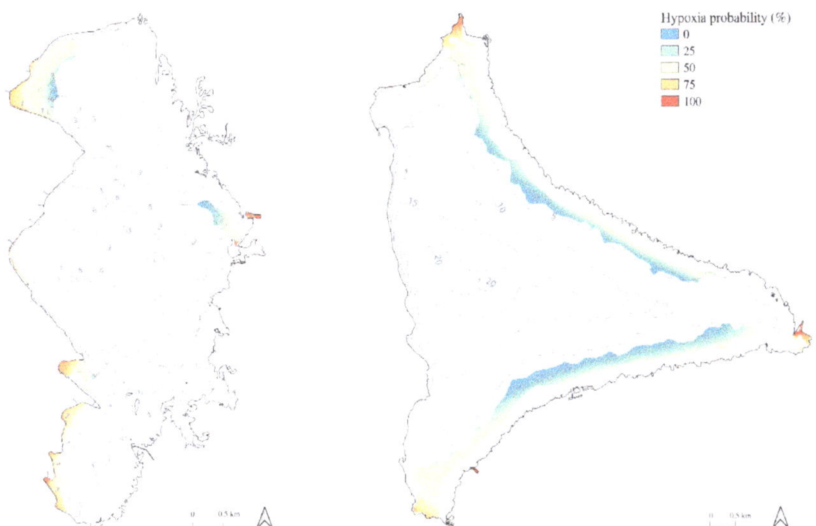

Figure 5. Hypoxia risk maps calculated for one year in densely vegetated areas of LAC Lake (on the **left**) and PAR Lake (on the **right**). Lakes bathymetry is also reported.

3. Discussion

In vegetated stands, diel variations of inorganic compounds typically reflect plants photosynthetic activity, with the lowest dissolved carbon and nitrogen concentrations measured in the water at late afternoon, corresponding to nutrients depletion by plants uptake, then an accumulation during the night, with a peak just before dawn. At the same time, DO and pH follow an exactly inverse pattern. In our study, the described nycthemeral shape was detectable only at some sites and mostly during summer. At other sites, heterotrophic activity, stimulated by temperature increase during summer and autumn, exceeded net oxygen release during the day, that resulting in hypoxia/anoxia events and buildup of CO_2, CH_4 and NH_4^+ in the water column. This observation is recurrent in dense stands formed by invasive macrophytes, where the sedimentation of organic matter generates an elevated benthic BOD during the period of senescence of plants; this implies a permanent DO deficit [15,31,32]. In dense hydrophyte stands, DO input from the atmosphere can be limited to the surficial layer of the water column, as long stems constitute a physical barrier, as floating-leaved macrophytes do [13]. In our case, the occurrence of a vertical "plant wall" at the external boundaries of vegetated stands may also lead to the annihilation of the horizontal flow of nutrients and DO from the pelagic to the littoral zones [22].

Hypoxic events and inorganic compounds buildup can be however contrasted by the wind action, which may induce local turbulent mixing and reaeration even within dense submerged canopies [14,18]. Coherently, some of the diel variations measured in our study showed a flattened shape, with constant values along the 24 h-cycle. On one hand, elevated DO values during the night could be attributable to convective mixing due to air temperature nightly decrease [17,33]. On the other hand, the maintaining of constant DO values along a diel cycle may be an indicator of stationary wind conditions and turbulent mixing; this supposition is supported by the second part of our study. Seasonal campaigns at wind-sheltered and -exposed sites showed that, ecosystem functioning was not ascribable to the plant presence/absence or to the seasonal biomass variation only. Indeed, DO and CO_2 saturation at wind-exposed sites hovered at about 100% all year round, indicating that wind-driven diffusion continuously outreached net production and consumption within the water column, even in invaded areas of the lake. Overall results show thus that the presence of invasive hydrophytes does not systematically promote water hypoxia, if local wind conditions allow an efficient water mixing by wind.

When considering the whole dataset, only DO values measured in LAC Lake during summer resulted in being dependent on plant density; moreover, vegetated stands in this lake mainly developed at sheltered sites [34]. Prevailing winds oriented from the northwest created low hydrodynamic conditions, because of the natural barrage formed by sand dunes [35]. Elevated plant biomass matching with shallow depths in wind-sheltered areas seemed to generate favorable conditions for water hypoxia, a phenomenon exacerbated by an elevated turnover of biomass during summer. In contrast, extremely elevated biomass measured in PAR Lake, largely exceeding values reported until now for *Egeria* spp. invaded sites [3,31], did not generate an extreme DO deficit even at wind-sheltered sites. The difference between the two lakes is evident also from a thermic point of view: at LAC Lake, a previous study had showed that water temperature measured in vegetated stands was significantly lower than that measured in plant-free areas, irrespective of the season [19]. The present study on PAR Lake shows instead that no significant difference exists between vegetated and plant-free areas, irrespective of the season (Figure S7). As for the DO and CO_2, the divergence in temperature results among the two lakes could be due to the different size of the lake, the second being larger and permitting fetch length—and thus, water mixing—to be more important.

The hypoxia risk map shows that elevated hypoxia probability is associated with wind-sheltered areas of the lakes, and that oxygenation shortage can affect a large total surface of several tens of hectares. Hypoxia risk is at its maximum in both lakes at enclosed and wind-sheltered areas, like small marinas and public boat launches, which are known to be important drivers of aquatic plant spread [36,37]. On the other hand, large surfaces of the lake invaded by elevated plant densities would not be affected by hypoxia and would thus not necessitate intervention. The hypoxia risk map we produced represents a preliminary and concrete tool, coupling field measurements and modelling, which can reduce plant management costs, as it indicates precisely where invasive plants constitute a problem for ecosystem functioning. A similar approach providing reproducible management tools has been recently published [38], that coupling lake depth or bathymetry to anoxia probability in the hypolimnion of deep lakes. Our model should be, however, calibrated site-specifically, because the intrinsic sedimentary features and the trophic status of the lake could affect the magnitude of hypoxia level and nutrients flux. The two lakes we studied presented different DOC values, sedimentary OM content and resulted in very different concentrations of CO_2 and CH_4. Also, due to a different fetch length, the reaeration strongly varied even within comparable wind velocity. A possible improvement of our method could have been to introduce the local bathymetry in the model. Indeed, waves induce vertical upward forces acting on the water column movements and sediment resuspension [34]; furthermore, wind-induced circulation in nearshore zones appears to be crucial in littoral plant-free areas [24]. We can expect an increase of wind effect on water mixing in shallow zones due to orbital movements translated to the lake bottom. Nevertheless, SAV also reduce waves action and current velocities within beds [39,40]. Future modelling works should thus focus on integrating vegetation in the photic region to better define how the cross-shore water circulation works. Another possible improvement in the future would be the use of automatic oxygen probes, in order to obtain a finer resolution scale of diel and seasonal variations, and perfect the calculation of hypoxia risk probability on a long temporal scale.

Our results highlight the need to consider local hydrodynamics in lake management decisions. Wind exposure should be used for spatially organizing management plans and prioritizing zones where invasive biomass control actions are needed. Mapping hypoxia risk in densely vegetated stands is a promising tool for the management of invasive hydrophytes in shallow lakes.

4. Materials and Methods

4.1. Study Area

Lacanau Lake and Parentis-Biscarrosse Lake are shallow lakes located in the southern Atlantic coast of France. Those lakes are characterized by sandy acidic substrate and classed as oligo-mesotrophic (Lacanau, 16 km^2) and eutrophic (Parentis-Biscarrosse, 32 km^2). Within the two lakes, large submerged stands of *Egeria densa* Planch. and *Lagarosiphon major* (Ridl.) Moss develop between 1 and 7 m deep, with dense stands being preferentially located at shallow and wind-sheltered sites, or at deep and wind-exposed sites [34].

4.2. Field Campaigns

Between June 2013 and November 2015 at Lacanau Lake, seasonal 24 h-cycle campaigns were carried out at 15 sites. Sampling sites were homogeneously distributed within the largest invasive macrophyte stands of the lake [34]. Water was collected within plant canopy at depths ranging from 100 to 330 cm, with a frequency of four times a day (two samplings during the day, between 11 a.m. and 3 p.m.; two samplings during the night, between 9 p.m. and 6 a.m.). Water temperature (T, °C), pH, dissolved oxygen (DO, expressed as saturation %), dissolved carbon dioxide (CO_2, %), dissolved methane (CH_4, µM), nitrate (NO_3^-, µM), ammonium (NH_4^+, µM) and dissolved organic carbon (DOC, mg L^{-1}) were measured according methods reported in [19]. Finally, we tested the influence of the sampling time on the biogeochemistry of the water column by a two-way ANOVA with interactions among factors. The diel variation (two levels: day vs. night) and the season (three levels: spring vs. summer vs. autumn) were considered as fixed factors, while the sampling site (fifteen levels) was considered as a random factor.

Between January 2016 and January 2017 at Parentis-Biscarrosse Lake, seasonal sampling campaigns were carried out, during the day only, at vegetated (3 wind-sheltered and 3 wind-exposed sites) and at plant-free sites (3 wind-sheltered and 3 wind-exposed). The degree of wind exposure was estimated by previous modeling of wind exposure Keddy Index [41]. Water was collected within plant canopy at depths ranging from 150 to 300 cm, between 11 a.m. and 3 p.m. Water samples collection, treatment and analyses are the same as those adopted in Lacanau Lake and reported in [19]. Finally, we tested the influence of spatial exposure to wind on the biogeochemistry of the water column by a three-way ANOVA with interactions among factors. The degree of wind exposure (two levels: exposed vs. sheltered), the plant presence (two levels: vegetated vs. plant-free) and the season (four levels: winter vs. spring vs. summer vs. autumn) were considered as fixed factors, while the sampling site (twelve levels) was considered as a random factor.

Normal distribution (Shapiro–Wilk Test) and homoscedasticity (Levene's Test) were tested before running ANOVAs. Post hoc analyses were performed by Tukey's Honestly Significant Difference (HSD) test. Statistical analyses were performed with R Program [42]. Mean values are reported with their standard deviation.

Macrophytes sampling was carried out by rake for total biomass (g_{DW} m^{-2}) measurements, immediately after water samplings, as reported in [19]. Concomitantly, within the plant stands, sediment samples were collected by grabber, as described in [34], for sedimentary organic matter (OM, as loss of ignition, % LOI) measurements. In order to test the dependence of DO saturation on plant biomass and OM content, a linear mixed-effects model fit by maximum likelihood was performed on the whole dataset (DO measurements from Lacanau and Parentis-Biscarrosse Lakes), with the sampling site as a random factor.

4.3. Wind Exposure Calculations

Wind exposure was calculated according [41] for both lakes by using a fetch matrix (i.e., the distance over which waves can build up) obtained from lake open-water raster grid cells (resolution of 17 m) for each wind compass direction (10–360°, in 10° increments). Wind data (hourly and daily mean speed and direction) were provided by Météo France in Cap-Ferret (44°37′54″ N, 1°14′53″ O) and Biscarrosse (44°25′54″ N, 1°14′51″ O) weather stations for Lacanau and Parentis-Biscarrosse Lakes, respectively. It is possible to generate values which should be related to the effect of wind at a given point (here, a grid cell) by using fetch and wind velocity. For a given compass direction, one measure of exposure is the product of mean wind speed and direction and the percent frequency of the wind blowing in that direction.

In order to position wind-sheltered and -exposed sampling sites in Parentis-Biscarrosse Lake, daily mean wind speed and direction were used to build a wind exposure map during 1-year period (2014). One measure of exposure was calculated for each grid cell over 36 compass wind directions according to a fetch matrix. A cell's total exposure is given by the sum of values calculated for all of the compass directions during the 1-year period. Sampling sites were chosen within lake areas identified as low- or highly-exposed to wind action.

4.4. Coupling DO and Wind Exposure

Keddy Index was calculated for the 4 h before the exact timing of water sampling. Then, each DO value was coupled to the sum of Keddy Index values for this period. This is the duration estimated being necessary for the water mixing at low depth [17,43]. In order to test the dependence of DO saturation on wind exposure, a Chow test was performed to determine the presence of a structural break at some point of the data series [44]. We used the *sctest* function from the *strucchange* package in R software to perform a Chow test, which resulted in $F = 10.7$, p-value = 2.7×10^{-5}. The significance of the test indicates that a structural breakpoint is present in the regression. Or else, that two regression lines can fit the pattern in the data more effectively than a single regression line. Finally, we applied the *segmented* function in R to analyze segmented relationships in the regression, in order to obtain a breakpoint value.

4.5. Hypoxia Risk Map Production

We calculated 4 h-long Keddy Index values each day during one year (2014 and 2016 for Lacanau and Parentis-Biscarrosse Lakes, respectively) for each raster cell corresponding to densely vegetated areas of the lake, and presenting a biomass >50 g_{DW} m^{-2} [34]. Each 4 h-long period and each grid cell in which wind exposure was under the breakpoint value, indicating a high risk of hypoxia, was classified as "1", whereas 4 h-long period with low risk of hypoxia (>2.9) were classified as "0". The probability of hypoxia was expressed as the percentage (0–100%) of 4 h-long periods where wind exposure was below the hypoxia threshold during one year. Finally, this probability was reported on raster grid cells to map the hypoxia risk at densely vegetated areas scale.

Supplementary Materials: The following are available online at https://www.mdpi.com/article/10.3390/plants10071269/s1, Figure S1: CO_2 results from seasonal 24 h-cycle campaigns in Lacanau Lake; Figure S2: pH results from seasonal 24 h-cycle campaigns in Lacanau Lake; Figure S3: CH_4 results from seasonal 24 h-cycle campaigns in Lacanau Lake; Figure S4: NH_4^+ results from seasonal 24 h-cycle campaigns in Lacanau Lake; Figure S5: NO_3^- results from seasonal 24 h-cycle campaigns in Lacanau Lake; Figure S6: Water temperature results from seasonal 24 h-cycle campaigns in Lacanau Lake; Figure S7: Water temperature results from seasonal campaigns in Parentis-Biscarrosse Lake; Figure S8: pH results from seasonal campaigns in Parentis-Biscarrosse Lake; Figure S9: CO_2 results from seasonal campaigns in Parentis-Biscarrosse Lake; Figure S10: CH_4 results from seasonal campaigns in Parentis-Biscarrosse Lake; Figure S11: NH_4^+ results from seasonal campaigns in Parentis-Biscarrosse Lake; Figure S12: NO_3^- results from seasonal campaigns in Parentis-Biscarrosse Lake. In S7, S8, S9, S10, S11 and S12, Tukey's HSD results are not reported for the seasonality factor. *** indicates p-value < 0.001, ** p-value < 0.01, * p-value < 0.05.

Author Contributions: C.R., V.B. and J.T.-R. conceptualized the project and methodology, M.E. and G.J. contributed to the lab and the field work. C.R. prepared the original draft; V.B., J.T.-R., M.E. and G.J. reviewed and edited the final version. All authors have read and agreed to the published version of the manuscript.

Funding: This research was funded by AEAG (Agence de l'Eau Adour-Garonne), within the conventions #310330122, #310330085, and #310330109, by Irstea (Institut National de Recherche en Sciences et Technologies pour l'Environnement et l'Agriculture), by the Regional Council Nouvelle-Aquitaine (CCRRDT) and by the Regional Research and Development Program (PSDR) within the project AquaVIT (#2015-1R20603—00005190-00005191).

Institutional Review Board Statement: Not applicable.

Informed Consent Statement: Not applicable.

Data Availability Statement: The data presented in this study are available in the article or supplementary material.

Acknowledgments: We would like to thank SIAEBVELG and Police du Lac de Lacanau, for facilitating night activities in the field. We are also grateful to B. Delest, A. Moreira, and K. Madarassou for nutrients analyses and to D. Poirier (EPOC, University of Bordeaux) for assisting in CH_4 analyses. J.-L. Keller, S. Moreira, T. Huguet, G. Ducasse, A. Poli, H. Guerreiro, J. Chabanne, J. Vedrenne, I. Halgand, R. Le Barh, J. Gueguen, N. Dagens and Y. Bourguignon considerably helped during field and laboratory activities and we sincerely acknowledge them. We are grateful to T. Leboucher and S. Boutry for contributing to the implementation of the hypoxia risk map. We are grateful to students A. Poli, V. Mauro and N. Naulin for preliminary data analysis. We acknowledge Météo France for wind data supplying. We are grateful to H. Ulloa and another anonymous reviewer for the useful comments on the original draft of the manuscript.

Conflicts of Interest: The authors declare no conflict of interest.

References

1. Caraco, N.F.; Cole, J.J. Contrasting impacts of a native and alien macrophyte on dissolved oxygen in a large river. *Ecol. Appl.* **2002**, *12*, 1496–1509. [CrossRef]
2. Pokorný, J.; Rejmánková, E. Oxygen regime in a fishpond with duckweeds (Lemnaceae) and Ceratophyllum. *Aquat. Bot.* **1983**, *17*, 125–137. [CrossRef]
3. Pelicice, F.M.; Agostinho, A.A.; Thomaz, S.M. Fish assemblages associated with Egeria in a tropical reservoir: Investigating the effects of plant biomass and diel period. *Acta Oecologica* **2005**, *27*, 9–16. [CrossRef]
4. Sorrell, B.K.; Downes, M.T.; Stanger, C.L. Methanotrophic bacteria and their activity on submerged aquatic macrophytes. *Aquat. Bot.* **2002**, *72*, 107–119. [CrossRef]
5. Yoshida, N.; Iguchi, H.; Yurimoto, H.; Murakami, A.; Sakai, Y. Aquatic plant surface as a niche for methanotrophs. *Front. Microbiol.* **2014**, *5*, 30. [CrossRef]
6. Sand-Jensen, K.; Prahl, C.; Stokholm, H. Oxygen release from roots of submerged aquatic macrophytes. *Oikos* **1982**, *38*, 349–354. [CrossRef]
7. Sorrell, B.K.; Dromgoole, F.I. Oxygen transport in the submerged freshwater macrophyte Egeria densa Planch. I. Oxygen production, storage and release. *Aquat. Bot.* **1987**, *28*, 63–80. [CrossRef]
8. Ribaudo, C.; Bartoli, M.; Racchetti, E.; Longhi, D.; Viaroli, P. Seasonal fluxes of O_2, DIC and CH_4 in sediments with Vallisneria spiralis: Indications for radial oxygen loss. *Aquat. Bot.* **2011**, *94*, 134–142. [CrossRef]
9. Haroon, M.F.; Hu, S.; Shi, Y.; Imelfort, M.; Keller, J.; Hugenholtz, P.; Yuan, Z.; Tyson, G.W. Anaerobic oxidation of methane coupled to nitrate reduction in a novel archaeal lineage. *Nature* **2013**, *500*, 567–570. [CrossRef] [PubMed]
10. Reddy, K.R.; Tucker, J.C.; Debusk, W.F. The role of Egeria in removing nitrogen and phosphorus from nutrient enriched waters. *J. Aquat. Plant Manag.* **1987**, *25*, 14–19.
11. Drexler, J.Z.; Khanna, S.; Lacy, J.R. Carbon storage and sediment trapping by Egeria densa Planch, a globally invasive, freshwater macrophyte. *Sci. Total Environ.* **2020**, *755*, 142602. [CrossRef]
12. Barko, J.W.; Godshalk, G.L.; Carter, V.; Rybicki, N.B. *Effects of Submersed Aquatic Macrophytes on Physical and Chemical Properties of Surrounding Water*; Aquatic Plant Control Research Program Technical Report A-88-1 I; Depart Army, U.S. Army Corps of Engineers: Washington, DC, USA, 1988; p. 28.
13. Frodge, J.D.; Thomas, G.L.; Pauley, G.B. Effects of canopy formation by floating and submergent aquatic macrophytes on the water quality of two shallow Pacific Northwest lakes. *Aquat. Bot.* **1990**, *38*, 231–248. [CrossRef]
14. Sand-Jensen, K.; Andersen, M.R.; Martinsen, K.T.; Borum, J.; Kristensen, E.; Kragh, T. Shallow plant-dominated lakes–extreme environmental variability, carbon cycling and ecological species challenges. *Ann. Bot.* **2019**, *124*, 355–366. [CrossRef]
15. Moore, T.P.; Clearwater, S.J.; Duggan, I.C.; Collier, K.J. Invasive macrophytes induce context-specific effects on oxygen, pH, and temperature in a hydropeaking reservoir. *River Res. Appl.* **2020**, *36*, 1717–1729. [CrossRef]
16. Herb, W.R.; Stefan, H.G. Dynamics of vertical mixing in a shallow lake with submersed macrophytes. *Water Resour. Res.* **2005**, *41*. [CrossRef]
17. Coates, M.J.; Folkard, A.M. The effects of littoral zone vegetation on turbulent mixing in lakes. *Ecol. Model.* **2009**, *220*, 2714–2726. [CrossRef]
18. Andersen, M.R.; Sand-Jensen, K.; Woolway, R.I.; Jones, I.D. Profound daily vertical stratification and mixing in a small, shallow, wind-exposed lake with submerged macrophytes. *Aquat. Sci.* **2017**, *79*, 395–406. [CrossRef]
19. Ribaudo, C.; Tison-Rosebery, J.; Buquet, D.; Jan, G.; Jamoneau, A.; Abril, G.; Anschutz, P.; Bertrin, V. Invasive aquatic plants as ecosystem engineers in an Oligo-Mesotrophic shallow lake. *Front. Plant Sci.* **2018**, *9*, 1781. [CrossRef] [PubMed]
20. Vilas, M.P.; Marti, C.L.; Oldham, C.E.; Hipsey, M.R. Macrophyte-induced thermal stratification in a shallow urban lake promotes conditions suitable for nitrogen-fixing cyanobacteria. *Hydrobiologia* **2018**, *806*, 411–426. [CrossRef]
21. Zhang, X.; Nepf, H.M. Thermally driven exchange flow between open water and an aquatic canopy. *J. Fluid Mech.* **2009**, *632*, 227. [CrossRef]
22. Ulloa, H.N.; Constantinescu, G.; Chang, K.; Horna-Munoz, D.; Hames, O.; Wüest, A. Horizontal transport under wind-induced resonance in stratified waterbodies. *Phys. Rev. Fluids* **2020**, *5*, 054503. [CrossRef]
23. Lin, Y.T. Wind effect on diurnal thermally driven flow in vegetated nearshore of a lake. *Environ. Fluid Mech.* **2015**, *15*, 161–178. [CrossRef]
24. Ulloa, H.N.; Constantinescu, G.; Chang, K.; Horna-Munoz, D.; Steiner, O.S.; Bouffard, D.; Wüest, A. Hydrodynamics of a periodically wind-forced small and narrow stratified basin: A large-eddy simulation experiment. *Environ. Fluid Mech.* **2019**, *19*, 667–698. [CrossRef]
25. Torma, P.; Wu, C.H. Temperature and circulation dynamics in a small and shallow lake: Effects of weak stratification and littoral submerged macrophytes. *Water* **2019**, *11*, 128. [CrossRef]
26. Havel, J.E.; Kovalenko, K.E.; Thomaz, S.M.; Amalfitano, S.; Kats, L.B. Aquatic invasive species: Challenges for the future. *Hydrobiologia* **2015**, *750*, 147–170. [CrossRef]
27. Matzek, V.; Covino, J.; Funk, J.L.; Saunders, M. Closing the knowing–doing gap in invasive plant management: Accessibility and interdisciplinarity of scientific research. *Conserv. Lett.* **2014**, *7*, 208–215. [CrossRef]
28. Young, S.L.; Kettenring, K.M. The social-ecological system driving effective invasive plant management: Two case studies of non-native Phragmites. *J. Environ. Manag.* **2020**, *267*, 110612. [CrossRef] [PubMed]

29. Le Floch, S.; Ginelli, L. The victorious battles of the lost war against aquatic invasive plants. 'Fluid' categorisation and multiple forms of ordinary commitment. *Trans. Inst. Br. Geogr.* **2021**. [CrossRef]

30. Cole, J.J.; Pace, M.L.; Carpenter, S.R.; Kitchell, J.F. Persistence of net heterotrophy in lakes during nutrient addition and food web manipulations. *Limnol. Oceanogr.* **2000**, *45*, 1718–1730. [CrossRef]

31. Bini, L.M.; Thomaz, S.M.; Carvalho, P. Limnological effects of Egeria najas planchon (Hydrocharitaceae) in the arms of Itaipu Reservoir (Brazil, Paraguay). *Limnology* **2010**, *11*, 39–47. [CrossRef]

32. Wang, H.; Lu, J.; Wang, W.; Yang, L.; Yin, C. Methane fluxes from the littoral zone of hypereutrophic Taihu Lake, China. *J. Geophys. Res. Atmos.* **2006**, *111*. [CrossRef]

33. Brothers, S.; Kazanjian, G.; Köhler, J.; Scharfenberger, U.; Hilt, S. Convective mixing and high littoral production established systematic errors in the diel oxygen curves of a shallow, eutrophic lake. *Limnol. Oceanogr. Methods* **2017**, *15*, 429–435. [CrossRef]

34. Bertrin, V.; Boutry, S.; Jan, G.; Ducasse, G.; Grigoletto, F.; Ribaudo, C. Effects of wind-induced sediment resuspension on distribution and morphological traits of aquatic weeds in shallow lakes. *J. Limnol.* **2017**, *76*, 84–96. [CrossRef]

35. Tastet, J.P.; Lalanne, R.; Maurin, B.; Dubos, B. Geological and archaeological chronology of a late Holocene coastal enclosure: The Sanguinet lake (SW France). *Geoarchaeology* **2008**, *23*, 131–149. [CrossRef]

36. Tamayo, M.; Olden, J. Forecasting the vulnerability of lakes to aquatic plant invasions. *Invasive Plant Sci. Manag.* **2014**, *7*, 32–45. [CrossRef]

37. Bertrin, V.; Boutry, S.; Alard, D.; Haury, J.; Jan, G.; Moreira, S.; Ribaudo, C. Prediction of macrophyte distribution: The role of natural versus anthropogenic physical disturbances. *Appl. Veg. Sci.* **2018**, *21*, 395–410. [CrossRef]

38. Deeds, J.; Amirbahman, A.; Norton, S.A.; Suitor, D.G.; Bacon, L.C. Predicting anoxia in low-nutrient temperate lakes. *Ecol. Appl.* **2021**. [CrossRef]

39. Carper, G.L.; Bachmann, R.W. Wind resuspension of sediments in a prairie lake. *Can. J. Fish. Aquat. Sci.* **1984**, *41*, 1763–1767. [CrossRef]

40. Schutten, J.; Dainty, J.; Davy, A. Wave-induced hydraulic forces on submerged aquatic plants in shallow lakes. *Ann. Bot.* **2004**, *93*, 333–341. [CrossRef] [PubMed]

41. Keddy, P.A. Quantifying within-lake gradients of wave energy: Interrelationships of wave energy, substrate particle size and shoreline plants in axe lake, Ontario. *Aquat. Bot.* **1982**, *14*, 41–58. [CrossRef]

42. R Core Team. *R: A Language and Environment for Statistical Computing*; R Foundation for Statistical Computing: Vienna, Austria, 2021. Available online: https://www.R-project.org/ (accessed on 23 May 2021).

43. Jalil, A.; Li, Y.; Du, W.; Wang, W.; Wang, J.; Gao, X.; Khan, H.O.S.; Pan, B.; Acharya, K. The role of wind field induced flow velocities in destratification and hypoxia reduction at Meiling Bay of large shallow Lake Taihu, China. *Environ. Pollut.* **2018**, *232*, 591–602. [CrossRef] [PubMed]

44. Chow, G.C. Tests of equality between sets of coefficients in two linear regressions. *Econom. J. Econom. Soc.* **1960**, 591–605. [CrossRef]

Review

Re-Establishment Techniques and Transplantations of Charophytes to Support Threatened Species

Irmgard Blindow [1,*], Maria Carlsson [2] and Klaus van de Weyer [3]

[1] Biological Station of Hiddensee, University of Greifswald, D-18565 Kloster, Germany
[2] County Administration Jönköpings Län, Hamngatan 4, S-551 86 Jönköping, Sweden; Maria.K.Carlsson@lansstyrelsen.se
[3] Lanaplan, Lobbericher Str. 5, D-41334 Nettetal, Germany; klaus.vdweyer@lanaplan.de
* Correspondence: blindi@uni-greifswald.de

Abstract: Re-establishment of submerged macrophytes and especially charophyte vegetation is a common aim in lake management. If revegetation does not happen spontaneously, transplantations may be a suitable option. Only rarely have transplantations been used as a tool to support threatened submerged macrophytes and, to a much lesser extent, charophytes. Such actions have to consider species-specific life strategies. K-strategists mainly inhabit permanent habitats, are perennial, have low fertility and poor dispersal ability, but are strong competitors and often form dense vegetation. R-strategists are annual species, inhabit shallow water and/or temporary habitats, and are richly fertile. They disperse easily but are weak competitors. While K-strategists easily can be planted as green biomass taken from another site, rare R-strategists often must be reproduced in cultures before they can be planted on-site. In Sweden, several charophyte species are extremely rare and fail to (re)establish, though apparently suitable habitats are available. Limited dispersal and/or lack of diaspore reservoirs are probable explanations. Transplantations are planned to secure the occurrences of these species in the country. This contribution reviews the knowledge on life forms, dispersal, establishment, and transplantations of submerged macrophytes with focus on charophytes and gives recommendations for the Swedish project.

Keywords: *Chara*; *Nitella*; *Tolypella*; *Nitellopsis*; re-establishment; revegetation; nutrients; herbivory

Citation: Blindow, I.; Carlsson, M.; van de Weyer, K. Re-Establishment Techniques and Transplantations of Charophytes to Support Threatened Species. *Plants* **2021**, *10*, 1830. https://doi.org/10.3390/plants10091830

Academic Editor: Stefano Accoroni

Received: 2 July 2021
Accepted: 25 August 2021
Published: 3 September 2021

Publisher's Note: MDPI stays neutral with regard to jurisdictional claims in published maps and institutional affiliations.

1. Introduction

To protect threatened macrophyte species in Sweden, an action plan started during 2017. The main aim of this program is to build knowledge which is considered necessary before actions are taken (Zinko 2017 [1]). The program includes 10 charophyte species (*Chara filiformis*, *C. subspinosa*, *C. braunii*, *Nitellopsis obtusa*, *Nitella translucens*, *N. mucronata*, *N. gracilis*, *N. syncarpa*, *N. confervacea*, *Tolypella canadensis*) and five angiosperm species (*Potamogeton acutifolius*, *P. compressus*, *P. friesii*, *P. rutilus*, *P. trichoides*). The selected charophyte species are rare in Sweden, which is surprising considering a high number of sites which seem suitable. Lack of knowledge about their occurrence in the country was and is one possible reason. Therefore, intensive monitoring was the main activity of several former action plans for threatened charophytes (Blindow 2009a,b,c,d,e [2–6]) and is still one main activity of the ongoing program. Except for *Tolypella canadensis*, however, lack of knowledge does not sufficiently explain the low number of sites for rare species. Oospores of these species are expected to be very rare in the diaspore reservoirs of lakes and small water bodies, which may restrict them from spontaneous (re)establishments. Transplantations of these species are therefore a second main activity of the ongoing action plan.

Experience with transplantations (e.g., translocations, see IUCN 2013 [7]) to protect threatened charophytes is still very limited. Fortunately, a number of threatened aquatic macrophytes have already been transplanted successfully, and experiences from these projects may be transferred to charophytes. Moreover, there is extensive literature on re-establishment of submerged macrophytes for other purposes such as lake restorations because of the positive impact of these plants on lake ecosystems and water quality (Hilt et al., in press [8]), which can be achieved by direct establishment (plantations) and/or indirectly by improving the habitat conditions for this vegetation. Submerged macrophytes act as sediment traps, store nutrients, retard shore erosions, and reduce phytoplankton densities by excretion of allelopathic substances—impacts which all increase water clarity. Together with their associated epiphyton, they offer a well-structured habitat, food, and oxygen and thereby favor species richness and biomass of macroinvertebrates. Both plants and macroinvertebrates are important food sources for fish and waterfowl. The vegetation further serves as a predation refuge for zooplankton, macroinvertebrates, and fish fry (Hilt et al., 2017 [9]).

Establishment success is dependent on dispersal and fertility but also competition with other plants. These abilities vary considerably among different life forms and species of submerged macrophytes. Detailed knowledge of these properties is essential to enable successful establishment and transplantation of submerged macrophytes.

This paper consists of three different parts: a review of ecological characteristics and life strategies of macrophytes (Sections 2–4) is followed by a review of management techniques to promote submerged macrophytes (Sections 6–9). Both parts first summarize knowledge about submerged macrophytes generally and end more specifically in a review about charophytes. The third part (Section 10) describes the "Swedish example", which aims at protection and especially transplantations of threatened charophytes and is based on the experiences reviewed in the first two parts.

2. Dispersal, Fertility, and Hibernation

Submerged macrophytes (re)establish from vegetative parts and/or diaspores that are transported to the water body or are already present on the site. Wind transport of diaspores (anemochory) is common in emergent plants but unusual in submerged plants, which mainly use water (hydrochory) but also different animals (zoochory) as transport vectors. Exozoochorous transport of green parts or turions is restricted to short distances, often within the same catchment area (Lacoul and Freedman 2006 [10], Soons et al., 2008 [11], Bakker et al., 2013 [12]). To reach remote water bodies and distant catchment areas, endozoochorous transport by waterfowl is the product of a co-evolutionary process (Clausen et al., 2002 [13], Figuerola and Green 2002 [14], Santamaria 2002 [15]). This transport requires the production of hard-shelled diaspores, which withstand the gut passage and often show improved germination after this passage (Clausen et al., 2002 [13], Figuerola and Green 2002 [14], Santamaria 2002 [15]). Such diaspores also tolerate harsh environmental conditions such as drying and freezing and serve as hibernacles, especially in temporary water bodies (Bonis and Grillas 2002 [16], Green et al., 2002 [17]).

The same mechanisms are applied in charophytes. Oospores tolerate both drying and freezing. They were once assumed to be transported by wind (Bakker et al., 2013 [12]), but it is doubtful if this transport has any major importance. Mature oospores are small (ca. 180 μm to >1000 μm; Wood 1959 [18], Haas 1994 [19], Krause 1997 [20]) but specifically heavy. Oospores were earlier shown to be transported by means of waterfowl, probably over high distances (Proctor 1959 [21], 1962 [22]), and to germinate better after a passage through a waterfowl gut (Proctor 1968 [23], Brochet et al., 2010 [24], Figuerola et al., 2010 [25]).

Charophytes hibernate as green plants or by means of specific vegetative hibernacles (bulbils) or oospores. As in vascular plants, hibernation modes vary considerably among species but also within species dependent on conditions such as water depth (Wang et al., 2015 [26]). For example, *Chara aspera* can hibernate as a green plant in deeper permanent habitats by means of bulbils and oospores in shallow water or exclusively by means of oospores, especially in temporary habitats (Blindow and Schütte 2007 [27]). In this species, oospores are assumed to serve mainly as long-term diaspore reservoir because they can survive long time periods but only have low annual germination rates; in contrast, bulbils germinate almost completely during spring but can survive just a few years and therefore are assumed to serve short-term diaspore reservoir (van den Berg et al., 2001 [28]). Generally, charophytes use oospores for long distance dispersal and for reestablishment from sediments after disturbances, and bulbils are used to maintain local populations (de Winton and Clayton 1996 [29], van den Berg et al., 2001 [28], Bonis and Grillas 2002 [16], Asaeda et al., 2007 [30], Brochet et al., 2010 [24]). Charophytes use three different modes to form dense vegetation with high interspecific differences in the relative importance of these modes: (A) vegetatively from omnipotent node cells, which can successfully be dispersed by means of fragments containing at least one node (Skurzyński and Bociąg 2011 [31]), (B) vegetatively from bulbils (Asaeda et al., 2007 [30], Wang et al., 2015 [26]), or (C) by germination of oospores (Skurzyński and Bociąg 2009 [32]).

Oospores collected while still situated on the plants are often in primary dormancy, which is broken after the winter or if the oospores are exposed to low temperatures for a longer time period (stratification); contrarily, oospores taken from sediments can germinate immediately (Takatori and Imahori 1971 [33], Sederias and Colman 2007 [34], Skurzyński and Bociąg 2009 [32]). Such oospores, however, have far lower germination success than bulbils, as they are in a secondary dormancy, which prevents them from germinating under unsuitable conditions (Stross 1989 [35], Holzhausen et al., 2017 [36]). Species-specific conditions of temperature, redox potential, and light are required to break dormancy and initiate germination (Casanova and Brock 1996 [37], Bonis and Grillas 2002 [16], de Winton et al., 2004 [38], Kalin and Smith 2007 [39], Skurzyński and Bociąg 2009 [32], Holzhausen et al., 2017 [36]). Oospores of species from temporary water bodies germinate far better after having been dried before (Sabbatini et al., 1987 [40], Casanova and Brock 1990 [41], 1996 [37], de Winton et al., 2004 [38]).

3. Interspecific Competition

Along a eutrophication gradient, submerged macrophytes are the dominating primary producers at low to moderate nutrient loadings, while phytoplankton dominates in highly eutrophic conditions. A shift from macrophyte to phytoplankton dominance occurs at a certain nutrient-related critical turbidity. This shift can happen rapidly in shallow lakes, which were assumed to occur in two different alternative stable states (Scheffer et al., 1993 [42]).

More recently, three different states of primary producer dominance were postulated to occur during progressive eutrophication, a macrophyte-dominated state with bottom-dwellers, a second macrophyte-dominated state with tall macrophytes, and a phytoplankton-dominated turbid state (Verhofstad et al., 2017 [43]). While the bottom-dweller state, often characterized by dense charophyte vegetation, is assumed to be rather stable, the tall macrophyte state, dominated by various angiosperms, is characterized by somewhat higher turbidity and lower stability (Meijer 2000 [44], Hilt et al., 2018 [45], Blindow et al., 2016 [46], Phillips et al., 2016 [47]) and therefore was called the "crashing" state (Sayer et al., 2010 [48]). Vice versa, tall macrophytes are sometimes the first submerged vegetation to establish in a turbid lake and to increase light availability in the water column far enough to enable a subsequent establishment of charophytes (Meijer 2000 [44], van den Berg et al., 2001 [28], Hargeby et al., 2007 [49]). Additionally, feedback mechanisms are assumed to differ between the two macrophyte-dominated states. While the refuge function for zooplankton seems to be of major importance in the state dominated by tall macrophytes, dense charophyte vegetation stabilizes the clearwater state mainly due to reduction of sediment resuspension, nutrient accumulation, and favoring of macroinvertebrates (Blindow et al., 2014 [50]).

Dominance patterns and interspecific competition among these different life forms of submerged plants (Figure 1) are mainly determined and affected by access to light and inorganic carbon. "Bottom-dwellers", such as isoetids and charophytes, but also some low-growing vascular plants form more or less dense vegetation close to the sediments, which prevents their occurrence in deeper, turbid water and therefore restricts them to less eutrophic environments (Barko and Smart 1981 [51], Blindow 1992a [52]). Most isoetids are adapted to soft water conditions with low concentrations of inorganic carbon in the water column and have developed several adaptations to this deficiency, such as carbon dioxide uptake from sediments and CAM metabolism. Generally, they lack the ability to assimilate bicarbonate (Madsen and Sand-Jensen 1991 [53], Keeley 1998 [54], Smolders et al., 2002 [55]). Apart from several *Nitella* species growing in soft water environments, charophytes occur mainly in calcium-rich water with higher pH values and bicarbonate as the main form of inorganic carbon. Here, they are highly competitive due to their efficient bicarbonate assimilation (van den Berg et al., 2002 [56], Ray et al., 2003 [57]). Charophytes therefore dominate the submerged vegetation in many oligo- to mesotrophic calcium-rich lakes, which were therefore called "Chara-lakes" by Samuelsson (1925 [58]).

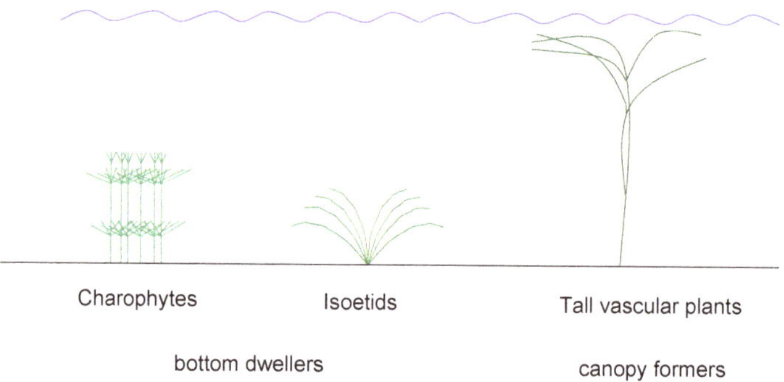

Figure 1. Different systematic groups and life forms of submerged plants, schematically.

Many vascular plants such as *Potamogeton* spp. and *Myriophyllum* spp. are tall and often form a canopy along the water surface, thus concentrating most of their photosynthetic biomass in regions with better light availability. These plants have a competitive advantage in turbid, more eutrophic environments, facilitated by often large hibernacles such as turions and tubers, which allow high growth rates during spring, even in turbid conditions (Blindow 1992a [52]). Most of these "canopy-formers" are able to assimilate bicarbonate but less efficiently than charophytes (van den Berg et al., 2002 [56]).

Experiments confirmed the different preferences observed in the field: charophytes are competitive at moderate nutrient concentrations, while tall angiosperms are superior competitors at higher nutrient conditions. van den Berg et al. (2002 [56]) demonstrated that the outcome of competition between *Chara aspera* and *Stuckenia pectinata* is dependent not only on light but also on bicarbonate availability. *Chara globularis* outcompeted *Myriophyllum spicatum* at low nutrient concentrations (Richter and Gross 2013 [59]). In another experiment, *C. globularis* developed far higher biomasses than angiosperms at low nutrient concentrations but far lower biomass at higher nutrient concentrations, while the growth rate of *Stuckenia pectinata* was not affected by the experimental condition (Bakker et al., 2010 [60]). In still another experiment, *Stuckenia pectinata* was outcompeted by charophytes at low nutrient concentrations, probably because of the efficient assimilation of nutrients and/or bicarbonate by the latter; in the same experiment, *Stuckenia pectinata* inhibited charophytes when it developed a "canopy", i.e., dense biomass close to the water surface (Hidding et al., 2010a [61]). In a system with experimental ponds, *Chara globularis* dominated at lower and *Elodea nuttallii* at higher nutrient concentrations (Bakker and Nolet 2014 [62]). In a newly created oligo- to mesotrophic lake dominated by charophytes, tall angiosperms were favored by the removal of *Chara* sp. and *Vaucheria* sp. in experimental plots (Vejříková et al., 2018 [63]).

4. Different Life Strategies in Charophytes

Among charophytes, both extreme R-strategists ("permanent pioneers") and extreme K-strategists with a strong impact on the whole ecosystem ("ecosystem engineers") can be identified (Schubert et al., 2018 [64]).

Typical R-strategists are annuals producing large quantities of oospores. These oospores are dispersed by waterfowl and can survive both drying and freezing and stay dormant for a long time, at least several decennia, in dry sediments (Krause 1997 [20], de Winton et al., 2000 [65], Rodrigo et al., 2015 [66]). In many newly created small water bodies, charophytes are the first submerged plants to establish but often disappear after several years due to competition of other, "late-coming" submerged plants (Casanova and Brock 1990 [41], Krause 1997 [20], Rodrigo et al., 2015 [66], Schubert et al., 2018 [64]). *Chara vulgaris*, *C. contraria*, *C. aspera*, and several *Nitella* species belong to these R-strategists, but most extreme are species such as *Tolypella intricata*, *T. glomerata*, and *Nitella capillaris*, which can also show up "spontaneously" in very small and temporal water bodies (see Figure 2). Already, Olsen (1944 [67]) and Hasslow (1931 [68]) mentioned their "meteoric" nature, while Allen (1950 [69]) and Fitzgerald (1985 [70]) called *Tolypella* spp. "vegetable comets". Oospores are most probably far more widespread than the sporadic records of these species, which only spend a very small part of their life cycle as green plants. Abundances are hard to estimate, which causes problems during red list assessments (Blindow 2009e [6]). In Sweden, *N. capillaris* was found in two small water bodies close to a former site more than 100 years after the last record of the species in the country (Blindow 2019 [71]).

Extreme K-strategists also belong to the charophyte group. Such species are perennial, produce only moderate numbers of oogonia, and therefore have a restricted ability to reach distant catchment areas. Under suitable conditions, however, they can form dense vegetation and outcompete other submerged macrophytes, acting as "nasty neighbors" (Figure 2). Because of their high biomasses, they act as "keystone organisms" in shallow water ecosystems and affect not only a number of physical and chemical factors but the whole food web structure (Hargeby et al., 1994 [72], Kufel and Kufel 2002 [73]). *Nitellopsis obtusa*, *Chara tomentosa*, *C. hispida*, and *C. subspinosa* belong to this group.

Figure 2. Different life strategies in charophytes: (**a**) *Nitella capillaris*, an extreme R-strategist, was re-discovered in this small water body near Kristianstad, about 100 years after the last record in the country. Photo by Bertil Möllerström. (**b**) The K-strategists *Chara subspinosa* and *C. tomentosa* form dense vegetation in Lake Levrasjön. Photo by Silke Oldorff.

5. (Re)establishment of Submerged Vegetation

(Re)establishment of submerged vegetation is therefore a major aim in many lake restorations projects. (Re)establishment can be achieved by improving the conditions for this vegetation and often without any plantations. Since some functions of this vegetation, such as increased habitat structure and substrate and predation refuge for smaller animals, are not dependent on living plants, even "plantations" of artificial plants have been applied in lake restorations (Schou et al., 2009 [74], Boll et al., 2012 [75], Balayla et al., 2017 [76], Jeppesen et al., 2017 [77]).

Sometimes, the opposite situation occurs, and "too dense" macrophytes are regarded as a nuisance. Dense vegetation clogs fishing nets and other fishing equipment, turbines, and other installations, impedes boat traffic and bathing, retards the water flow-through in channels, and causes high oxygen consumption during night (Jellyman et al., 2009 [78]).

Many publications investigate reasons for expansion and decline of submerged plants and deal with the restoration of this vegetation, including a strikingly high number of reviews. Bakker et al. (2013 [12]) summarized "case studies" of lake restorations which caused an expansion of submerged macrophytes, often combined with improved water clarity. Blindow et al. (2014 [50]) discussed differences in the feedback mechanisms between angiosperms and charophytes. Hussner et al. (2014 [79]) and Hilt et al. (2006 [80]) described the effect of single management measures on submerged macrophytes and gave detailed recommendations for macrophyte restoration. Phillips et al. (2016 [47]) discussed causes for the disappearance of submerged vegetation from shallow lakes and asked what we have learned during the past 40 years. van Katwijk et al. (2016 [81]) and Zhang et al. (2021 [82]) presented a global analysis of seagrass restoration projects. Jeppesen et al. (2017 [77]) treated the development of submerged vegetation after biomanipulations. Verhofstad et al. (2017 [43]) summarized the knowledge about the development of dense submerged vegetation after restorations, including the importance of sediments, light, and diaspore reservoirs in this process. Hilt et al. (2018 [45]) clarified the relationships between nutrient load and dominating vegetation type with and without biomanipulation. Two regional reviews summarized global experiences and case studies concerning transplantations of submerged macrophytes (van de Weyer et al., 2021 [83]) and submerged macrophytes with focus on charophytes (Blindow 2019 [71]). Finally, Rodrigo (2021 [84]) reviewed revegetation with submerged macrophytes including charophytes as a restoration tool for natural and constructed wetlands.

This extensive literature provides a good knowledge basis about which environmental conditions favor submerged macrophytes and shows that nutrient level and grazing pressure are the most important factors to be considered. High nutrient levels disfavor submerged plants because of poor water column light availability. A reduction of nutrient concentrations by means of (external) precipitation of phosphorus or by so-called "flushing" therefore has a positive impact on submerged vegetation (Meijer 2000 [44], van den Berg et al., 2001 [28]). Additionally, reduction of internal fertilization has generally a positive effect but may be combined with a risk of (mechanically) damaging the vegetation. Besides a decrease of overall nutrient concentrations, sediment removal reduces resuspension, allows a better anchorage of plants in the sediments, and exposes formerly covered seed banks but may reduce a major part of the diaspore reservoir. Covering of sediments reduces resuspension but also covers the seed banks and therefore can impede re-establishments. Oxidation of the sediment surface and (internal) phosphorus precipitation can be harmful due to mechanical disturbance and rapid pH changes (Hussner et al., 2014 [79]). Additionally, repeated mowing can favor submerged vegetation, as nutrients are removed and the ecosystem is maintained in a lower nutrient status (Kuiper et al., 2016 [85], see below).

High grazing pressure from fish, waterfowl, and crayfish can jeopardize the (re)establishment of submerged vegetation (van der Wal et al., 2013 [86], Hussner et al., 2014 [79]). Grazing pressure from fish and waterfowl is low in most natural lakes (Marklund et al., 2002 [87], Rip et al., 2006 [88]). Waterfowl can, however, have a major effect on density and species composition of submerged vegetation when present in high numbers (Søndergaard et al., 1996 [89], van Donk and Otte 1996 [90], Hilt et al., 2006 [80], van Onsem and Triest 2018 [91]). Especially high densities of herbivorous and benthivorous fish are harmful to submerged macrophytes (Hutorowicz and Dziedzic 2008 [92], Hussner et al., 2014 [79], Hilt et al., 2006 [80], Zinko 2017 [1]). During lake restoration, submerged vegetation has therefore often been fenced to avoid damage by grazing (Irfanullah and Moss 2004 [93], Hilt et al., 2006 [80], Hussner et al., 2014 [79], Jeppesen et al., 2017 [77]). Biomanipulation, e.g., the reduction of planktivorous/benthivouous fish or the implantation of piscivorous fish, favors submerged vegetation due to a reduction of mechanical damage and increase of zooplankton, which in turn reduces phytoplankton (Hussner et al., 2014 [79]). Spontaneous (re)establishment of submerged vegetation after biomanipulation has commonly been observed (Lauridsen et al., 1993 [94], van Donk and Otte 1996 [90], Fugl and Myssen 2007 [95], Sandby and Hansen 2007 [96], Verhofstad et al., 2017 [43],

Jeppesen et al., 2017 [77]). Vice versa, numerous plantations of submerged plants failed because of (often illegal) simultaneous carp implantations (see references below and in Table 1).

Table 1. Case studies for transplantations of charophytes, sorted country-wise. Methods specify, if plants are planted in pots, on textile mats, as green plant biomass, as oospores or as sediment containing oospores, and if areas were covered with sheets to impede competing species. Accompanying measures (Accomp): C—cutting of competing macrophytes; F—fish reduction; N—nutrient reduction; imp—implementation of *Anodonta* and *Salvelinus*, species assumed to favour submerged vegetation; Success/problems: + full success, ± some success, − no success of transplantations; C—competition; E—eutrophication; H—herbivory.

Site	Habitat	Method	Accomp	Charophyte Species Established	SUCCESS/PROBLEMS	Sources (No. of References)
Austria						
Mieminger Badesee	lake	sheets	C; N	*C. contraria*	not finished	[79]; A. La Rosée, pers. comm.
Canada						
Upper Link Lake	lake	green plants		*Nitella flexilis*	+	[97]
Germany						
Steinhöringer Badesee	lake	sheets; textile mats	imp	*C. globularis, C. papillosa*	±; H	[79]
Teichanlage Wielenbach	pond	textile mats		*C. globularis, C. contraria*	+; C	[79,98]
Bachtelweiher	lake	sheets; textile mats	F	*C. globularis, C. contraria*	−; E	[79,98]
Unterer Inselsee	lake	textile mats		*C. globularis, C. contraria*	±; E	[98]
Lake Phoenix	lake	green plants; sediment	N	*C. globularis, C. contraria, C. vulgaris*	+; C	[99–101]; own data
Baldeneysee	lake	green plants; sediment		*C. globularis, C. hispida, Nitellopsis obtusa*	±; C	[102]
Blücher-Park-Weiher	lake	green plants; sediment	N	*C. globularis, C. contraria, C. vulgaris, C. hispida, Nitellopsis obtusa*	+	[103]
Weißenstädter See	lake	green plants; textile mats	F	*Nitella flexilis*	−; H	[79]
Buchreuther Weiher	lake	sheets;		*C. globularis*		[80]
Wuckersee	lake	sediment	N	different *Chara* spp.	+	A. Hussner, pers. comm.; R. Mauersberger, pers. comm.
Behlendorfer See	lake	green plants		*C. subspinosa, C. contraria, Nitellopsis obtusa*	+	[104]
Baarer Kiesgrube	gravel pit	textile mats		*C. contraria*		[80]
Kiesgrube am Reeser See	gravel pit	green plants; textile mats		*C. contraria*	−	[79]
The Netherlands						
various lakes	lake	green plants.; sediment		charophytes	±	[79]
New Zealand						
Lake Rotoroa	lake	green plants; precultures	F	charophytes	±; H	[78,105]
Lake Rotomanuka	lake	pots		charophytes	−; H, C	[78]
Spain						
Albufera de València	lagoon	pots; precultures	N	*C. hispida, C. baltica, C. vulgaris, Nitella hyalina*	±; H	[106–108]

Table 1. *Cont.*

Site	Habitat	Method	Accomp	Charophyte Species Established	SUCCESS/PROBLEMS	Sources (No. of References)
Sweden						
Tinnerbäcken	ponds	green plants		*C. globularis, C. virgata, Nitella flexilis, N. opaca*	+	1
Forsmark	ponds	green plants		*C. globularis, C. virgata*	+	1
Växjö lakes	lakes	pots	F	*Nitella flexilis* vel *opaca*	+	[109,110]
Switzerland						
Action Plan	ponds	precultures		*Nitella hyalina*	+	[111]; A. Schwarzer, pers. comm.
USA						
Lake Susan, Minnesota	lake		F	*Chara vulgaris*	±	[112]
Lake Cooper, Texas	lake	oospores		*Chara vulgaris*	−; H, dessication	[113]
El Dorado Lake, Kansas	lake	oospores		*Chara vulgaris*	−; H	[114]

Grazing pressure differs highly among different plant species. Thus, the highly "palatable" *Stuckenia pectinata* was favored by protection against grazing, while *Myriophyllum spicatum* grew better in open, unprotected plots (Vejřiková et al., 2018 [63]). Grazing effects also interact with nutrient conditions. An experimental study showed that grazing pressure was higher at higher nutrient concentration, which was explained by higher plant palatability (Bakker and Nolet 2014 [62]). Verhofstad et al. (2017 [43]) described the intricate interactions among nutrients, fish, and macrophyte composition: high densities of herbivorous fish or waterfowl give rise to a lake ecosystem without submerged vegetation but with dominance of phytoplankton. Biomanipulation can cause a re-establishment of submerged vegetation with dominance of bottom-dwellers at lower nutrient conditions and tall species at high nutrient concentrations, the latter of which can be replaced by phytoplankton if nutrient loading increases further.

Moreover, water level and water level fluctuations have a high impact on submerged vegetation (Mäemets et al., 2018 [115]). In large, wind-exposed lakes, sediment resuspension can cause high turbidities, which can prevent (re)establishment of submerged vegetation, even if nutrient concentrations are rather low (Schutten et al., 2005 [116]). Artificial islands, enclosures, and other protecting installations have been applied to locally reduce resuspension and allow an establishment of macrophytes (Hussner et al., 2014 [79]). Restoration success can be substantially improved if several measures are combined (Kozak and Gołdyn 2016 [117]).

In a number of countries, lake brownification is increasing due to multiple mechanisms such as land use, climate change, and a return to less acidification (Temnerud et al., 2014 [118]). Higher water color causes reduced growth rates of submerged macrophytes (Reitsema et al., 2020 [119]), including charophytes (Choudhury et al., 2019 [120]).

Even under favorable conditions, (re)establishment of macrophytes may fail because of lack of diaspores. Diaspore banks should therefore be investigated before lake restorations to estimate the potential for re-establishments (Rodrigo and Alonso-Guillen 2013 [121], Hussner et al., 2014 [79], Holzhausen et al., 2017 [36]). A shift of macrophyte species composition is often observed after successful lake restorations and is explained by the large differences in numbers and longevity of diaspores among these species (Bakker et al., 2013 [12]). The composition of diaspores often differs widely from the composition of the actual vegetation. Densities of charophyte oospores can exceed several 10,000 m^{-2} of lake sediment, while the densities of the (far larger) angiosperm diaspores are several orders of magnitude lower (de Winton et al., 2000 [65], van den Berg et al., 2001 [28], Steinhardt and Selig 2007 [122], 2009 [123], Blindow et al., 2016 [46], Verhofstad et al., 2017 [43], Holzhausen et al., 2017 [36]). In germination experiments with freshwater sediments, charophytes developed higher germling densities (van Onsem and Triest 2018 [91]), while angiosperm germling densities were higher in experiments with brackish water sediments (Blindow et al., 2016 [46]).

Restorations of nutrient-rich lakes sometimes aim at favoring angiosperms such as *Stuckenia pectinata*, which are well adapted to higher turbidity (Coffey 2001 [124], Jellyman et al., 2009 [78]). Often, however, charophyte vegetation is preferred before tall macrophytes (Moss and van Donk 1990 [125]). Charophytes form dense vegetation with high biodiversity and a high biomass per lake surface unit and have therefore a stronger impact on phytoplankton and light availability than angiosperms. The share of rare species is high. Many species are winter-green or have a long growth period, which gives a more permanent effect on phytoplankton and light. Finally, these "bottom-dwellers" do not hamper bathing and boating as much as tall macrophytes which reach up to the water surface (Blindow 1992b [126], van den Berg et al., 1998 [127], Coops et al., 2002 [128], Kufel and Kufel 2002 [73], Bakker et al., 2013 [12], Blindow et al., 2014 [50], Hussner et al., 2014 [79], Verhofstad et al., 2017 [43], Zinko 2017 [1]).

6. Transplantations of Submerged Vegetation

"Direct" establishment of submerged macrophytes by means of transplantations (e.g., translocations, see IUCN 2013 [7]) has been applied during lake restorations, often combined with other measures such as nutrient reduction and biomanipulation (Hussner et al., 2014 [79]) but also in running water to increase habitat quality (Riis et al., 2009 [129]). Once established, submerged vegetation contributes to the stabilization of a clearwater state and therefore causes a more sustainable effect of lake restorations. Transplantations have also been applied to increase the biodiversity of aquatic macrophytes (Muller et al., 2013 [130], Rodrigo and Carabal 2020 [108]) and to create habitats for fish (Slagle and Allen 2008 [131], Fleming et al., 2011 [132]). Transplantations are time consuming (Jeppesen et al., 2017 [77]) and can be successful only if environmental conditions are suitable for submerged macrophytes (e.g., Hussner et al., 2014 [79], Hilt et al., 2006 [80], van de Weyer et al., 2021 [83]). Time and money are wasted if the warning given by Bakker et al. (2013 [12]) is not considered: "Subsequently one should wonder why macrophytes are not spontaneously returning to the restored water body. This may indicate that growing conditions are still not good enough and in that case transplanting will be unsuccessful".

Transplantations may be a suitable option if submerged plants do not (re)establish spontaneously in spite of suitable ecological conditions, which indicates that sufficient diaspores of native species are lacking. Based on experiences from a number of case studies, Hussner et al. (2014 [79]), Hilt et al. (2006 [80]), and van de Weyer et al. (2021 [83]) gave detailed recommendations regarding conditions and how such transplantations should be performed. Project aims should be defined, necessary permits from owners and nature conservation authorities should be obtained, threat factors should be reduced, ecological conditions and the colonization potential should be investigated, suitable plantation areas and methods as well as suitable species and donor sites should be selected, and, finally, experiences should thoroughly be documented (see Figure 3).

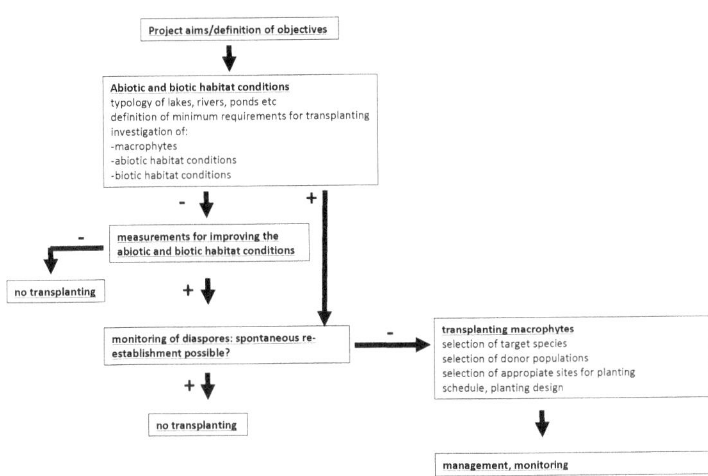

Figure 3. Checklist for re-establishments of submerged vegetation. From van de Weyer et al. (2021 [83]), modified.

Knowledge about which conditions and procedures favor submerged vegetation and which influences should be avoided is therefore essential. Data on nutrients, light, depth profile, sediment structure, exposition, as well as occurrence and abundance of herbivorous animals such as fish, crayfish, and waterfowl should be available if transplanting is considered (Grodowitz et al., 2009 [133], Hussner et al., 2014 [79]). Exceedingly high nutrient concentrations and/or high densities of cyprinid fish or grass carp are the main reasons for failures (see references in Table 1).

Project aims, environmental conditions, and colonization ability are factors to be considered when suitable species are selected for transplantations. Hussner et al. (2014 [79]) presented a list of species suitable for transplantations in Central European lakes and recommended transplantation of *Chara* spp. in alkaline, calcium-rich lakes. Vice versa, Jellyman et al. (2009 [78]) advised against plantations of species adapted to low nutrient conditions such as charophytes in eutrophicated lakes and recommended the use of *Stuckenia pectinata* for such environments. In China, *Vallisneria natans* is often planted, which is relatively tolerant against eutrophication (Li et al., 2008 [134]), but transplantations of this species fail at high fish densities and elevated nutrient concentrations, especially when both effects are combined (Gu et al., 2018 [135]). Rodrigo and Carabal (2020 [108]) recommended transplantation of *Myriophyllum spicatum*, *Stuckenia pectinata*, and *C. vulgaris*, as these species are widely available, easy to cultivate, and in experiments turned out to be rather grazing-resistant, while species such as *Ceratophyllum demersum*, *Nitella hyalina*, and *Tolypella glomerata* could be established once a vegetation cover has developed to increase biodiversity.

There are various techniques to plant aquatic macrophytes. The plants can be taken directly from a suitable donor site or transplanted after pre-culture. Green plants or plant parts, tubers, and rhizomes can be transferred to the target site. In laboratory experiments, some submerged plants such as *Myriophyllum spicatum* could easily be established from fragments, while, in other species such as *Potamogeton pusillus*, only few fragments survived after plantation (Barrat-Segretain et al., 1998 [136], 1999 [137], Vári 2013 [138]). Different kinds of substrates have been used, preferably decomposable ones, such as jute mats, wood, wool, or decomposable pots (Rott 2005 [139], Hoffmann et al., 2013 [140], Hussner et al., 2014 [79], van de Weyer et al., 2021 [83]). Substrates and techniques differ considerably in costs and especially in labor input. Establishment success, however, seems generally to be less dependent on substrate type and planting technique but is severely jeopardized by unsuitable conditions such as strong currents, unconsolidated sediments, and low light availability. Sediments also should have a sufficiently high share of organic material and may not contain toxic substances. Protection against grazing is especially important as long as plant biomasses and expansion on the target site are low (Lauridsen et al., 1993 [94], Irfannulah and Moss 2004 [93], Hilt et al., 2006 [80], Moore et al., 2010 [141], Jeppesen et al., 2017 [77], Rohal et al., 2021 [142], van de Weyer et al., 2021 [83]; Figure 4).

Figure 4. Dense charophyte vegetation (*Chara subspinosa, C. tomentosa*) inside grazing protections, Lake Wucker, Germany. Photo by Klaus van de Weyer.

Transplantations often start with so-called "founder colonies". These plantations, usually in protected exclosures, can be increased in the following years until the plants can expand by themselves and outside of the enclosures in the lake (Smart et al., 1998 [143], Smart and Dick 1999 [144], Jellyman et al., 2009 [78], Hussner et al., 2014 [79]). A sufficiently high share of the lake surface (around 30%) should be shallow enough to allow establishment by submerged vegetation (Jeppesen et al., 2017 [77]). In smaller lakes, the total area has been planted (van de Weyer et al., 2014 [99]) after a complete fish removal (see also Moss et al., 1996 [145]). Seagrass investigations demonstrate the advantages to transplant large intact patches rather than dispersed plots (Zhang et al., 2021 [82]).

Few attempts to (re)establish submerged macrophytes have been made in warmer regions, where this vegetation often is seen as a nuisance, except for China, where submerged plants have been planted in large quantities during lake restorations (Jeppesen et al., 2017 [77]). In smaller lakes, plantations were often successful when protected against herbivorous fish but failed in some cases due to expansion of floating-leaved plants (Chen et al., 2009 [146], Jeppesen et al., 2017 [77]).

7. Transplantations of Charophytes

Charophytes are rather commonly selected for transplantations for various reasons. Most common are transplantations connected to lake restorations. A number of charophyte species form dense and sometimes winter-green vegetation, which can store substantial quantities of nutrients and has a stronger and more sustainable impact on water quality than water angiosperms (Blindow 1992b [126], Kufel and Kufel 2002 [73], Blindow et al., 2014 [50]). In most cases, a mix of different species is planted with dominance of common species. *Chara contraria, C. globularis, C. papillosa, C. vulgaris,* and *Nitella mucronata* are recommended, but especially large species which can form dense vegetation such as *Chara tomentosa* and *Nitellopsis obtusa* (Hussner et al., 2014 [79]). Charophytes are more sensible against eutrophication than other submerged macrophytes. While waterfowl often prefer angiosperms before charophytes (Hidding et al., 2010b [147], Langhelle et al., 1996 [148]), crayfish prefer charophytes before angiosperms (Nyström and Strand 1996 [149], Zinko 2017 [1]). Zinko (2017 [1]) therefore advised never to implement crayfish in habitats with threatened macrophytes.

All available case studies on transplantations of charophytes are described in Table 1. For these transplantations, green plants, preferably protected by enclosures, and/or sediments rich in oospores were used. A number of these projects failed, often due to (sometimes illegal) fish implantations or nutrient loadings.

Other transplantation projects prefer charophytes, as they are bottom-dwellers and therefore are less disturbing for various activities such as boating and swimming than tall macrophytes (Hilt et al., 2006 [80]); they also provide valuable habitats for fish (Dick et al., 2004 [113], Dick and Smart 2004 [114]). A mixture of aquatic macrophytes including charophytes is sometimes transplanted to increase biodiversity (Rodrigo and Carabal 2020 [108], Rodrigo 2021 [84]; see Figure 5). Charophytes were also transplanted as agents to accumulate radioactive substances ("biological polishing"; Smith and Kalin 1992 [97]). Rarely, threatened charophytes are transplanted as a measure to protect these species (see below).

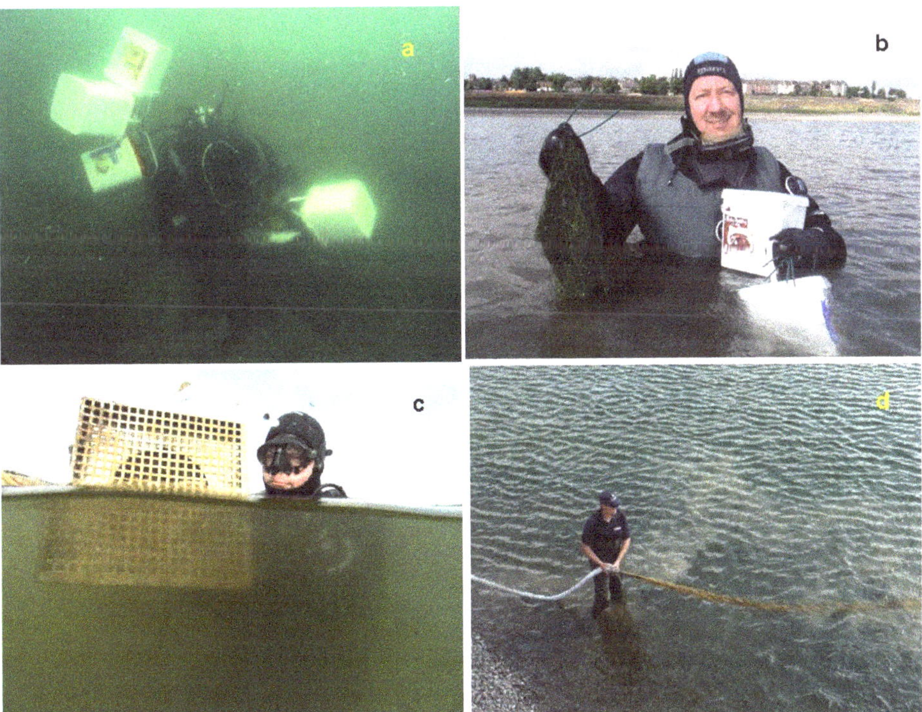

Figure 5. Lake Phoenix, Germany. (**a**): charophytes (green plants) are collected by divers in the donor lake, (**b**): planting of charophytes in L. Phoenix, (**c**): collection of water and sediment containing oospores by divers using a pump in the donor lake, (**d**): implementation of donor lake water and sediment in L. Phoenix. Photos by Klaus van de Weyer.

8. Transplantations of Threatened Aquatic Vascular Plants

While there are a number of experiences with both indirect and direct establishments (transplantations), plantations aiming at the protection of threatened species (e.g., population restorations, see IUCN 2013 [7]) have given rise to different kinds of projects as well as a new field of research (Seddon et al., 2007 [150], Jeppesen et al., 2017 [77]). Prior to transplantations, the presence of viable diaspores should be investigated in the transplantation site (Bakker et al., 2013 [12], Verhofstad et al., 2017 [43], Holzhausen et al., 2017 [36]). If an establishment from the present diaspore reservoir is not possible, transplantations may be a suitable option to support the regional population. Therefore, necessary permits and potentially negative consequences such as damage of the donor original population, gene pool contaminations, and introduction of neophytic species attached to the donor plant material have to be considered (Barett and Kohn 1991 [151], Foster Huenneke 1991 [152], Hussner et al., 2014 [79], Holzhausen et al., 2017 [36]).

There are few guidelines or recommendations for transplantations of rare aquatic plants. Guidelines for transplantations of rare terrestrial plants were developed in several countries such as Germany (Sukopp and Trautmann 1981 [153]), the USA. (Falk et al., 1996 [154]), and Sweden (Wetterin 2008 [155]). The IUCN (2013 [7]) provided guidelines for transplantations (translocations) of rare animals and plants. These publications agree in their main points:

- A species should be transplanted only if it does not establish spontaneously;
- Laws have to be followed and necessary permits must be obtained;
- Species may only be planted within their (recent or historic) distribution area;
- Donor plants should be obtained from a site close by and be genetically similar to the original population;
- The donor population may not be damaged;
- Transplantation sites must correspond to the species' environmental demands;
- All transplantations have to be monitored and documented scientifically over a longer time period;
- Protection and appropriate management of the transplantation site has to be guaranteed.

Falk et al. (1996 [154]) warned for failures: "A replacement population can be established only if the original causes of decline have been eliminated".

There are some experiences with transplantations of rare aquatic vascular plants. Among isoetids, the endemic *Isoetes malinverniana* was successfully transplanted in Italian small water bodies (Abeli et al., 2017 [156]). Transplantations of *Littorella uniflora, Isoetes lacustris, Lobelia dortmanna* succeeded in German lakes, especially if the plants were protected against grazing (Lenzewski 2019 [157]).

Both plant fragments and tubers of rare *Potamogeton* species were successfully transplanted in the UK (*P. compressus*; Markwell and Halls 2008 [158]), the USA (*P. amplifolius*; Storch et al., 1986 [159]), and Sweden (*P. acutifolius, P. compressus, P. trichoides*; Nilsson 2017 [160], Reuterskiöld 2017 [161], Zinko 2017 [1]).

Schwarzer and Wolff (2005 [162]) used both living plants and sporangia for the reestablishment of *Salvinia natans* in Germany. Ibars and Estrelles (2012 [163]) described the successful transplantation of soil spore banks to recover a lost population of *Marsilea quadrifolia* in Spain.

9. Transplantations of Threatened Charophytes

Indirect and direct establishment of charophyte vegetation have been part of a number of restoration projects (see above and Table 1). These experiences provide extensive knowledge about suitable environmental conditions for charophytes (Stewart 2008 [164]), which is an important prerequisite for successful transplantations (se Bakker et al., 2013 [12]). Together with transplantations of other threatened aquatic macrophytes (see above), these activities provide knowledge essential for what were, up to now, hardly applied transplantations of threatened charophytes. Bakker et al. (2013 [12]) and Jeppesen et al. (2017 [77]) mention the need for transplantations of threatened submerged macrophytes including charophytes to maintain biodiversity. For Swedish wetlands, Ekologgruppen (2009 [165]) recommended transplantations of threatened charophytes such as *Chara papillosa, Nitella gracilis*, and *N. mucronata*. Becker (2014 [166]), however, did not include transplantations among the numerous actions suggested to protect threatened charophytes in Germany.

According to our knowledge, the Swiss action plan for *Nitella hyalina* was the first time a threatened charophyte species was planted aiming to re-establish the species in its (former) Swiss distribution area (Schwarzer 2017 [111]). Fresh plant material was collected in France during 2017 and pre-cultured outdoors. These pre-cultures were successful. The plants hibernated and produced richly fertile biomass during 2018, when *Nitella hyalina* was planted in suitable sites close to Lake Zürich. During the following years, the species was stable in six out of 10 sites and expanded in these sites (see Figure 6). Plantations in additional sites are planned (A. Schwarzer, pers. comm.).

Figure 6. Transplantation of *Nitella hyalina* in Switzerland. (**a**) Precultivation in different tanks in a garden. (**b**) Target site during 2019. Transplanted *N. hyalina* (red circle) within vegetation consisting of different *Chara* species. (**c,d**) Target site during 2020. (**d**) Some *N. hyalina* had hibernated (red circle); establishment of *N. hyalina* outside of the original plantation is indicated by red arrows. Photos by A. Schwarzer.

Both fresh plant material and oospores can be used for transplantations depending on the life strategy of the species in question.

9.1. Establishment from Shoot Fragments

Many species can easily be established from shoot fragments. Shoot apices containing at least two nodes are used with the lowest node pushed down into the sediment. Node cells are omnipotent (Skurzyński and Bociąg 2011 [31]) and, in most cases, readily develop rhizoids and new growth. For such precultures, glass beakers with low nutrient water (tape water or water from the donor site) can be used, and sediments with a moderately high organic content provide nutrients. Sediment from the donor site eventually mixed with sand is often most suitable.

A number of charophyte species from temperate regions have been cultured from shoot fragments, in most cases successfully (see Table 2). Bociąg and Rekowska (2012 [167]) cultivated shoot fragments successfully from a number of species. Thereby, *Chara globularis* had the highest growth rates, followed by *C. subspinosa*; the lowest rates were found in *C. tomentosa* and *C. aspera*.

Table 2. References for successful culture of single charophyte species from shoot fragments. Swedish program species are shown in bold.

Species	Sources
Chara aculeolata	V. Krautkrämer, pers. comm.
Chara aspera	Blindow et al. (2003 [168]); Bociąg and Rekowska (2012 [167]); V. Krautkrämer, pers. comm.; M. Rodrigo, pers. comm.; own data
Chara baltica	Wüstenberg et al. (2011 [169]); A. Holzhausen, pers. comm.; own data
Chara canescens	A. Holzhausen, pers. comm.; V. Krautkrämer, pers. comm.; M. Rodrigo, pers. comm.
Chara contraria	A. Holzhausen, pers. comm.; V. Krautkrämer, pers. comm.; own data
Chara globularis	Bakker et al. (2010 [60]); Bociąg and Rekowska (2012 [167]); Richter & Gross (2013 [59]); V. Krautkrämer, pers. comm.; own data
Chara hispida	Wüstenberg et al. (2011 [169]); Rodrigo et al. (2017 [170]); V. Krautkrämer, pers. comm.; M. Rodrigo, pers. comm.; own data
Chara horrida	Own data
Chara papillosa	Own data
Chara subspinosa	Bociąg and Rekowska (2012 [167]); A. Holzhausen, pers. comm; own data
Chara tomentosa	Wüstenberg et al. (2011 [169]); Bociąg and Rekowska (2012 [167]); A. Holzhausen, pers. comm
Chara virgata	Own data
Chara vulgaris	Rodrigo et al. (2017 [170]); A. Holzhausen, pers. comm
Lamprothamnium papillosum	M. Rodrigo, pers. comm.
Nitella gracilis	M. Rodrigo, pers. comm.
Nitella hyalina	M. Rodrigo, pers. comm.
Nitella mucronata	V. Krautkrämer, pers. comm.
Nitella opaca	V. Krautkrämer, pers. comm.
Nitella tenuissima	V. Krautkrämer, pers. comm
Nitella translucens	Own data

Most *Chara* spp. can easily be cultured, often for many years, but generally, cultivation seems to be more difficult for species without cortex such as *Nitella* spp. and *Nitellopsis obtusa* (A. Holzhausen, pers. comm.). Species without cortex and long internodes such as *Nitellopsis obtusa* and *Nitella translucens* were cultured for physiological experiments, either in outdoor ponds or (more frequently) in the laboratory, but growth rates were not published for such cultures. *Nitellopsis obtusa* was transferred from the field to aquaria with tape water or site water in room temperature and under lamps and thus kept alive until the start of the experiments (Kurtyka et al., 2011 [171], Kisnieriene et al., 2012 [172]). In a laboratory of the University of Valencia, Spain, a number of charophyte species are kept in culture in small pots containing a sand/sediment substrate mixture, which are placed in larger beakers with tape water (Rodrigo et al., 2017 [170], Rodrigo 2021 [84]).

Alternatively, outdoor mesocosms (V. Krautkrämer, pers. comm., Richter and Gross 2013 [59]) or experimental ponds (Bakker et al., 2010 [60]) have been used to culture charophytes. Krautkrämer (pers. comm.) successfully used plastic containers containing different kinds of sediment and tap or site water for such cultures.

A new culture method was developed by Wüstenberg et al. (2011 [169]). Charophyte shoot fragments are planted in sand enriched with K_3PO_4 and covered with pure sand without nutrient addition. The overlying water consists of a nutrient solution without phosphorus. Enclosed in a polyethylen membrane, a bicarbonate reservoir provides a permanent supply of inorganic carbon. The advantage of this method is that growth rates of microalgae are kept low, while the charophytes can take up phosphorus from the sediment. Growth rate of charophytes are very high in such cultures.

9.2. Establishment from Oospores

Some charophyte species cannot be established from shoot fragments (see above). Especially, annual species with rich oospore production can be easier to establish from oospores. Establishment from oospores is complicated by the generally low germination success (see above) and the demand for species-specific germination conditions. Oospores of *Chara globularis* only grow at low redox potential (Forsberg 1965 [173]), while other species do not share this requirement (A. Holzhausen, pers. comm.). Germination has sometimes failed in autoclaved sediments and has been successful only if the sediment contained a certain organic share (A. Holzhausen, pers. comm.). A number of European species have high germination success at 15 °C but not at 20 °C (A. Holzhausen, pers. comm.). Temperature is probably acting as an indicator for the most suitable season (spring) for germination, while summer temperatures indicate that it is too late. Some species such as *Nitella furcata* and *Chara zeylanica*, however, only germinate during a so-called "germination window" during spring, which seems to open independently of temperature (Sokol and Stross 1986 [174], Stross 1989 [35]). Additionally, the presence of toxic substances can inhibit germination, as shown for *Chara hispida* in the presence of microcystin (Rojo et al., 2013 [175]). $Fe_2(SO_4)_3$, which sometimes is used to immobilize phosphorus in lake restoration, was shown to inhibit charophyte oospore germination (Rybak et al., 2017 [176]). Oospore germination of both *Chara* sp. and *Nitella* sp. was reduced by high concentrations of Cu (Kelly et al., 2012 [177]), and oospores of *Chara vulgaris* showed lower germination after exposure to high concentrations of Ni (Kalin and Smith 2007 [39]), sulfide, or Fe^{2+} (Sederias and Colman 2009 [178]).

Generally, oospores should be stratified, and sediments should be dried and provided with a certain share of organic matter before germination experiments are started. The specific germination demands of the species in question must be known, such as light (Holzhausen et al., 2017 [36]). The viability of oospores collected from sediments should be investigated. The so-called "crash tests" give a first indication: viable oospores show a "resistance to crushing" when pressed. Additionally, triphenyltetrazoliumchloride (TTC) staining is a good indicator for viability (Holzhausen et al., 2017 [36]).

9.3. Precultures

Charophyte species which do not form dense vegetation but occur as single plants on their sites often have to be precultured to obtain sufficient biomass for transplantations. Many species can easily be reproduced in larger or smaller containers with suitable sediments and water (see above), eventually with transplantations to other containers. The plants can be cultured indoors with artificial light or outdoors in larger containers or mesocosms. The latter alternative is assumed to be more promising, as the plants already are adapted to the on-site climate when transferred to their target sites. A good example is the Swiss Action Plan for *Nitella hyalina* with precultures in a market garden, which were bought by the canton of Zürich to culture aquatic macrophytes (Schwarzer 2017 [111]; Schwarzer, pers. comm.).

9.4. Accompanying Techniques

9.4.1. eDNA Analyses

eDNA analyses of water samples are already widely applied to detect a large range of aquatic organisms (see reviews by Thomsen and Willerslev 2018 [179] and Ruppert et al., 2019 [180]). In Sweden, eDNA analyses have successfully been applied for several years with the focus on fish, mussels, and crayfish (Bohman 2018 [181], von Proschwitz and Wengström 2021 [182]). Aquatic plants are, however, largely under-represented in such analyses compared to aquatic animals (Thomsen and Willerslev 2018 [179]). In a Canadian investigation, eDNA analyses identified more species belonging to the genera of *Potamogeton* and *Zannichellia* than "traditional" methods (Kuzmina et al., 2018 [183]). Muha et al. (2018 [184]) detected invasive aquatic plants by means of eDNA analysis.

The method has not yet been tested systematically for charophytes but seems promising. Charophytes are assumed to release larger DNA quantities than vascular plants. When damaged by, e.g., grazing, the content of the large internode cell, which contains a high number of nuculid and chloroplasts, is released into the water column. Some first investigations confirmed that charophytes are easily detected in water samples. Thereby, markers using both nucleus and chloroplast genes are applied (Nowak, pers. comm.).

Diaspore investigations are important if transplantations of rare species are considered in sites where these species are absent in the vegetation. Such plantations should be avoided if viable oospores still are present in the sediment. Instead, re-establishment from the site's "own" diaspores should be promoted (Bakker et al., 2013 [12], Verhofstad et al., 2017 [43], Zinko 2017 [1], Holzhausen et al., 2017 [36]). "Classical" diaspore reservoir investigations are suitable to quantify and determine oospores and to check their viability (Holzhausen 2017 [36]) but are labor-intensive and connected with a high risk of missing rare species. eDNA analyses of sediment samples are less expensive and may be more suitable to detect rare species in the diaspore reservoir, especially *Nitella* spp. and *Tolypella* spp. Species belonging to these genera have often high oospore production (see below), and species-specific primers already exist (P. Nowak, pers. comm.). Thereby, sediment samples down to 10 cm could be analyzed, which corresponds to the layer containing viable oospores (van Onsem and Triest 2018 [91]). In terrestrial habitats, eDNA analyses have already been applied to identify diaspores in soil samples (Fahner et al., 2016 [185]).

9.4.2. Harvesting

Harvesting of submerged vegetation is a very old technique traditionally applied to fertilize arable fields and still used for this purpose in many countries (Roger and Watanabe 1984 [186]). Recently, the technique was recommended to achieve a complete phosphorus recycling (Quilliam et al., 2015 [187]).

The effects of cutting on submerged vegetation and interactions among different macrophytes were calculated in several models such as CHARISMA (Van Nes et al., 2002 [188], 2003 [189]), SAGA (Hootsmans 1999 [190]), and, more recently, PCLake (Kuiper et al., 2017 [85]). Practically, cutting has been applied to remove vegetation which is regarded as "obstacles" around bathing places but also to eliminate "undesired" macrophyte species in lake restoration projects. The harvested biomass has to be removed to prevent nutrient release and oxygen consumption, leaving the major part of the lake's submerged vegetation intact in order to uphold the clearwater feedback mechanisms (Hussner et al., 2014 [79], Hilt et al., 2006 [80]). Harvesting is labor- and cost-intensive, increasingly so as many plant species can rapidly regenerate (Abernethy et al., 1996 [191]). Experiences from a number of case-studies showed highly variable and even contradictory whole-ecosystem effects (Engel 1990 [192], Nichols and Lathrop 1994 [193], Barrat-Segretain and Amoros 1996 [136], Morris et al., 2003 [194], Bal et al., 2006 [195]; Morris et al., 2006 [196]).

Generally, macrophyte cutting seems to favor charophytes, which is explained by the removal of shading tall macrophytes and was therefore recommended as a method to favor rare charophytes (Zinko 2017 [1]). In a Polish lake, *Nitella mucronata* increased after macrophyte harvesting, while tall macrophytes, especially *Elodea canadensis*, decreased (Lawniczak-Malinska and Achtenberg 2018 [197]). Similarly, *Nitella mucronata* increased in a Swedish lake after cutting of floating-leaved plants (Kyrkander and Örnborg 2015 [198]).

9.4.3. Indicator Species

To select suitable habitats for rare charophytes, which are poor competitors, other bottom-dwellers with similar habitat characteristics such as *Nitella* spp., *Chara globularis*, *C. virgata*, isoetids, *Pilularia globulifera*, and *Elatine hexandra* could function as "indicator species" (Zinko 2017 [1]).

10. The Swedish Example: How to Protect Rare Charophytes

Based on knowledge and practical experience, recommendations are here given for the protection of threatened charophyte species included in the actual action plan in Sweden (Zinko 2017 [1]). For more detailed information about ecology, dispersal mechanisms, competitive strength, life strategy, number of sites, red list status, and trends, see Blindow (2009a–d [2–5]), Zinko (2017 [1]), SLU Artdatabanken (2020 [199]), and Artportalen (https://www.artportalen.se, accessed on 3 September 2021).

In Sweden, a general strategy for transplantations of native threatened aquatic species was implemented (Wetterin 2008 [155]). On a regional level, the county administration of Östergötland developed a strategy for cultivation and translocations of threatened species (Antonsson 2012 [200]). A national strategy for translocations of aquatic plants and animals is in a state of preparation. For red-listed species (which is the case for all program species), permits may be necessary for transplantations according to the national environmental law (Miljöbalken 12 kap 6§).

The 10 program species differ widely in rareness/number of sites and especially life strategies. Consequently, different actions with different priorities are recommended to secure the species within the country (Table 3). Survey is recommended for some species, either by "classical" methods and/or by means of eDNA of sediment or water samples. Transplantations are recommended for species which are assumed to be hampered from expansion because of rareness and lack of oospores in the diaspore reservoirs, not lack of suitable sites. Some species are highly competitive (K-strategists) and can form dense and extensive biomass once they have reached a new site but have only restricted dispersal abilities. Biomass of such species can be collected from donor sites without jeopardizing the population. To test suitable techniques, these species should first be transplanted on-site. Prior to transplantations to new sites, the occurrence of rare species which potentially could be outcompeted by the "newcomers" has to be investigated, and the transplantation material has to be checked for contamination with undesired species such as neophytes. Species with only low biomass on their actual sites, mainly weak competitors (R-strategists), may have to be precultured. Methods have not been tested for any of these species, but the method developed for *Nitella hyalina* has been very successful (see Table 1) and could be applied. Cutting tall macrophyte vegetation may additionally support the establishment of these weak competitors, and indicator species may help to identify suitable sites. In an initial stage, all transplants need to be protected against grazing and be followed by a detailed monitoring and, if necessary, actions to improve water quality and reduce herbivorous/benthivorous fish.

Table 3. Number of sites (records after 2000), life strategy, and recommended actions for the 10 charophyte species included in the Swedish action plan for threatend macrophytes (Zinko 2017 [1]). Strategy: r = r strategist. k = k strategist. int = intermediate. eDNA: specified, if analysis of sediment (sed.) and/or water samples is recommended. Transplantations (Tr.), direct and/or after precultivation (precult.): 1 = high priority; 2 = lower priority. ? = strategy may deviate in the Swedish populations. Cutting: Harvesting of tall macrophytes to improve establishment. Indicator: Indicator species are used to identify suitable habitats. For further explanations, see text.

Species	No of Sites	Strategy	Survey	eDNA	Tr. Direct	Tr. Precult.	Cutting	Indicator	Comment
Chara filiformis	1	int			1		x		
Chara subspinosa	16	k			2				highly competitive
Nitellopsis obtusa	17	k			2				highly competitive
Nitella translucens	6	k	x	water	1				
Nitella mucronata	about 50	int					(x)		no actions recommended
Nitella gracilis	20	r		sed.		1	x	x	
Nitella syncarpa	2	r		sed.	1	1	x	x	
Nitella confervacea		r	x	sed.	1?	1	x	x	
Chara braunii	3 *	int?			2	2			
Tolypella canadensis	6	k?	x	water					stable population?

* freshwater sites only.

Swedish authorities, similar to authorities in other countries, also include taxa with a doubtful taxonomic rank in conservational efforts. Consequently, both *C. filiformis* and *C. subspinosa* were included in the recent action plan to protect threatened macrophyte species (Zinko 2017 [1]), though they can genetically not be separated from *C. contraria* and *C. hispida*, respectively (Nowak et al., 2016 [201], Nowak, pers. comm.). The reason for this decision is that, similar to other taxonomic groups, the selection of species in charophytes is "man-made" rather than corresponding to the biological species concept. Genetic analyses are of limited support in, e.g., the so-called "Hartmania complex" within the genus of *Chara* (which includes *C. subspinosa* and *C. hispida*), because of generally close clustering of all taxa belonging to this group (see Nowak et al., 2016 [201]).

Chara filiformis has been described as annual and perennial (Migula 1897 [202], Olsen 1944 [67]). Life cycle is poorly known, but reproduction by both fragmentation and bulbils has been observed (Migula 1897 [202], Teppke 2014 [203]). The species can form dense monospecific vegetation but grows also associated with other plants, mainly other charophytes (Blindow 2009a [2], Teppke 2014 [203], Brzozowski et al., 2018 [204]). The competitive abilities and the dispersal mechanisms of the species are rather unknown. The species is typical for calcium-rich lakes. *Chara filiformis* can easily be kept in culture for a long time (Olsen 1944 [67], A. Holzhausen, pers. comm.). In experiments, oospores only germinated at low light (Holzhausen et al., 2017 [36]). The species was successfully transplanted in a German pond (R. Mauersberger, pers. comm.).

Lake Levrasjön in Scania is its only Swedish site. It was found for the first time during 1860 and seems since then to have occurred in the lake (Wahlstedt 1862 [205], Hasslow 1931 [68], Blindow 2009a [2], own observations). The species should be transplanted to other calcium-rich lakes close to Lake Levrasjön, preferably as green plants after a test of transplantations in Lake Levrasjön. Cutting of tall macrophytes is recommended in Lake Levrasjön to stabilize the occurrence of *C. filiformis* in the lake.

Chara subspinosa and *Nitellopsis obtusa* belong to the most "extreme" K-strategists among charophytes. Both species form dense vegetation and are highly competitive but are commonly sterile and assumed to be poor colonizers (Pereya-Ramos 1981 [206], Pełechaty 2005 [207], Langangen 2007 [208], Blindow 1992b [126], Schubert et al., 2014 [209]). This is especially the case for the dioecious *N. obtusa*, which in some lakes is represented by only one sex (Krause 1997 [20], Blindow 2009a [2], Kabus 2014 [210]). Both species occur in permanent habitats, most commonly in calcium-rich lakes, and are perennial. While *C. subspinosa* mainly hibernates green, *N. obtusa* hibernates by means of bulbils and as green plants in deeper water (Pereyra-Ramos 1981 [206], Hargeby 1990 [211], Skurzynski and Bociąg 2011 [31], Kabus 2014 [210], Cahill 2017 [212]).

C. subspinosa can easily be cultured in the laboratory (see Table 2), but is rarely fertile (Bociąg and Rekowska 2012 [167]). Skurzyński and Bociąg (2009, 2011 [31,32]) succeeded in cultivating the species from oospores, though only a low share (5%) of oospores germinated at 18 °C, and no germination was observed at 5 °C. Sediment redox potential did not affect germination, while the germination was retarded in the dark. Holzhausen (pers. comm.) observed higher germination success of oospores taken from sediments at 15 °C at high light intensities. Oospore implantations in lake restoration projects were discussed (Skurzyński & Bociąg 2009 [32]). Transplantations of fresh biomass were already successfully applied in Lake Behlendorfer See, Germany (Meis et al., 2018 [104]). In Lake Wuckersee, Germany, *C. subspinosa* established spontaneously in enclosures protected against cyprinid fish, showing that these fish are a serious threat factor (A. Hussner, pers. comm).

Bulbils of *N. obtusa* germinated readily at both high and low light conditions, while oospore germination failed. Cultivation of green plants was successful in natural sediments but not sand and less easily than *Chara* spp. (Holzhausen et al., 2017 [36], A. Holzhausen, pers. comm.). Krautkrämer (pers. comm.) failed in culturing the species. For physiological experiments, the species was kept in laboratory cultures for longer times, but no information on growth rates was given (Kurtyka et al., 2011 [171], Kisnieriene et al., 2012 [172]).

In Sweden, *Chara subspinosa* and *N. obtusa* occur in 16 and 17 sites, respectively, all of them calcium-rich lakes (see Figure 2). *C. subspinosa* is difficult to investigate, as it is hard to distinguish from *C. hispida*. *C. subspinosa* and *N. obtusa* have disappeared from a number of their former sites, probably because of eutrophication (Kyrkander 2007 [213], Zinko 2017 [1], Herbst et al., 2018 [214], Artportalen: accessed 7 May 2021). In *N. obtusa*, however, this decline was compensated by the colonization of new sites during the extension of the distribution areas to northern regions (Blindow 2009a [2]).

Transplantations are recommended to secure the occurrence of both species in the country and to counteract their assumed poor dispersal abilities, preferably on sites where they disappeared before, given that the on-site conditions are favorable. Preculture is not necessary, as dense vegetation is present on the actual sites (Kyrkander 2007 [213], Zinko 2017 [1], own observations). Lake Krankesjön in southern Sweden shifted to a clearwater state during the 1980s, and charophytes expanded (Hargeby et al., 1994 [72]). *C. subspinosa* was observed for the first time during 1995 (Blindow 2009a [2]); *Nitellopsis obtusa* was observed during 2009 (Artportalen: accessed 7 May 2021). Both species have since then expanded, thereby reducing the former dense vegetation of *Chara tomentosa* (own observations). Additionally, in North America, where *Nitellopsis obtusa* is an invasive plant, it has outcompeted other submerged macrophytes (Brainard and Schulz 2017 [215], Cahill 2017 [212]). Because of the high competitive strength of *C. subspinosa* and *N. obtusa*, there is a certain risk that other submerged macrophytes are outcompeted after plantations of (one of) these target species. A detailed investigation of submerged vegetation, including a search for rare species, is therefore necessary before transplantations (Zinko 2017 [1]). Both species are, however, especially suitable for transplantations in the context of lake restorations because of their ability to form dense vegetation. They could be planted in enclosures in their former site Lakes Ringsjöarna combined with other measures to improve water quality. This question is already discussed by the local administration (Richard Nilsson, Ringsjöns vattenråd, Höörs kommun, pers. comm.), especially as the water quality of the lakes has recently improved (Ekologigruppen Ekoplan AB 2019 [216]).

Nitella translucens hibernates as green plant (Wahlstedt 1875 [217], Migula 1897 [202], van Raam 1998 [218], Becker and Doege 2014 [219]). It is often fertile, but also sterile plants are commonly found (Becker and Doege 2014 [219], Blindow 2009b [3]). Nothing seems to be known about the dispersal abilities of this species. *N. translucens* can form dense monospecific vegetation (Becker and Doege 2014 [219]), which indicates strong vegetative reproduction and a rather high competitive ability. The species occurs in calcium-poor, oligo- to mesotrophic water, often with high contents of humic substances and sediments consisting of dy, and prefers subneutral to neutral pH (Bruinsma 2007 [220], Becker and Doege 2014 [219]). In Sweden, the species was characterized as typical for forests lakes (Zinko 2017 [1]). Nothing seems to be known about oospore germination (A. Holzhausen, pers. comm.). The species was cultured in the laboratory in small plastic containers with nutrient solution and artificial light at a pH of 5.5 and temperatures between 21 and 24 °C (Cruz-Mireles and Ortega-Blake 1991 [221]). Spanswick (1972 [222]) and Spanswick and Miller (1977 [223]) also cultured the species in the laboratory.

In Sweden, *Nitella translucens* occurs in six actual sites in the southern part of the country and has disappeared from five (Artportalen: accessed 7 May 2021). There may be a rather high number of unknown sites (Zinko 2017 [1], Å. Widgren, pers. comm.). Apart from field investigations, possibly supported by eDNA analyses (P. Nowak, pers. comm.), transplantations are planned for some of the species' former sites if water quality seems appropriate and after a test of plantations within one of its actual sites. On some actual sites, biomass seems to be sufficient for plantations, which erases the need for precultures. A pilot study with transplantations within Lake Älmtasjön, one of the actual sites, is planned for the summer of 2021 (Å. Widgren, pers. comm.).

Nitella mucronata is both annual and perennial with hibernation as a green plant (Wahlstedt 1875 [217], Migula 1897 [202], Olsen 1944 [67], Forsberg 1960 [224]). Little is known about the dispersal abilities of the species. Both fertile and sterile plants are common (Olsen 1944 [67], Korsch 2014a [225]). The species can form monospecific vegetation and is therefore assumed to be a rather good competitor (Blindow 2009b [3]). It occurs in a broad range of habitats such as lakes, small water bodies, and running water in both calcium-rich water and soft water, ranging from oligotrophic to eutrophic conditions with varying conductivities, and it seems to be less sensible against eutrophication than many other charophytes (Simons and Nat 1996 [226], Doege et al., 2014 [227], Korsch 2014a [225]). In the laboratory, oospores only germinated at high light, not at low light conditions (Holzhausen et al., 2017 [36]). The species can rather easily be kept in culture (V. Krautkrämer, pers. comm.) but less easily than many *Chara* spp. (A. Holzhausen, pers. comm.).

Intensive field investigations during the former action plan (Blindow 2009b [3]) increased the number of known sites in Sweden to around 50 (Artportalen: accessed 5 May 2021). Plantations seem promising and have been successful in Lake Phoenix, Germany (see Figure 5; Table 1), but are not considered necessary to secure the species' occurrence in Sweden. Plantations could, however, be applied during lake restorations (Zinko 2017 [1]). Cutting of tall macrophytes is recommended to favor the species on its recent sites.

Nitella gracilis, *N. syncarpa*, and *N. confervacea* are three small, slender charophyte species. They have an annual life cycle and only hibernate occasionally as green plants in deeper water (Wahlstedt 1875 [205], Hasslow 1931 [68], Krause 1997 [20], Korsch 2014b [228], Korte et al., 2014 [229], Pätzold et al., 2014 [230]). All three species are typical pioneer plants. They are often richly fertile with probably good dispersal abilities and often colonize newly created water bodies but can disappear soon because of competition from other plants and only rarely form monospecific vegetation (Wahlstedt 1875 [205], Du Rietz 1945 [231], Dahlgren 1953 [232], Koistinen 2003 [233], Blindow 2009b [3], Korsch 2014b [228]).

Nitella gracilis mainly occurs in small water bodies including temporary ponds, ditches, and pools and prefers oligo- to mesotrophic, calcium-poor, and shallow water (Doege et al., 2014 [227], Korsch 2014b [228]). In Sweden, it has about 20 actual sites. It has disappeared from several former sites (Kyrkander 2007 [213], Thuresson 2019 [234], Artportalen: accessed 7 May 2021) and is threatened by both eutrophication and acidification (Becker 2014 [166]). The species has been found in small water bodies, oligotrophic, even acidified lakes and brackish water with low salinities, down to more than 5 m depth (Artportalen: accessed 14 October 2018, Thuresson 2019 [234]). Nothing seems to be known about conditions for oospore germination (A. Holzhausen, pers. comm.). The species was successfully cultivated in the laboratory (M. Rodrigo, pers. comm.).

Nitella syncarpa occurs in lakes and small water bodies, including temporary ones, in subneutral to alkaline water and under oligo- to eutrophic conditions, mainly in shallow water, occasionally down to 8 m depth (Vesić et al., 2011 [235], Korte et al., 2014 [229]). Nothing seems to be known about conditions for oospore germination (A. Holzhausen, pers. comm.). Zherelova (1989a,b [236,237]) probably cultivated the species in the laboratory but did not specify any methods. In Sweden, *N. syncarpa* only occurs in two recent sites and seems to have disappeared from a number of its former sites (Blindow 2009b [3], Artportalen: accessed 7 May 2021). The occurrence on one of its recent sites is threatened by eutrophication (Kyrkander and Örnborg 2012 [238]). The species is one of the most threatened charophytes in Sweden, and actions to secure its occurrence in the country have a high priority (see Table 3).

Nitella confervacea occurs in small water bodies, including temporary ones, and in oligo- to mesotrophic, occasionally even eutrophic lakes, in hard and soft water, mainly in shallow water but occasionally several meters deep (Vesić et al., 2011 [235], Doege et al., 2014 [227], Pätzold et al., 2014 [230], Zinko 2017 [1]). Nothing seems to be known about oospore germination or culture conditions (A. Holzhausen, pers. comm.). The species is known from 10 actual Swedish sites and has disappeared from a number of its former sites (Artportalen: accessed 7 May 2021, Thuresson 2019 [234]).

Transplantations seem important to secure all three species in Sweden. As they only have low biomasses on the actual sites, precultivation is probably necessary. *N. confervacea* has rather high biomass in Lake Möckeln (own observations), which potentially can be used for a direct transfer. In Lake Limsjön, the biomass of *N. syncarpa* is rather large (Kyrkander and Örnborg 2012 [238]) and, therefore, removal of part of this population for transplantations was suggested (Zinko 2017 [1]). As the three species are typical pioneer plants, transplantations should not be focused on former sites but on suitable habitats within their recent distribution area, such as lake shores with sparse vegetation and newly created small water bodies (Zinko 2017 [1]). Indicator species may help selecting such habitats. Cutting of taller macrophytes could support the establishment. The species may be over-looked on many sites. Especially *N. confervacea* is hard to find because of its small size and risk of confusion with *Nitella wahlbergiana*, which is rather abundant in the country (Langangen 2007 [208], Zinko 2017 [1]). Resting oospores may be far more common than green plants and could be tracked by means of eDNA.

Chara braunii is mainly annual and hibernates by means of oospores but occasionally also as a green plant (Wahlstedt 1864 [239], Migula 1897 [202], Langangen 1974 [240], Franke and Doege 2014 [241]). The species is richly fertile and has been assumed to have good dispersal ability (Migula 1897 [202], Krause 1997 [20], Langangen et al., 2002 [242], Zhakova 2003 [243], Franke and Doege 2014 [241], Blindow 2009c [4]). It has been characterized as a poor competitor (Migula 1897 [202], Krause and Walter 1985 [244]) but can dominate in sites where competing vegetation is erased during winter, such as fish ponds that fall dry during winter (Krause and Walter 1985 [244]). The species occurs mainly in small water bodies but also in permanent habitats such as springs (Krause 1997 [20]) and even in the deep water zones of larger lakes down to 33 m (Blindow et al., 2018 [245]). It can be found in oligotrophic to eutrophic conditions, hard and soft water, and freshwater and brackish water. Mass development in a fish pond which was dried and frozen during winter (Migula 1897 [202]) indicates that oospores not only survive drying and freezing but that germination may be stimulated by such conditions. Schmidt et al. (1996 [246]) characterized *C. braunii* as a "permanent pioneer" in fish ponds. In its Swedish Bothnian Bay sites, the species occurs in a depth of 0.1 to 0.7 m (Artportalen: accessed 14 October 2018), where ice action during winter is strong, and any hibernation as green plants is hardly possible (Idestam-Almqvist 2000 [247]).

The species has often been cultured. In Japan, it was kept outdoors in containers with tape water and a sand/soil mixture (Amirnia et al., 2019 [248]). Imahori and Iwasa (1965 [249]) and Sato et al. (2014 [250]) obtained axenic cultures after surface sterilization of oospores with sodiumhypochloride (see Forsberg 1965 [173]) in containers with a sand/soil mixture, distilled water, and artificial light at 23 °C. The cultivation method developed by Wüstenberg et al. (2011 [169]) was successfully applied at the University of Marburg, Germany (S. Rensing, pers. comm.). Foissner et al. (1996 [251]) and Schmölzer et al. (2011 [252]) described successful cultivation and high growth rates in aquaria containing a peat/sand mixture and distilled water with artificial light at around 20 °C. Cultures failed, however, at the University of Valencia, Spain (M. Rodrigo, pers. comm.).

In Sweden, the species occurs in around 20 actual sites in the Bothnian Bay (Pekkari 1953 [253], Tolstoy & Österlund 2003 [254], Artportalen: accessed 7 May 2021). For long time, these brackish water sites were the only ones known in the country after the species disappeared from two former freshwater sites probably because of eutrophication (Blindow 2009c [4]). During 2018 and 2019, *C. braunii* was detected in three larger freshwater lakes, one of which (Lake Finjasjön) was heavily eutrophicated (Artportalen: accessed 7 May 2021). Freshwater and brackish water occurrences are highly separate from each other not only geographically but also ecologically. While brackish water plants are typical R-strategists, hibernation, reproduction, and competitive behaviors of the freshwater plants are largely unknown. The genetic diversity of *C. braunii* is unusually large, indicating that the species may consist of several taxonomic clusters (P. Nowak, pers. comm.).

Transplantations are not planned for the Bothnian Bay, as the occurrence in this area is assumed to be secured, but are recommended to support the occurrence in freshwater. Transplantations from one of the two freshwater lakes to suitable sites close by are eventually considered after preculture if the on-site biomass is too limited

Tolypella canadensis is an arctic charophyte with a circumpolar distribution (Romanov and Kopyrina 2016 [255]) and low on-site temperatures throughout (Langangen 1993 [256], Romanov and Kopyrina 2016 [255]). In Scandinavia, both fertile and sterile plants have been found. Oospores sometimes seem not to ripen before the end of the short growing period (Langangen 1993 [256], Langangen and Blindow 1995 [257]). The species is perennial and hibernates as green plants or by means of bulbils (Romanov and Kopyrina 2016 [255]). Nothing is known about its dispersal abilities or its competitive abilities, but it has often been found in dense monospecific vegetation (Langangen 1993 [256], Krause 1997 [20], Artportalen: accessed 14 October 2018). The species has been found in lakes and slowly running water; it prefers deeper water and soft water conditions with low Ca concentrations and neutral pH (Langangen 1993 [256], Langangen and Blindow 1995 [257], Romanov & Kopyrina 2016 [255]). Nothing is known about oospore germination (A. Holzhausen, pers. comm.). In a culture experiments, the plants died when exposed to temperatures exceeding 15 °C (Langangen 1993 [256]).

In Sweden, there are six actual sites, all in the county of Norrbotten (Artportalen: accessed 7 May 2021). During field investigation, the species was relocated most of its former sites (Pettersson et al., 2008 [258], Blindow 2009d [4], Zinko 2017 [1], Artportalen: accessed 8 October 2018). The occurrence in Sweden seems to be secure despite the low number of sites known. The species is assumed to have been widely overlooked, as field investigations in this part of the country are difficult and expensive. eDNA analyses of water samples have been successfully tested (P. Nowak, pers. comm.) and can help to reduce the costs for these investigations.

11. Final Remarks

The Swedish Action Plan (Zinko 2017 [1]) is an ambitious project. The extensive literature reviewed in this paper shows that successful re-establishment and transplantation has to consider life strategies, which vary considerably among charophytes, and that management techniques have to be adapted to the different species and life strategies. Existing experiences on re-establishments and transplantations of charophytes provide a sound basis for the transplantations planned. Especially, the successful transplantation of *Nitella hyalina* in Switzerland is most promising. Starting this action plan, Sweden has taken a pioneer roll in the protection of threatened charophytes. A thorough documentation of the results and the experiences is of outermost importance.

Author Contributions: I.B. and M.C. conceptualized the project. M.C. was responsible for the project administration. I.B. and K.v.d.W. collected data and literature sources. I.B. prepared the original draft; I.B., M.C. and K.v.d.W. reviewed and edited the final version. All authors have read and agreed to the published version of the manuscript.

Funding: The publication was financially supported by the County Administration of Jönköping, Sweden, and the University of Greifswald, Germany.

Institutional Review Board Statement: Not applicable.

Informed Consent Statement: Not applicable.

Acknowledgments: The following persons gave various kinds of information and/or provided valuable literature sources, which is gratefully acknowledged: Roland Bengtsson, Vincent Bertrin, Aurelie Boissezon, John Bruinsma, Michelle Casanova, J. Delay, Ulrike Hamann, Andreas Hedrén, Sabine Hilt, Anja Holzhausen, Andreas Hussner, Erik Jeppesen, Chrystal Kelly, Volker Krautkrämer, Tina Kyrkander, Alexandra La Rosée (nee Hösch), Agnieszka Ławniczak-Malińska, Rüdiger Mauersberger, Emile Nat, Richard Nilsson, Petra Nowak, Mariusz Pełechaty, Stefan Rensing, Maria Rodrigo, Arno Schwarzer, Mikael Svensson, Nick Stewart, Mats Thuresson, Hans-Christoph Vahle, Åke Widgren. Bertil Möllerström, Silke Oldorff, and Arno Schwarzer kindly provided photo material. The comments of two anonymous reviewers considerably improved this paper.

Conflicts of Interest: The authors declare no conflict of interest.

References

1. Zinko, U. Kunskapsuppbyggande program—15 hotade makrofytarter i permanenta vatten. *Havs-och Vattenmyndighetens Rapp.* **2017**, *6*.
2. Blindow, I. *Åtgärdsprogram för hotade kransalger: Arter i kalkrika sjöar 2008–2011. Trådsträfse (Chara filiformis), spretsträfse (Chara rudis), stjärnslinke (Nitellopsis obtusa)*; Report 5848; Naturvårdsverket: Stockholm, Sweden, 2009.
3. Blindow, I. *Åtgärdsprogram för hotade kransalger: Slinke-arter i sjöar och småvatten 2008–2011. Grovslinke (Nitella translucens), uddslinke (Nitella mucronata), spädslinke (Nitella gracilis), höstslinke (Nitella syncarpa), dvärgslinke (Nitella confervacea)*; Report 5850; Naturvårdsverket: Stockholm, Sweden, 2009.
4. Blindow, I. *Åtgärdsprogram för hotade kransalger: Tuvsträfse och barklöst sträfse 2008–2011. Tuvsträfse (Chara connivens), barklöst sträfse (Chara braunii)*; Report 5851; Naturvårdsverket: Stockholm, Sweden, 2009.
5. Blindow, I. *Åtgärdsprogram för hotade kransalger: Fjällrufse 2008–2011. (Tolypella canadensis)*; Report 5852; Naturvårdsverket: Stockholm, Sweden, 2009.
6. Blindow, I. *Åtgärdsprogram för hotade kransalger: Arter i småvatten/periodiska vatten 2008–2011. Vårslinke (Nitella capillaris), uddrufse (Tolypella intricata), trubbrufse (Tolypella glomerata)*; Report 5849; Naturvårdsverket: Stockholm, Sweden, 2009.
7. IUCN. *Guidelines for Reintroductions and Other Conservation Translocations*; IUCN: Gland, Switzerland, 2013.
8. Hilt, S.; Vermaat, J.E.; van de Weyer, K. *Macrophytes. Encyclopedia of Inland Waters*, 2nd ed.; Subsection 4: Lakes as Home for Life; Elsevier: Amsterdam, The Netherlands, in press.
9. Hilt, S.; Brothers, S.; Jeppesen, E.; Veraart, A.; Kosten, S. Translating regime shifts in shallow lakes into changes in ecosystem functions and services. *BioScience* **2017**, *67*, 928–936. [CrossRef]
10. Lacoul, P.; Freedman, B. Environmental influences on aquatic plants in freshwater ecosystems. *Environ. Rev.* **2006**, *14*, 89–136. [CrossRef]
11. Soons, M.B.; Van Der Vlugt, C.; Van Lith, B.; Heil, G.W.; Klaassen, M. Small seed size increases the potential for dispersal of wetland plants by ducks. *J. Ecol.* **2008**, *96*, 619–627. [CrossRef]
12. Bakker, E.S.; Sarneel, J.; Gulati, R.D.; Liu, Z.; van Donk, E. Restoring macrophyte diversity in shallow temperate lakes: Biotic versus abiotic constraints. *Hydrobiologia* **2013**, *710*, 23–37. [CrossRef]
13. Clausen, P.; Nolet, B.; Fox, A.; Klaassen, M. Long-distance endozoochorous dispersal of submerged macrophyte seeds by migratory waterbirds in northern Europe—A critical review of possibilities and limitations. *Acta Oecologica* **2002**, *23*, 191–203. [CrossRef]
14. Figuerola, J.; Green, A.J. Dispersal of aquatic organisms by waterbirds: A review of past research and priorities for future studies. *Freshw. Biol.* **2002**, *47*, 483–494. [CrossRef]
15. Santamaria, L. Why are most aquatic plants widely distributed? Dispersal, clonal growth and small-scale heterogeneity in a stressful environment. *Acta Oecologica* **2002**, *23*, 137–154. [CrossRef]
16. Bonis, A.; Grillas, P. Deposition, germination and spatio-temporal patterns of charophyte propagule banks: A review. *Aquat. Bot.* **2002**, *72*, 235–248. [CrossRef]
17. Green, A.J.; Figuerola, J.; Sánchez, M.I. Implications of waterbird ecology for the dispersal of aquatic organisms. *Acta Oecologica* **2002**, *23*, 177–189. [CrossRef]
18. Wood, R.D. Gametangial constants of extant Charophyta for use in micropaleobotany. *J. Paleontol.* **1959**, *33*, 186–194.

19. Haas, J.N. First identification key for charophyte oospores from central Europe. *Eur. J. Phycol.* **1994**, *29*, 227–235. [CrossRef]
20. Krause, W. *Charales (Charophycae). Süsswasserflora von Mitteleuropa, Band 18*; Gustav Fischer: Verlag Jena, Germany, 1997.
21. Proctor, V.W. Dispersal of fresh-water algae by migratory water birds. *Science* **1959**, *130*, 623–624. [CrossRef] [PubMed]
22. Proctor, V.W. Viability of *Chara* oospores taken from migratory water birds. *Ecology* **1962**, *43*, 528–529. [CrossRef]
23. Proctor, V.W. Long-distance dispersal of seeds by retention in digestive tract of birds. *Science* **1968**, *160*, 321–322. [CrossRef]
24. Brochet, A.L.; Guillemain, M.; Fritz, H.; Gauthier-Clerc, M.; Green, A.J. Plant dispersal by teal (*Anas crecca*) in the Camargue: Duck guts are more important than their feet. *Freshw. Biol.* **2010**, *55*, 1262–1273. [CrossRef]
25. Figuerola, J.; Charalambidou, I.; Santamaria, L.; Green, A.J. Internal dispersal of seeds by waterfowl: Effect of seed size on gut passage time and germination patterns. *Naturwissenschaften* **2010**, *97*, 555–565. [CrossRef] [PubMed]
26. Wang, H.; Liu, C.; Yu, D. Morphological and reproductive differences among three charophyte species in response to variation in water depth. *Aquat. Biol.* **2015**, *24*, 91–100. [CrossRef]
27. Blindow, I.; Schütte, M. Elongation and mat formation of *Chara aspera* under different light and salinity conditions. *Hydrobiologia* **2007**, *584*, 69–76. [CrossRef]
28. Van den Berg, M.S.; Coops, H.; Simons, J. Propagule bank buildup of *Chara aspera* and its significance for colonization of a shallow lake. *Hydrobiologia* **2001**, *462*, 9–17. [CrossRef]
29. De Winton, M.D.; Clayton, J.S. The impact of invasive submerged weed species on seed banks in lake sediments. *Aquat. Bot.* **1996**, *53*, 31–45. [CrossRef]
30. Asaeda, T.; Rajapakse, L.; Sanderson, B. Morphological and reproductive acclimations to growth of two charophyte species in shallow and deep water. *Aquat. Bot.* **2007**, *86*, 393–401. [CrossRef]
31. Skurzyński, P.; Bociąg, K. Vegetative propagation of *Chara rudis* (Characeae, Chlorophyta). *Phycologia* **2011**, *50*, 194–201. [CrossRef]
32. Skurzyński, P.; Bociąg, K. The effect of environmental conditions on the germination of *Chara rudis* oospores (Characeae, Chlorophyta). *Charophytes* **2009**, *1*, 61–67.
33. Takatori, S.; Imahori, K. Light reactions in the control of oospore germination of *Chara delicatula*. *Phycologia* **1971**, *10*, 221–228. [CrossRef]
34. Sederias, J.; Colman, B. The interaction of light and low temperature on breaking the dormancy of *Chara vulgaris* oospores. *Aquat. Bot.* **2007**, *87*, 229–234. [CrossRef]
35. Stross, R.G. The temporal window of germination in oospores of *Chara* (Charophyceae) following primary dormancy in the laboratory. *New Phytol.* **1989**, *113*, 491–495. [CrossRef]
36. Holzhausen, A.; Porsche, C.; Schubert, H. Viability assessment and estimation of the germination potential of charophyte oospores: Testing for site and species specificity. *Bot. Lett.* **2017**, *165*, 147–158. [CrossRef]
37. Casanova, M.T.; Brock, M.A. Can oospore germination patterns explain charophyte distribution in permanent and temporary wetlands? *Aquat. Bot.* **1996**, *54*, 297–312. [CrossRef]
38. De Winton, M.D.; Casanova, M.T.; Clayton, J.S. Charophyte germination and establishment under low irradiance. *Aquat. Bot.* **2004**, *79*, 175–187. [CrossRef]
39. Kalin, M.; Smith, M.P. Germination of *Chara vulgaris* and *Nitella flexilis* oospores: What are the relevant factors triggering germination? *Aquat. Bot.* **2007**, *87*, 235–241. [CrossRef]
40. Sabbatini, M.R.; Argüello, J.A.; Fernández, O.A.; Bottini, R.A. Dormancy and growth inhibitor levels in oospores of *Chara contraria* A. Braun ex Kütz. (Charophyta). *Aquat. Bot.* **1987**, *28*, 189–194. [CrossRef]
41. Casanova, M.T.; Brock, M.A. Charophyte germination and establishment from the seed bank of an Australian temporary lake. *Aquat. Bot.* **1990**, *36*, 247–254. [CrossRef]
42. Scheffer, M.; Hosper, S.H.; Meijer, M.-L.; Moss, B.; Jeppesen, E. Alternative equilibria in shallow lakes. *Trends Evol. Ecol.* **1993**, *8*, 275–279. [CrossRef]
43. Verhofstad, M.; Núñez, M.A.; Reichman, E.; van Donk, E.; Lamers, L.; Bakker, E. Mass development of monospecific submerged macrophyte vegetation after the restoration of shallow lakes: Roles of light, sediment nutrient levels, and propagule density. *Aquat. Bot.* **2017**, *141*, 29–38. [CrossRef]
44. Meijer, M.L. Biomanipulation in the Netherlands—15 Years of Experience. Ph.D. Thesis, Wageningen University, Wageningen, The Netherlands, 2000.
45. Hilt, S.; Nuñez, M.M.A.; Bakker, E.; Blindow, I.; Davidson, T.; Gillefalk, M.; Hansson, L.-A.; Janse, J.H.; Janssen, A.; Jeppesen, E.; et al. Response of submerged macrophyte communities to external and internal restoration measures in north temperate shallow lakes. *Front. Plant Sci.* **2018**, *9*, 194. [CrossRef] [PubMed]
46. Blindow, I.; Dietrich, J.; Möllmann, N.; Schubert, H. Growth, photosynthesis and fertility of *Chara aspera* under different light and salinity conditions. *Aquat. Bot.* **2003**, *76*, 213–234. [CrossRef]
47. Phillips, G.L.; Willby, N.; Moss, B. Submerged macrophyte decline in shallow lakes: What have we learnt in the last forty years? *Aquat. Bot.* **2016**, *135*, 37–45. [CrossRef]
48. Sayer, C.D.; Davidson, T.A.; Jones, J.I. Seasonal dynamics of macrophytes and phytoplankton in shallow lakes: A eutrophication-driven pathway from plants to plankton? *Freshw. Biol.* **2010**, *55*, 500–513. [CrossRef]
49. Hargeby, A.; Blindow, I.; Andersson, G. Long-term patterns of shifts between clear and turbid states in Lake Krankesjön and Lake Tåkern. *Ecosystems* **2007**, *10*, 29–36. [CrossRef]

50. Blindow, I.; Dahlke, S.; Dewart, A.; Flügge, S.; Hendreschke, M.; Kerkow, A.; Meyer, J. Composition and diaspore reservoir of submerged macrophytes in a shallow brackish water bay of the southern Baltic Sea–influence of eutrophication and climate. *Hydrobiologia* **2016**, *778*, 121–136. [CrossRef]
51. Barko, J.W.; Smart, R.M. Comparative influences of light and temperature on the growth and metabolism of selected submersed freshwater macrophytes. *Ecol. Monogr.* **1981**, *51*, 219–236. [CrossRef]
52. Blindow, I. Decline of charophytes during eutrophication: Comparison with angiosperms. *Freshw. Biol.* **1992**, *28*, 9–14. [CrossRef]
53. Madsen, T.V.; Sand-Jensen, K. Photosynthetic carbon assimilation in aquatic macrophytes. *Aquat. Bot.* **1991**, *41*, 5–40. [CrossRef]
54. Keeley, J.E. CAM photosynthesis in submerged aquatic plants. *Bot. Rev.* **1998**, *64*, 121–175. [CrossRef]
55. Smolders, A.; Lucassen, E.; Roelofs, J. The isoetid environment: Biogeochemistry and threats. *Aquat. Bot.* **2002**, *73*, 325–350. [CrossRef]
56. Van den Berg, M.S.; Coops, H.; Simons, J.; Pilon, J. A comparative study of the use of inorganic carbon resources by *Chara aspera* and *Potamogeton pectinatus*. *Aquat. Bot.* **2002**, *72*, 219–233. [CrossRef]
57. Ray, S.; Klenell, M.; Choo, K.-S.; Pedersén, M.; Snoeijs, P. Carbon acquisition mechanisms in *Chara tomentosa*. *Aquat. Bot.* **2003**, *76*, 141–154. [CrossRef]
58. Samuelsson, G. Untersuchungen über die höhere Wasserflora von Dalarne. *Sv. Växtsoc. Sällsk. Handl.* **1925**, *9*, 1–31.
59. Richter, D.; Gross, E.M. *Chara* can outcompete *Myriophyllum* under low phosphorus supply. *Aquat. Sci.* **2013**, *75*, 457–467. [CrossRef]
60. Bakker, E.; van Donk, E.; Declerck, S.; Helmsing, N.; Hidding, B.; Nolet, B. Effect of macrophyte community composition and nutrient enrichment on plant biomass and algal blooms. *Basic Appl. Ecol.* **2010**, *11*, 432–439. [CrossRef]
61. Hidding, B.; Brederveld, R.J.; Nolet, B.A. How a bottom-dweller beats the canopy: Inhibition of an aquatic weed (*Potamogeton pectinatus*) by macroalgae (*Chara* spp.). *Freshw. Biol.* **2010**, *55*, 1758–1768. [CrossRef]
62. Bakker, E.S.; Nolet, B.A. Experimental evidence for enhanced top-down control of freshwater macrophytes with nutrient enrichment. *Oecologia* **2014**, *176*, 825–836. [CrossRef]
63. Vejříková, I.; Vejřík, L.; Lepš, J.; Kočvara, L.; Sajdlová, Z.; Čtvrtlíková, M.; Peterka, J. Impact of herbivory and competition on lake ecosystem structure: Underwater experimental manipulation. *Sci. Rep.* **2018**, *8*, 12130. [CrossRef]
64. Schubert, H.; Blindow, I.; Bueno, N.C.; Casanova, M.T.; Pełechaty, M.; Pukacz, A. Ecology of charophytes—Permanent pioneers and ecosystem engineers. *Perspect. Phycol.* **2018**, *5*, 61–74. [CrossRef]
65. De Winton, M.D.; Clayton, J.S.; Champion, P.D. Seedling emergence from seed banks of 15 New Zealand lakes with contrasting vegetation histories. *Aquat. Bot.* **2000**, *66*, 181–194. [CrossRef]
66. Rodrigo, M.A.; Rojo, C.; Segura, M.; Alonso-Guillén, J.L.; Martín, M.; Vera, P. The role of charophytes in a Mediterranean pond created for restoration purposes. *Aquat. Bot.* **2015**, *120*, 101–111. [CrossRef]
67. Olsen, S. *Danish Charophyta, Chorological, Ecological and Biological Investigations*; Det Kongelige Danske Videnskabernes Selskab, Biologiske Skrifter: Copenhagen, Denmark, 1944; Volume 3, pp. 1–240.
68. Hasslow, O.J. Sveriges characeer. *Bot. Not.* **1931**, *1931*, 63–136.
69. Allen, G.O. British Stoneworts (Charophyta). In *The Haslemere Natural History Society*; T. Bungle & Co.: Arbroath, UK, 1950.
70. Fitzgerald, R. 'A vegetable comet' *Tolypella prolifera* appears again at Amberly Wild Brooks. *BSBI News*, August 1985; 32–33.
71. Blindow, I. *Litteraturstudie om Kransalger*; Report; County Administration Jönköpings Län: Jönköping, Sweden, 2019; Volume 20, pp. 1–84.
72. Hargeby, A.; Andersson, G.; Blindow, I.; Johansson, S. Trophic web structure in a shallow eutrophic lake during a dominance shift from phytoplankton to submerged macrophytes. *Hydrobiologia* **1994**, *279*, 83–90. [CrossRef]
73. Kufel, L.; Kufel, I. Chara beds acting as nutrient sinks in shallow lakes—A review. *Aquat. Bot.* **2002**, *72*, 249–260. [CrossRef]
74. Schou, M.O.; Risholt, C.; Lauridsen, T.; Søndergaard, M.; Grønkjær, P.; Jeppesen, E. Restoring lakes by using artificial plant beds: Habitat selection of zooplankton in a clear and a turbid shallow lake. *Freshw. Biol.* **2009**, *54*, 1520–1531. [CrossRef]
75. Boll, T.; Balayla, D.; Andersen, F.; Jeppesen, E. Can artificial plant beds be used to enhance macroinvertebrate food resources for perch (*Perca fluviatilis* L.) during the initial phase of lake restoration by cyprinid removal? *Hydrobiologia* **2011**, *679*, 175–186. [CrossRef]
76. Balayla, D.; Boll, T.; Trochine, C.; Jeppesen, E. Could artificial plant beds favour microcrustaceans during biomanipulation of eutrophic shallow lakes? *Hydrobiologia* **2017**, *802*, 221–233. [CrossRef]
77. Jeppesen, E.; Søndergaard, M.; Lauridsen, T.L.; Davidson, T.A.; Liu, Z.; Mazzeo, N.; Trochine, C.; Özkan, K.; Jensen, H.S.; Trolle, D.; et al. Biomanipulation as a restoration tool to combat eutrophication: Recent advances and future challenges. *Adv. Ecol. Res.* **2017**, *47*, 411–488. [CrossRef]
78. Jellyman, D.; Walsh, J.; de Winton, M.; Sutherland, D. *A Review of the Potential to Re-Establish Macrophyte Beds in Te Waihora (Lake Ellesmere)*; Report No. R09/38; Environment Canterbury, Regional Council: Christchurch, New Zealand, 2009; ISBN 978-1-86937-968-1.
79. Hussner, A.; Gross, E.M.; van de Weyer, K.; Hilt, S. Handlungsempfehlung zur Abschätzung der Chancen einer Wiederansiedlung von Wasserpflanzen bei der Restaurierung von Flachseen Deutschlands. Arbeitskreis "Flachseen" der Deutschen Gesellschaft für Limnologie e.V. (editor). 2014. Available online: http://www.dgl-ev.de (accessed on 4 May 2021).

80. Hilt, S.; Gross, E.M.; Hupfer, M.; Morscheid, H.; Mählmann, J.; Melzer, A.; Poltz, J.; Sandrock, S.; Scharf, E.-M.; Schneider, S.; et al. Restoration of submerged vegetation in shallow eutrophic lakes–A guideline and state of the art in Germany. *Limnologica* **2006**, *36*, 155–171. [CrossRef]

81. Van Katwijk, M.M.; Thorhaug, A.; Marbà, N.; Orth, R.J.; Duarte, C.M.; Kendrick, G.A.; Althuizen, I.; Balestri, E.; Bernard, G.; Cambridge, M.L.; et al. Global analysis of seagrass restoration: The importance of large-scale planting. *J. Appl. Ecol.* **2015**, *53*, 567–578. [CrossRef]

82. Zhang, Y.S.; Gittman, R.K.; Donaher, S.E.; Trackenberg, S.N.; van der Heide, T.; Silliman, B.R. Inclusion of Intra- and Interspecific Facilitation Expands the Theoretical Framework for Seagrass Restoration. *Front. Mar. Sci.* **2021**, *8*. [CrossRef]

83. Van de Weyer, K.; Meis, S.; Stuhr, J. *Entwicklung eines Handlungsleitfadens für die Ansiedlung von aquatischen Makrophyten in schleswig-holsteinischen Seen*; Landesamt für Landwirtschaft, Umwelt und ländliche Räume des Landes Schleswig-Holstein: Kiel, Germany, 2021.

84. Rodrigo, M. Wetland restoration with hydrophytes: A review. *Plants* **2021**, *10*, 1035. [CrossRef] [PubMed]

85. Kuiper, J.J.; Verhofstad, M.J.J.M.; Louwers, E.L.M.; Bakker, E.S.; Brederveld, R.J.; Van Gerven, L.P.A.; Janssen, A.B.G.; De Klein, J.J.M.; Mooij, W.M. Mowing submerged macrophytes in shallow lakes with alternative stable states: Battling the good guys? *Environ. Manag.* **2017**, *59*, 619–634. [CrossRef]

86. Van der Wal, J.E.M.; Dorenbosch, M.; Immers, A.K.; Forteza, C.V.; Geurts, J.J.M.; Peeters, E.T.H.M.; Koese, B.; Bakker, E.S. Invasive crayfish threaten the development of submerged macrophytes in lake restoration. *PLoS ONE* **2013**, *8*, e78579. [CrossRef]

87. Marklund, O.; Sandsten, H.; Hansson, L.-A.; Blindow, I. Effects of waterfowl and fish on submerged vegetation and macroinvertebrates. *Freshw. Biol.* **2002**, *47*, 2049–2059. [CrossRef]

88. Rip, W.; Rawee, N.; de Jong, A. Alternation between clear, high-vegetation and turbid, low-vegetation states in a shallow lake: The role of birds. *Aquat. Bot.* **2006**, *85*, 184–190. [CrossRef]

89. Søndergaard, M.; Bruun, L.; Lauridsen, T.; Jeppesen, E.; Vindbæk Madsen, T. The impact of grazing waterfowl on submerged macrophytes: In situ experiments in a shal-low eutrophic lake. *Aquat. Bot.* **1996**, *53*, 73–84. [CrossRef]

90. Van Donk, E.; Otte, A. Effects of grazing by fish and waterfowl on the biomass and species composition of submerged macrophytes. *Hydrobiologia* **1996**, *340*, 285–290. [CrossRef]

91. Van Onsem, S.; Triest, L. Turbidity, waterfowl herbivory, and propagule banks shape submerged aquatic vegetation in ponds. *Front. Plant Sci.* **2018**, *9*, 1514. [CrossRef]

92. Hutorowicz, A.; Dziedzic, J. Long-term changes in macrophyte vegetation after reduction of fish stock in a shallow lake. *Aquat. Bot.* **2007**, *88*, 265–272. [CrossRef]

93. Irfanullah, H.; Moss, B. Factors influencing the return of submerged plants to a clear-water, shallow temperate lake. *Aquat. Bot.* **2004**, *80*, 177–191. [CrossRef]

94. Lauridsen, T.; Jeppesen, E.; Andersen, F.Ø. Colonization of submerged macrophytes in shallow fish manipulated Lake Væng: Impact of sediment composition and birds grazing. *Aquat. Bot.* **1993**, *46*, 1–15. [CrossRef]

95. Fugl, K.; Myssen, P.M. Lake Maribo Søndersø. In *Sørestaurering i Danmark. Report from NERI No. 636.2007*; Liboriussen, L., Søndergaard, M., Jeppesen, E., Eds.; Danmarks Miljøundersøgelser Aarhus Universitet: Aarhus, Danmark, 2007.

96. Sandby, K.S.; Hansen, J. Lake Arreskov. In *Sørestaurering i Danmark. Report from NERI No. 636.2007*; Liboriussen, L., Søndergaard, M., Jeppesen, E., Eds.; Danmarks Miljøundersøgelser Aarhus Universitet: Aarhus, Danmark, 2007.

97. Smith, M.P.; Kalin, M. *Ecological Engineering and the Chara Process Applied to the Rabbit Lake Drainage Basin*; Final Report for Cameco, Boojum Technical Reports; Laurentian University: Greater Sudbury, ON, Canada, 1992.

98. Mählmann, J.; Arnold, R.; Herrmann, L.; Morscheid, H.; Mattukat, F. Künstliche Wiederbesiedlung von submersen Makrophyten in Standgewässern mit Hilfe eines textilen Vegetationstragsystems. *Rostock. Meeresbiol. Beiträge* **2006**, *15*, 133–145.

99. Van de Weyer, K.; Sümer, G.; Hueppe, H.; Petruck, A. Das Konzept PHOENIX See: Nachhaltiges Management von Makrophyten-Massenentwicklungen durch eine Kombination nährstoffarmer Standortbedingungen und Bepflanzung mit Armleuchteralgen. *Korresp. Wasserwirtsch.* **2014**, *2014*, 23–27.

100. Van de Weyer, K.; Sümer, G.; Meis, S. Erfahrungen mit unterschiedlichen Sohlbelegungsmaterialien zum Management von Makrophyten-Massenentwicklungen im PHOENIX See (Dortmund). *Korresp. Wasserwirtsch.* **2016**, *2016*, 353–356.

101. Van de Weyer, K.; Meis, S.; Sümer, G. Entwicklung von Flora und Vegetation im PHOENIX See (Dortmund)—Fünf Jahre nach Anpflanzungen mit Armleuchteralgen. *Rostocker Meeresbiol. Beiträge* **2017**, *27*, 7–18.

102. Lanaplan. Test-Bepflanzungen im Baldeneysee mit Armleuchteralgen im Jahr 2020 im Rahmen des Projektes ELODEA II—Erprobung und Bewertung innovativer Methoden zur Kontrolle des Makrophytenwachstums in den Ruhrstauseen. Unpublished report. 2020.

103. Lanaplan. Sanierung Blücherpark-Weiher in Köln-Bepflanzungen mit Armleuchteralgen im Jahr 2020. Unpublished report. 2020.

104. Meis, S.; van de Weyer, K.; Siepen, F.; Stuhr, J.; Hamann, U. WRRL-Maßnahme zur Verbesserung der Wasserpflanzenvegetation am Behlendorfer See (Schleswig-Holstein). In *Erweiterte Zusammenfassungen der Jahrestagung 2018 (Kamp-Lintfort)*; Deutsche Gesellschaft für Limnologie (DGL): Essen, Germany, 2019; p. 181.

105. Dugdale, T.M.; Hicks, B.J.; De Winton, M.; Taumoepeau, A. Fish exclosures versus intensive fishing to restore charophytes in a shallow New Zealand lake. *Aquat. Conserv. Mar. Freshw. Ecosyst.* **2006**, *16*, 193–202. [CrossRef]

106. Rodrigo, M.; Rojo, C.; Alonso-Guillén, J.L.; Vera, P. Restoration of two small Mediterranean lagoons: The dynamics of submerged macrophytes and factors that affect the success of revegetation. *Ecol. Eng.* **2013**, *54*, 1–15. [CrossRef]

107. Alonso-Guillén, J.L. Charophytes in Restoration of Aquatic Ecosystems. A Study Case within Albufera de València Natural Park. Ph.D. Thesis, Universitat de Valencia, Valencia, Spain, 2011.

108. Rodrigo, M.A.; Carabal, N. Selecting submerged macrophyte species for replanting in Mediterranean eutrophic wetlands. *Glob. Ecol. Conserv.* **2020**, *24*, e01349. [CrossRef]

109. ALcontrol AB, Hushållningssällskapet I Halland. *Vattenväxter i Växjösjön och Södra Bergundasjön*; Final Report; Växjö kommun: Växjö, Sweden, 2017.

110. Växjö Kommun. *Reduktionsfiske i Växjösjöarna*; Final report; Council of Växjö: Växjö, Sweden, 2018.

111. Schwarzer, A. *Aktionsplan Vielästige Glanzleuchteralge (Nitella hyalina (DC.) C. Agardh)*; Baudirektion Kanton Zürich, Amt für Landschaft und Natur: Zürich, Switzerland, 2017.

112. Knopik, J.M.; Newman, R.M. Transplanting aquatic macrophytes to restore the littoral community of a eutrophic lake after the removal of common carp. *Lake Reserv. Manag.* **2018**, *34*, 365–375. [CrossRef]

113. Dick, G.O.; Smart, M.; Smith, J.K. *Aquatic Vegetation Restauration in Cooper Lake, Texas: A Case Study*; U.S. Army Corps of Engineers: Washington, DC, USA, 2004.

114. Dick, G.O.; Smart, M. *Aquatic Vegetation Restauration in El Dorado Lake, Kansas: A Case Study*; Final report; U.S. Army Corps of Engineers: Washington, DC, USA, 2004.

115. Mäemets, H.; Laugaste, R.; Palmik, K.; Haldna, M. Response of primary producers to water level fluctuations of Lake Peipsi. *Proc. Estonian Acad. Sci.* **2018**, *67*, 231. [CrossRef]

116. Schutten, J.; Dainty, J.; Davy, A.J. Root anchorage and its significance for submerged plants in shallow lakes. *J. Ecol.* **2005**, *93*, 556–571. [CrossRef]

117. Kozak, A.; Gołdyn, R. Macrophyte response to the protection and restoration measures of four water bodies. *Int. Rev. Hydrobiol.* **2016**, *101*, 160–172. [CrossRef]

118. Temnerud, J.; Hytteborn, J.K.; Futter, M.N.; Köhler, S.J. Evaluating common drivers for color, iron and organic carbon in Swedish watercourses. *AMBIO* **2014**, *43*, 30–44. [CrossRef] [PubMed]

119. Reitsema, R.E.; Wolters, J.-W.; Preiner, S.; Meire, P.; Hein, T.; De Boeck, G.; Blust, R.; Schoelynck, J. Response of submerged macrophyte growth, morphology, chlorophyll content and nutrient stoichiometry to increased flow velocity and elevated CO_2 and dissolved organic carbon concentrations. *Front. Environ. Sci.* **2020**, *11*. [CrossRef]

120. Choudhury, M.I.; Urrutia-Cordero, P.; Zhang, H.; Ekvall, M.K.; Medeiros, L.R.; Hansson, L.-A. Charophytes collapse beyond a critical warming and brownification threshold in shallow lake systems. *Sci. Total. Environ.* **2019**, *661*, 148–154. [CrossRef]

121. Rodrigo, M.; Alonso-Guillén, J.L. Assessing the potential of Albufera de València Lagoon sediments for the restoration of charophyte meadows. *Ecol. Eng.* **2013**, *60*, 445–452. [CrossRef]

122. Steinhardt, T.; Selig, U. Spatial distribution patterns and relationship between recent vegetation and diaspore bank of a brackish coastal lagoon on the southern Baltic Sea. *Estuar. Coast. Shelf Sci.* **2007**, *74*, 205–214. [CrossRef]

123. Steinhardt, T.; Selig, U. Comparison of recent vegetation and diaspore banks along abiotic gradients in brackish coastal lagoons. *Aquat. Bot.* **2009**, *91*, 20–26. [CrossRef]

124. Coffey, W.W. The Feasibility of Submerged Macrophyte Re-Establishment in Kaituna Lagoon, Lake Ellesmere (Te Waihora). Master's Thesis, Lincoln University, Christchurch, New Zealand, 2001.

125. Moss, B.; van Donk, E. Engineering and biological approaches to the restoration from eutrophication of shallow lakes in which aquatic plant communities are important components. *Hydrobiologia* **1990**, *200*, 367–377. [CrossRef]

126. Blindow, I. Long- and short-term dynamics of submerged macrophytes in two shallow eutrophic lakes. *Freshw. Biol.* **1992**, *28*, 15–27. [CrossRef]

127. Van den Berg, M.S.; Coops, H.; Meijer, M.-L.; Scheffer, M.; Simons, J. Clear water associated with a dense Chara vegetation in the shallow and turbid Lake Veluwemeer, The Netherlands. In *The Structuring Role of Submerged Macrophytes in Lakes*; Jeppesen, E., Søndergaard, M., Søndergaard, M., Christoffersen, K., Eds.; Springer: New York, NY, USA, 1998; pp. 339–352.

128. Coops, H.; Van Nes, E.; van den Berg, M.V.D.; Butijn, G. Promoting low-canopy macrophytes to compromise conservation and recreational navigation in a shallow lake. *Aquat. Ecol.* **2002**, *36*, 483–492. [CrossRef]

129. Riis, T.; Schultz, R.; Olsen, H.-M.; Katborg, C.K. Transplanting macrophytes to rehabilitate streams: Experience and recommendations. *Aquat. Ecol.* **2009**, *43*, 935–942. [CrossRef]

130. Muller, I.; Buisson, E.; Mouronval, J.-B.; Mesléard, F. Temporary wetland restoration after rice cultivation: Is soil transfer required for aquatic plant colonization? *Knowl. Manag. Aquat. Ecosyst.* **2013**, *411*, 3. [CrossRef]

131. Slagle, Z.; Allen, M.S. Should we plant macrophytes? Restored habitat use by the fish community of Lake Apopka, Florida. *Lake Reserv. Manag.* **2018**, *34*, 296–305. [CrossRef]

132. Fleming, J.P.; Madsen, J.D.; Dibble, E.D. Macrophyte re-establishment for fish habitat in Little Bear Creek Reservoir, Alabama, USA. *J. Freshw. Ecol.* **2011**, *26*, 105–114. [CrossRef]

133. Grodowitz, M.J.; Smart, M.; Dick, D.O.; Stokes, J.A.; Snow, J. Development of a Multi-Attribute Utility Analysis Model for Selecting Aquatic Plant Restoration Sites in Reservoirs. ERDC/TN APCRP-EA-21. 2009. Available online: https://www.researchgate.net/publication/235059861_Development_of_a_Multi-Attribute_Utility_Analysis_Model_for_Selecting_Aquatic_Plant_Restoration_Sites_in_Reservoirs (accessed on 7 May 2021).

134. Li, C. *Eutrophication and Restoration of Huizhou West Lake*; Guangdong Science Press: Guangzhou, China, 2009; p. 168.

135. Gu, J.; He, H.; Jin, H.; Yu, J.; Jeppesen, E.; Nairn, R.W.; Li, K. Synergistic negative effects of small-sized benthivorous fish and nitrogen loading on the growth of submerged macrophytes–Relevance for shallow lake restoration. *Sci. Total Environ.* **2018**, *610–611*, 1572–1580. [CrossRef]

136. Barrat-Segretain, M.-H.; Bornette, G.; Hering-Vilas-Bôas, A. Comparative abilities of vegetative regeneration among aquatic plants growing in disturbed habitats. *Aquat. Bot.* **1998**, *60*, 201–211. [CrossRef]

137. Barrat-Segretain, M.H.; Henry, C.P.; Bornette, G. Regeneration and colonisation of aquatic plant fragments in relation to the disturbance frequency of their habitats. *Arch. Hydrobiol.* **1999**, *145*, 111–127. [CrossRef]

138. Vari, A. Colonisation by fragments in six common aquatic macrophyte species. *Fundam. Appl. Limnol.* **2013**, *183*, 15–26. [CrossRef]

139. Rott, T. Gewässerbelastung durch Cyanobakterien–Ergebnisbericht über den Einsatz von Makrophyten in der Restaurierung eines Flachsees. *Deutsche Ges. Limnologie-Tag.* **2004**, *2005*, 496–500.

140. Hoffmann, M.A.; González, A.B.; Raeder, U.; Melzer, A. Experimental weed control of *Najas marina* ssp. *intermedia* and *Elodea nuttallii* in lakes using biodegradable jute matting. *J. Limnol.* **2013**, *72*, e39. [CrossRef]

141. Moore, K.A.; Shields, E.C.; Jarvis, J.C. The role of habitat and herbivory on the restoration of tidal freshwater submerged aquatic vegetation populations. *Restor. Ecol.* **2008**, *18*, 596–604. [CrossRef]

142. Rohal, C.; Reynolds, L.; Adams, C.; Martin, C.; Latimer, E.; Walsh, S.; Slater, J. Biological and practical tradeoffs in planting techniques for submerged aquatic vegetation. *Aquat. Bot.* **2020**, *170*, 103347. [CrossRef]

143. Smart, R.M.; Dick, G.O.; Doyle, R.D. Techniques for establishing native aquatic plants. *J. Aquat. Plant Manag.* **1998**, *36*, 44–49.

144. Smart, R.M.; Dick, G.O. *Propagation and Establishment of Aquatic Plants: A Handbook for Ecosystem Restoration Projects*; U.S. Army Corps of Engineers: Washington, DC, USA, 1999. [CrossRef]

145. Moss, B.; Madgwick, J.; Phillips, G.L. *A Guide to the Restoration of Nutrient-Enriched Shallow Lakes*; Broads Authority & Environment Agency: Norwich, UK, 1996.

146. Chen, K.-N.; Bao, C.-H.; Zhou, W.-P. Ecological restoration in eutrophic Lake Wuli: A large enclosure experiment. *Ecol. Eng.* **2009**, *35*, 1646–1655. [CrossRef]

147. Hidding, B.; Bakker, E.; Keuper, F.; De Boer, T.; De Vries, P.; Nolet, B. Differences in tolerance of pondweeds and charophytes to vertebrate herbivores in a shallow Baltic estuary. *Aquat. Bot.* **2010**, *93*, 123–128. [CrossRef]

148. Langhelle, A.; Lundgren, F.; Marklund, O. Var söker Tåkerns simfåglar efter föda? *Vingspegeln* **1996**, *15*, 58–66.

149. Nyström, P.; Strand, J. Grazing by a native and an exotic crayfish on aquatic macrophytes. *Freshw. Biol.* **1996**, *36*, 673–682. [CrossRef]

150. Seddon, P.J.; Armstrong, D.P.; Maloney, R.F. Developing the Science of Reintroduction Biology. *Conserv. Biol.* **2007**, *21*, 303–312. [CrossRef]

151. Barett, S.C.H.; Kohn, J.R. Genetic and evolutionary consequences of small population size in plants: Implications for conservation. In *Genetics and Conservation of Rare Plants*; Falk, D.A., Holsinger, K.E., Eds.; Oxford University Press: New York, NY, USA, 1991; pp. 3–30.

152. Foster Huenneke, L. Ecological implications of genetic variation in plant populations. In *Genetics and Conservation of Rare Plants*; Falk, D.A., Holsinger, K.E., Eds.; Oxford University Press: New York, NY, USA, 1991; pp. 31–44.

153. Sukopp, H.; Trautmann, W. Ausbringung von Wildpflanzen. *Nat. und Landsch.* **1981**, *56*, 368–369.

154. Falk, D.A.; Millar, C.I.; Olwell, M. *Restoring Diversity. Strategies for Reintroduction of Endangered Plants*; Island Press: Washington, DC, USA; Covelo, CA, USA, 1996.

155. Wetterin, M. *Utsättning av vilda växt- och Djurarter i Nature*; Dnr 401-3708-08 Nl; Swedish Environmental Protection Agency: Stockholm, Sweden, 2008.

156. Abeli, T.; Cauzzi, P.; Rossi, G.; Pistoja, F.; Mucciarelli, M. A gleam of hope for the critically endangered *Isoëtes malinverniana*: Use of small-scale translocations to guide conservation planning. *Aquat. Conserv. Mar. Freshw. Ecosyst.* **2017**, *28*, 501–505. [CrossRef]

157. Lenzewski, N. *Massnahmen zur Förderung und Entwicklung der Strandlingsrasen in schleswig-holsteinischen Seen*; Unpublished Report; Landesamt für Landwirtschaft, Umwelt und ländliche Räume des Landes Schleswig-Holstein: Kiel, Germany, 2019.

158. Markwell, H.J.; Halls, J. Translocation of a nationally scarce aquatic plant, grass-wrack pondweed *Potamogeton compressus*, at South Walsham Marshes, Norfolk, England. *Conserv. Evidence* **2008**, *5*, 69–73.

159. Storch, T.A.; Winter, J.D.; Neff, C. The employment of macrophyte transplantating techniques to establish *Potamogeton amplifolius* beds in Chautauqua Lake, New York. *Lake Reserv. Manag.* **1986**, *2*, 263–266. [CrossRef]

160. Nilsson, E. Spets-och knölnate. Utsättning i befintliga och nygrävda vatten. In Proceedings of the Symposium on Lake Restoration, Linköping, Sweden, 10–11 October 2017.

161. Reuterskiöld, E. Aktiv etablering av sällsynta natearter i anlagda våtmarker. In Proceedings of the Symposium on Lake Restoration, Linköping, Sweden, 10–11 October 2017.

162. Schwarzer, A.; Wolff, P. Der Gemeine Schwimmfarn (*Salvinia natans* [L.] ALL.) am Oberrhein. Ökologische Untersuchungen und Ansiedlungsmaßnahmen für eine hoch-gradig gefährdete Wasserpflanze. *Nat. Landsch. Bad.-Württ* **2005**, *75*, 333–360.

163. Ibars, A.M.; Estrelles, E. Recent development in ex situ and in situ conservation of ferns. *Fern Gaz.* **2012**, *19*, 67–86.

164. Stewart, N. *Creating Gravel Pit Ponds and Lakes for Stoneworts*; © Pond Conservation: Oxford, UK, 2008.

165. Ekologgruppen. *Aktiv etablering av sällsynta våtmarksarter i anlagda våtmarker och dammar*; Dnr 2009; 25-10989/07; Jordbruksverket, Försöks- och Utvecklingsprojekt (FoU): Landskrona, Sweden, 2009.

166. Becker, R. Gefährdung und Schutz von Characeen. In *Armleuchteralgen. Die Characeen Deutschlands*; Arbeitsgruppe Characeen Deutschlands, Ed.; Springer Spektrum: Berlin, Germany, 2014; pp. 149–192.
167. Bociąg, K.; Rekowska, E. Are stoneworts (Characeae) clonal plants? *Aquat. Bot.* **2012**, *100*, 25–34. [CrossRef]
168. Blindow, I.; Hargeby, A.; Hilt, S. Facilitation of clear-water conditions in shallow lakes by macrophytes: Differences between charophyte and angiosperm dominance. *Hydrobiologia* **2014**, *737*, 99–110. [CrossRef]
169. Wüstenberg, A.; Pörs, Y.; Ehwald, R. Culturing of stoneworts and submersed angiosperms with phosphate uptake exclusively from an artificial sediment. *Freshw. Biol.* **2011**, *56*, 1531–1539. [CrossRef]
170. Rodrigo, M.A.; Puche, E.; Rojo, C. On the tolerance of charophytes to high-nitrate concentrations. *Chem. Ecol.* **2017**, *34*, 22–42. [CrossRef]
171. Kurtyka, R.; Burdach, Z.; Karcz, W. Effect of cadmium and lead on the membrane potential and photoelectric reaction of *Nitellopsis obtusa* cells. *Gen. Physiol. Biophys.* **2011**, *30*, 52–58. [CrossRef]
172. Kisnieriene, V.; Ditchenko, T.; Kudryashov, A.; Sakalauskas, V.; Yurin, V.; Rukšėnas, O. The effect of acetylcholine on Characeae K+ channels at rest and during action potential generation. *Open Life Sci.* **2012**, *7*, 1066–1075. [CrossRef]
173. Forsberg, C. Sterile Germination of Oospores of Chara and Seeds of *Najas marina*. *Physiol. Plant.* **1965**, *18*, 128–137. [CrossRef]
174. Sokol, R.C.; Stross, R.G. Annual germination window in oospores of *Nitella furcata* (Charophyceae). *J. Phycol.* **1986**, *22*, 403–406. [CrossRef]
175. Rojo, C.; Segura, M.; Cortés, F.; Rodrigo, M.A. Allelopathic effects of microcystin-LR on the germination, growth and metabolism of five charophyte species and one submerged angiosperm. *Aquat. Toxicol.* **2013**, *144–145*, 1–10. [CrossRef]
176. Rybak, M.; Joniak, T.; Gąbka, M.; Sobczyński, T. The inhibition of growth and oospores production in *Chara hispida* L. as an effect of iron sulphate addition: Conclusions for the use of iron coagulants in lake restoration. *Ecol. Eng.* **2017**, *105*, 1–6. [CrossRef]
177. Kelly, C.L.; Hofstra, D.E.; de Winton, M.; Hamilton, D.P. Charophyte germination responses to herbicide application. *J. Aquat. Plant Manag.* **2012**, *50*, 150–154.
178. Sederias, J.; Colman, B. Inhibition of *Chara vulgaris* oospore germination by sulfidic sediments. *Aquat. Bot.* **2009**, *91*, 273–278. [CrossRef]
179. Thomsen, P.F.; Willerslev, E. Environmental DNA—An emerging tool in conservation for monitoring past and present biodiversity. *Biol. Conserv.* **2015**, *183*, 4–18. [CrossRef]
180. Ruppert, K.M.; Kline, R.J.; Rahman, S. Past, present, and future perspectives of environmental DNA (eDNA) metabarcoding: A systematic review in methods, monitoring, and applications of global eDNA. *Glob. Ecol. Conserv.* **2019**, *17*, e00547. [CrossRef]
181. Bohman, P. *eDNA i en droppe vatten. Vattenprovtagning av DNA från fisk, kräftor och musslor—En Kunskapssammanställning*; Sveriges lantbruksuniversitet Aqua Reports: Uppsala, Sweden, 2018.
182. Von Proschwitz, T.; Wengström, N. Zoogeography, ecology, and conservation status of the large freshwater mussels in Sweden. *Hydrobiologia* **2021**, *848*, 2869–2890. [CrossRef]
183. Kuzmina, M.L.; Braukmann, T.; Zakharov, E.V. Finding the pond through the weeds: EDNA reveals underestimated diversity of pondweeds. *Appl. Plant Sci.* **2018**, *6*, e01155. [CrossRef]
184. Muha, T.P.; Skukan, R.; Borrell, Y.J.; Rico, J.M.; Garcia de Leaniz, C.; Garcia-Vazquez, E.; Consuegra, S. Contrasting seasonal and spatial distribution of native and invasive Codium seaweed revealed by targeting species-specific eDNA. *Ecol. Evol.* **2019**, *9*, 8567–8579. [CrossRef]
185. Fahner, N.A.; Shokralla, S.; Baird, D.J.; Hajibabaei, M. Large-Scale Monitoring of Plants through Environmental DNA Metabarcoding of Soil: Recovery, Resolution, and Annotation of Four DNA Markers. *PLoS ONE* **2016**, *11*, e0157505. [CrossRef]
186. Roger, P.A.; Watanabe, I. *Algae and Aquatic Weeds as Source of Organic Matter and Plant Nutrients for Wetland Rice. I: Organic Matter and Rice*; The International Rice Research Institute: Los Banos, Philippines, 1984; pp. 147–168.
187. Quilliam, R.S.; van Niekerk, M.; Chadwick, D.R.; Cross, P.; Hanley, N.; Jones, D.L.; Vinten, A.J.A.; Willby, N.; Oliver, D.M. Can macrophyte harvesting from eutrophic water close the loop on nutrient loss from agricultural land? *J. Environ. Manag.* **2015**, *152*, 210–217. [CrossRef] [PubMed]
188. Van Nes, E.H.; Scheffer, M.; Van den Berg, M.S.; Coops, H. Aquatic macrophytes: Restore, eradicate or is there a compromise? *Aquat. Bot.* **2002**, *72*, 387–403. [CrossRef]
189. Van Nes, E.H.; Scheffer, M.; Berg, M.S.V.D.; Coops, H. Charisma: A spatial explicit simulation model of submerged macrophytes. *Ecol. Model.* **2003**, *159*, 103–116. [CrossRef]
190. Hootsmans, M.J.M. Modelling *Potamogeton pectinatus*: For better or for worse. *Hydrobiologia* **1999**, *415*, 7–11. [CrossRef]
191. Abernethy, V.J.; Sabbatini, M.R.; Murphy, K.J. Response of *Elodea canadensis* Michx, and *Myriophyllum spicatum* L. to shade, cutting and competition in experimental culture. *Hydrobiologia* **1996**, *340*, 219–224. [CrossRef]
192. Engel, S. Ecological impacts of harvesting macrophytes in Halverson Lake, Wisconsin. *J. Aquat. Plant Manag.* **1990**, *28*, 41–45.
193. Nichols, S.; Lathrop, R.C. Impact of harvesting on aquatic plant communities in Lake Wingra, Wisconsin. *J. Aquat. Plant Manag.* **1994**, *32*, 33–36.
194. Morris, K.; Boon, P.I.; Bailey, P.C.; Hughes, L. Alternative stable states in the aquatic vegetation of shallow urban lakes. I. Effects of plant harvesting and low-level nutrient enrichment. *Mar. Freshw. Res.* **2003**, *54*, 185–200. [CrossRef]
195. Bal, K.D.; Van Belleghem, S.; Deckere, E.; Meire, P. The re-growth capacity of sago pondweed following mechanical cutting. *J. Aquat. Plant Manag.* **2006**, *44*, 139–141.

196. Morris, K.; Bailey, P.C.E.; Boon, P.I.; Hughes, L. Effects of plant harvesting and nutrient enrichment on phytoplankton community structure in a shallow urban lake. *Hydrobiologia* **2006**, *571*, 77–91. [CrossRef]
197. Lawniczak-Malińska, A.E.; Achtenberg, K. On the use of macrophytes to maintain functionality of overgrown lowland lakes. *Ecol. Eng.* **2018**, *113*, 52–60. [CrossRef]
198. Kyrkander, T.; Örnborg, J. Metodstudie rörande bekämpning av gul näckros, *Nuphar lutea*, i naturreservatet i Asköviken-Tidö. *Länsstyrelsen i Västmanlands län Rapp.* **2015**, *1*, 28.
199. SLU Artdatabanken. Rödlistade arter i Sverige 2020. SLU, Uppsala. 2020. Available online: www.artdatabanken.se/globalassets/ew/subw/artd/2.-var-verksamhet/publikationer/31.-rodlista-2020/rodlista-2020 (accessed on 8 August 2021).
200. Antonsson, K. *Policy för odling och utsättning av rödlistade arter i Östergötlands län. 2012–2018*; Dnr. 511-5818-1000-001; Länsstyrelsen i Östergötlands län: Linköping, Sweden, 2012.
201. Nowak, P.; Schubert, H.; Schaible, R. Molecular evaluation of the validity of the morphological characters of three Swedish Chara sections: Chara, Grovesia, and Desvauxia (Charales, Charophyceae). *Aquat. Bot.* **2016**, *134*, 113–119. [CrossRef]
202. Migula, W. Die Characeen. In *Rabenhorst, Kryptogamenflora. Die Characeen Deutschlands, Österreichs und der Schweiz*, 2nd ed.; Eduard Kummer: Leipzig, Germany, 1897; Volume 5.
203. Teppke, M. *Chara filiformis*. In *Armleuchteralgen. Die Characeen Deutschlands*; Arbeitsgruppe Characeen Deutschlands, Ed.; Springer Spektrum: Berlin, Germany, 2014; pp. 292–299.
204. Brzozowski, M.; Pełechaty, M.; Pietruczuk, K. Co-occurrence of the charophyte *Lychnothamnus barbatus* with higher trophy submerged macrophyte indicators. *Aquat. Bot.* **2018**, *151*, 51–55. [CrossRef]
205. Wahlstedt, L.J. Bidrag till kännedomen om de skandinaviska arterna af växtfamiljen Characeae. Ph.D. Thesis, University of Lund, Lund, Sweden, 1862.
206. Pereyra-Ramos, E. The ecological role of Characeae in the lake littoral. *Ecol. Pol.* **1981**, *29*, 167–209.
207. Pełechaty, M. Does spatially varied phytolittoral vegetation with significant contribution of charophytes cause spatial and temporal heterogeneity of physical-chemical properties of the pelagic waters of a tachymictic lake? *Pol. J. Environ. Stud.* **2005**, *14*, 63–73.
208. Langangen, A. *Charophytes of the Nordic Countries*; Saeculum ANS: Oslo, Norway, 2007.
209. Schubert, H.; Blindow, I.; van de Weyer, K. *Chara subspinosa*. In *Armleuchteralgen. Die Characeen Deutschlands*; Arbeitsgruppe Characeen Deutschlands, Ed.; Springer Spektrum: Berlin, Germany, 2014; pp. 345–354.
210. Kabus, T. *Nitellopsis obtusa*. In *Armleuchteralgen. Die Characeen Deutschlands*; Arbeitsgruppe Characeen Deutschlands, Ed.; Springer Spektrum: Berlin, Germany, 2014; pp. 505–514.
211. Hargeby, A. Macrophyte associated invertebrates and the effect of habitat permanence. *Oikos* **1990**, *57*, 338–346. [CrossRef]
212. Cahill, B.C. *State of Michigan's Status and Strategy for Starry Stonewort (Nitellopsis obtusa (Desv. in Loisel.) J. Groves) Management*; Technical Report; Central Michigan University: Mount Pleasant, MI, USA, 2017.
213. Kyrkander, T. Inventering av kransalger i sötvatten 2007. *Länsstyrelsen Västra Götalands Län Rapp.* **2007**, *91*, 72.
214. Herbst, A.; Henningsen, L.; Schubert, H.; Blindow, I. Encrustations and element composition of charophytes from fresh or brackish water sites–habitat- or species-specific differences? *Aquat. Bot.* **2018**, *148*, 29–34. [CrossRef]
215. Brainard, A.S.; Schulz, K.L. Impacts of the cryptic macroalgal invader, *Nitellopsis obtusa*, on macrophyte communities. *Freshw. Sci.* **2017**, *36*, 55–62. [CrossRef]
216. Ekologigruppen Ekoplan AB. *Makrofytinventering i Ringsjön 2019*; Lund, Sweden, 2019. Available online: https://www.ringsjon.se/wp-content/uploads/2020/04/Makrofytinventering-i-Ringsjon-2019_web.pdf (accessed on 3 September 2021).
217. Wahlstedt, L.J. *Monografi öfver Sveriges och Norges Characeer*; Boktryckeri: Christianstad, Sweden, 1875.
218. Van Raam, J.C. *Handboek Kranswieren*; Charaboek: Hilversum, The Netherlands, 1998; p. 200.
219. Becker, R.; Doege, A. *Nitella translucens*. In *Armleuchteralgen. Die Characeen Deutschlands*; Arbeitsgruppe Characeen Deutschlands, Ed.; Springer Spektrum: Berlin, Germany, 2014; pp. 493–505.
220. Bruinsma, J. Waterplanten in poelen langs de Tongelreep bij de Achelse Kluis. *Gorteria* **2007**, *32*, 111–123.
221. Cruz-Mireles, R.M.; Ortega-Blake, I. Effect of Na_3VO_4 on the P State of *Nitella translucens*. *Plant Physiol.* **1991**, *96*, 91–97. [CrossRef] [PubMed]
222. Spanswick, R. Evidence for an electrogenic ion pump in *Nitella translucens*. I. The effects of pH, K+, Na+, light and temperature on the membrane potential and resistance. *Biochim. Biophys. Acta (BBA)—Biomembr.* **1972**, *288*, 73–89. [CrossRef]
223. Spanswick, R.M.; Miller, A.G. Measurement of the Cytoplasmic pH in *Nitella translucens*. Comparison of values obtained by microelectrode and weak acid methods. *Plant Physiol.* **1977**, *59*, 664–666. [CrossRef] [PubMed]
224. Forsberg, C. Subaquatic macrovegetation in Ösbysjön, Djursholm. *Acta Oecologia Scand.* **1960**, *11*, 183–199. [CrossRef]
225. Korsch, H. *Nitella mucronata*. In *Armleuchteralgen. Die Characeen Deutschlands*; Arbeitsgruppe Characeen Deutschlands, Ed.; Springer Spektrum: Berlin, Germany, 2014; pp. 455–463.
226. Simons, J.; Nat, E. Past and present distribution of stoneworts (Characeae) in The Netherlands. *Hydrobiologia* **1996**, *340*, 127–135. [CrossRef]
227. Doege, A.; van de Weyer, K.; Becker, R.; Schubert, H. Bioindikation mit Characeen. In *Armleuchteralgen, Die Characeen Deutschlands*; Arbeitsgruppe Characeen Deutschlands, Ed.; Springer Spektrum: Berlin, Germany, 2014; pp. 97–138.
228. Korsch, H. *Nitella gracilis*. In *Armleuchteralgen. Die Characeen Deutschlands*; Arbeitsgruppe Characeen Deutschlands, Ed.; Springer Spektrum: Berlin, Germany, 2014; pp. 435–443.

229. Korte, E.; Pätzold, F.; Doege, A. *Nitella syncarpa. Die Characeen Deutschlands*; Arbeitsgruppe Characeen Deutschlands, Ed.; Springer Spektrum: Berlin, Germany, 2014; pp. 477–485.
230. Pätzold, F.; Korte, E.; Blindow, I. *Nitella confervacea*. In *Armleuchteralgen. Die Characeen Deutschlands*; Arbeitsgruppe Characeen Deutschlands, Ed.; Springer Spektrum: Berlin, Germany, 2014; pp. 413–420.
231. Du Rietz, G.E. Nitella Nordstedtiana i två uppländska sjöar. *Sven. Bot. Tidskr.* **1945**, *39*, 83–94.
232. Dahlgren, L. Nya lokaler för Nitella Nordstedtiana och Equisetum variegatum i Södermanland. *Bot. Not.* **1953**, *1953*, 142–143.
233. Koistinen, M. *Nitella confervacea* (Breb.) A. Braun ex Leonh. 1863. In *Charophytes of the Baltic Sea. The Baltic Marine Biologists Publication No. 19*; Schubert, H., Blindow, I., Eds.; Gantner Verlag: Ruggell, Liechtenstein, 2003; pp. 168–173.
234. Thuresson, M. Inventeringar av hotade makrofyter 2018. Kransalger i Skären och Malmsjön samt styvnate i Sparren. Länsstyrelsen Stockholm, Stockholm, Sweden. *Fakta* **2019**, 1. Available online: https://www.lansstyrelsen.se/download/18.2c30d6f167c5e8e7 c0d653/1546949171502/Inventeringar%20av%20hotade%20makrofyter%202018.pdf (accessed on 3 September 2021).
235. Vesić, A.; Blaženčić, J.; Stanković, M. Charophytes (Charophyta) in the Zasavica special nature reserve. *Arch. Biol. Sci.* **2011**, *63*, 883–888. [CrossRef]
236. Zherelova, O.M. Activation of choride channels in the plasmalemma of *Nitella syncarpa* by inositol 1,4,5-trisphosphate. *FEBS Lett.* **1989**, *249*, 105–107. [CrossRef]
237. Zherelova, O.M. Protein kinase S is involved in Ca2+ channels in plasmalemma of *Nitella syncarpa*. *FEBS Lett.* **1989**, *242*, 330–332. [CrossRef]
238. Kyrkander, T.; Örnborg, J. Kransalger i Dalarnas län. Inventeringar 2008–2010. *Länsstyrelsen i Dalarnas län Rapp.* **2012**, 8. Available online: https://www.lansstyrelsen.se/download/18.746760b71768421ad5520679/1611829785021/2012-08_Kransalger%20i% 20Dalarna_inventeringar%202008-2010.pdf (accessed on 3 September 2021).
239. Wahlstedt, L.J. Om characeernas knoppar och öfvervintring. Thesis, University of Lund, Lund, Sweden, 1864. Available online: https://www.google.co.uk/books/edition/Om_Characeernas_knoppar_och_%C3%B6fvervintri/16QVAAAAYAAJ? hl=en&gbpv=0 (accessed on 3 September 2021).
240. Langangen, A. Ecology and distribution of Norwegian charophytes. *Norw. J. Bot.* **1974**, *21*, 31–52.
241. Franke, T.; Doege, A. *Chara braunii*. In *Armleuchteralgen. Die Characeen Deutschlands*; Arbeitsgruppe Characeen Deutschlands, Ed.; Springer Spektrum: Berlin, Germany, 2014; pp. 253–261.
242. Langangen, A.; Koistinen, M.; Blindow, I. The charophytes of Finland. *Memo. Soc. Fauna Flora Fenn.* **2002**, *78*, 17–48.
243. Zhakova, L.V. *Chara braunii* C.C. Gmel. 1826. In *The Baltic Marine Biologists Publication No. 19*; Schubert, H., Blindow, I., Eds.; Charophytes of the Baltic Sea; Gantner Verlag: Ruggel, Liechtenstein, 2003; pp. 64–69.
244. Krause, W.; Walter, E. Die Characeen der Teiche in Oberfranken. *Ber. Bayer. Bot. Ges.* **1985**, *65*, 51–58.
245. Blindow, I.; Marquardt, R.; Schories, D.; Schubert, H. Charophytes of Chile—Taxonomy. 1. Chareae. *Nova Hedwig.* **2018**, *107*, 1–47. [CrossRef]
246. Schmidt, D.; Garniel, A.; Geissler, U.; Gutowski, A.; Kies, L.; Krause, W.; Melzer, A.; Samietz, R.; Schütz, W.; van de Weyer, K.; et al. Rote Liste der Armleuchteralgen (Charophyceae) Deutschlands. 2. *Fassung. Schr. für Veg.* **1996**, *28*, 547–576.
247. Idestam-Almqvist, J. Dynamics of submerged aquatic vegetation on shallow soft bottoms in the Baltic Sea. *J. Vegegation Sci.* **2000**, *11*, 425–432. [CrossRef]
248. Amirnia, S.; Asaeda, T.; Takeuchi, C.; Kaneko, Y. Manganese-mediated immobilization of arsenic by calcifying macro-algae, Chara braunii. *Sci. Total. Environ.* **2018**, *646*, 661–669. [CrossRef]
249. Imahori, K.; Iwasa, K. Pure culture and chemical regulation of the growth of charophytes. *Phycologia* **1965**, *4*, 127–134. [CrossRef]
250. Sato, M.; Sakayama, H.; Sato, M.; Ito, M.; Sekimoto, H. Characterization of sexual reproductive processes in *Chara braunii* (Charales, Charophyceae). *Phycol. Res.* **2014**, *62*, 214–221. [CrossRef]
251. Foissner, I.; Lichtscheidl, I.K.; Wasteneys, G.O. Actin-based vesicle dynamics and exocytosis during wound wall formation in Characean internodal cells. *Cell Motil. Cytoskelet.* **1996**, *35*, 35–48. [CrossRef]
252. Schmölzer, P.M.; Höftberger, M.; Foissner, I. Plasma Membrane Domains Participate in pH Banding of *Chara* Internodal Cells. *Plant Cell Physiol.* **2011**, *52*, 1274–1288. [CrossRef] [PubMed]
253. Pekkari, S. Fynd av *Chara braunii* vid Haparanda. *Bot. Not.* **1953**, *1953*, 73–77.
254. Tolstoy, A.; Österlund, K. *Alger vid Sveriges Östersjökust. En Fotoflora*; ArtDatabanken: Uppsala, Sweden, 2003.
255. Romanov, R.; Kopyrina, L.I. *Tolypella canadensis* (Charales, Charophyceae) in Asia: Final evidence of its circumpolar distribution. *Nova Hedwig.* **2016**, *102*, 423–427. [CrossRef]
256. Langangen, A. *Tolypella canadensis*, a charophyte new to the European flora. *Cryptogam. Algol.* **1993**, *14*, 221–231.
257. Langangen, A.; Blindow, I. Kransalgen *Tolypella canadensis* Sawa i Scandinavia. *Polarflokken* **1995**, *19*, 131–137.
258. Pettersson, M.; Edin, R.; Johansson, L. *Inventering av fjällrufse, Tolypella canadensis, i fem sjöar i Norrbottens län, 2007*; Report 7; County Administration: Norrlands Län, Sweden, 2008.

Review

Wetland Restoration with Hydrophytes: A Review

Maria A. Rodrigo 

Integrative Ecology Group, Cavanilles Institute for Biodiversity and Evolutionary Biology, University of Valencia, Catedrático José Beltrán 2, Paterna, 46980 Valencia, Spain; maria.a.rodrigo@uv.es

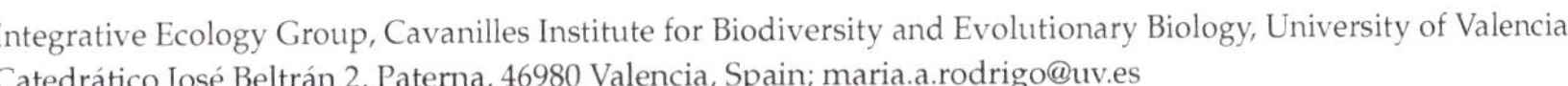

Abstract: Restoration cases with hydrophytes (those which develop all their vital functions inside the water or very close to the water surface, e.g., flowering) are less abundant compared to those using emergent plants. Here, I synthesize the latest knowledge in wetland restoration based on revegetation with hydrophytes and stress common challenges and potential solutions. The review mainly focusses on natural wetlands but also includes information about naturalized constructed wetlands, which nowadays are being used not only to improve water quality but also to increase biodiversity. Available publications, peer-reviewed and any public domain, from the last 20 years, were reviewed. Several countries developed pilot case-studies and field-scale projects with more or less success, the large-scale ones being less frequent. Using floating species is less generalized than submerged species. Sediment transfer is more adequate for temporary wetlands. Hydrophyte revegetation as a restoration tool could be improved by selecting suitable wetlands, increasing focus on species biology and ecology, choosing the suitable propagation and revegetation techniques (seeding, planting). The clear negative factors which prevent the revegetation success (herbivory, microalgae, filamentous green algae, water and sediment composition) have to be considered. Policy-making and wetland restoration practices must more effectively integrate the information already known, particularly under future climatic scenarios.

Keywords: revegetation; submerged macrophytes; floating macrophytes; aquatic phanerogams; charophytes; seeding; planting; transplanting; sediment transfer; natural wetlands; constructed wetlands

Citation: Rodrigo, M.A. Wetland Restoration with Hydrophytes: A Review. *Plants* **2021**, *10*, 1035. https://doi.org/10.3390/plants10061035

Academic Editor: Angelo Troia

Received: 28 March 2021
Accepted: 19 May 2021
Published: 21 May 2021

Publisher's Note: MDPI stays neutral with regard to jurisdictional claims in published maps and institutional affiliations.

1. Introduction

The term "wetland" broadly spans various types of water bodies, including seagrass meadows, coastal marshes (salt, brackish and freshwater tidal), forested wetlands (riparian, floodplain, bottomland hardwood, mangroves, etc.), and inland freshwater and saline wetlands (emergent wetlands, sedge meadows, wet prairies, fens, vascular plants in bogs, and temporary or seasonal wetlands, such as vernal pools and mudflats). Wetlands in the world provide essential ecosystem functions and services [1]: support biodiversity for conservation, improve water quality for downstream waters, combat sea-level rise, protect coastlines, mitigate the effects of flooding, drought and climate change, and provide habitat for recreation and other activities [2–5]. However, historically wetlands have been heavily impacted by humans, resulting in a loss of more than half of the wetlands globally, with significant impacts and risks to wildlife, humans and economies. Therefore, the restoration of this type of habitats is a must for the welfare of humanity. In March 2019, the United Nations declared 2021–2030 the Decade of Ecosystem Restoration [6] and thus, integral wetland restoration must be considered within these priorities and efforts.

One of the active strategies for aquatic ecosystem restoration has been traditionally seedling or planting macrophytes [7–9]. However, the majority of these study cases focusses on emergent aquatic plants, such as *Typha* spp., *Juncus* spp., *Phragmites australis*, etc. [10–12]. Restoration cases with hydrophytes, understanding them as the aquatic "plants" in a strict sense (that is, those which develop all their vital functions inside the water or very close to the water surface as the case of flowering; thus, they live submerged or floating in the water) are less abundant. Among other reasons, it is because working with hydrophytes is much more challenging compared to emergent plants. Moreover, much of this information is broadly scattered throughout the peer-reviewed and grey literature. Hence, there are no synthetic comprehensive reviews for restoration of wetlands based on hydrophytes. Here, I synthesize the latest knowledge in wetland restoration based on revegetation with hydrophytes and stress common challenges and potential solutions. Within hydrophytes, phanerogams, macroalgae and aquatic pteridophytes and bryophytes (mainly mosses and liverworts) can be considered. However, this review is restricted to submerged and floating phanerogams in continental wetlands, macroalgae, such as charophytes, or seagrasses for coastal wetlands. The conservation consensus is clear: "the protection of intact undisturbed environments is the only real solution to conservation of natural communities". However, what to do with those wetlands already affected? Revegetation is not the perfect solution for the conservation of wetlands. This is because many times it is not able to perpetuate species at risk, nor maintaining complex natural communities, but less is nothing. Therefore, this review mainly focusses on natural wetlands but also includes some information about more or less naturalized constructed wetlands, which nowadays are being used not only to improve water quality [13] but also to increase biodiversity and recover other ecosystem services such as carbon sequestration [14–17]. Available publications, both peer-reviewed and any public domain, from the last 20 years, have been reviewed (although I might have missed some cases). Personal expertise is also provided. I address two scales of interventions: (i) outdoor experimental approaches and (ii) large-scale actions in the field. I focus on issues related to the different approaches used (seedling, planting, transferring sediment, etc.). I also discuss the most common hydrophyte species used in restoration, the factors affecting revegetation and stress the challenges to evaluate the success of revegetation.

2. Natural and Constructed Wetlands

The ideal situation would be, of course, the preservation of natural wetlands whenever possible and that the reconstruction should be considered only as a last resort [18]. This is sometimes not possible because natural wetlands have disappeared or are severely degraded, and constructed wetlands (CWs) are implemented. CWs are artificial wetlands designed to intercept wastewater and remove a wide range of pollutants before discharge into natural water bodies. Surface-flow CWs are similar to natural marshes as they tend to occupy shallow channels and basins through which water flows at low velocities above and within the substrate. They mimic natural wetland ecosystems that combine physical, chemical, and biological processes to purify the water quality in more-controlled and efficient ways [19]. On the other hand, wetland restoration aims to restore lost biodiversity and to provide ecosystem services, such as flood-peak reduction and water-quality improvement, for instance, through phytoremediation. A successful restoration project may need to consider incorporating different wet environments (e.g., ponds, shallow lagoons, wet meadows, etc.), possibly combining areas for phytoremediation with areas of low nutrient content. In the last two decades, there has been an increasing trend to implement CWs in protected areas, such as national or natural parks in all continents (e.g., in Europa: Italy [20,21], Spain [16,22], Poland [23]; America [24,25]; Asia [26]; Africa [27]; Oceania [28], etc.). Some authors stress that wetland restoration must be prioritized over the creation of artificial wetlands, because, even when intended for conservation, they may not provide an adequate replacement of, for example, waterbird-supported functions [29]. However, other authors indicate that the biodiversity of constructed wetlands for wastewater treatment can be enhanced through proper design and management [30]. Moreover, CWs may serve as experimental pilot areas where treatments and procedures for revegetation to be further applied in wider natural but degraded wetlands can be tested [16].

3. Leader Countries in Wetland Revegetation with Hydrophytes

Since almost all countries in the world have their wetlands affected by pollution and many other problems, a large part of them has attempted to restore wetlands. However, in countries, where there are many large lakes, restoration has focused on them. In countries, where the scarcity of large continental water bodies is the normal situation, such as countries in the Mediterranean, with a semiarid climate, wetlands take a special relevance. The United States of America (USA) is the first country in the ranking of records obtained in a search in the "Web of Science" (WoS) about "wetland restoration" by world countries (more than 3000 records), followed by China (more than 1170 records) (Figure 1a). Australia, Canada, United Kingdom, and The Netherlands showed between 419 and 215 records. Germany, Spain, Africa (mostly South Africa), France, India, Mexico, Poland, Korea and Finland yielded between 196 and 100 records. Japan, New Zealand, Brazil, Italy, Denmark, Sweden and Belgium showed between 95 and 61 records. The rest of the countries showed less than 60 records.

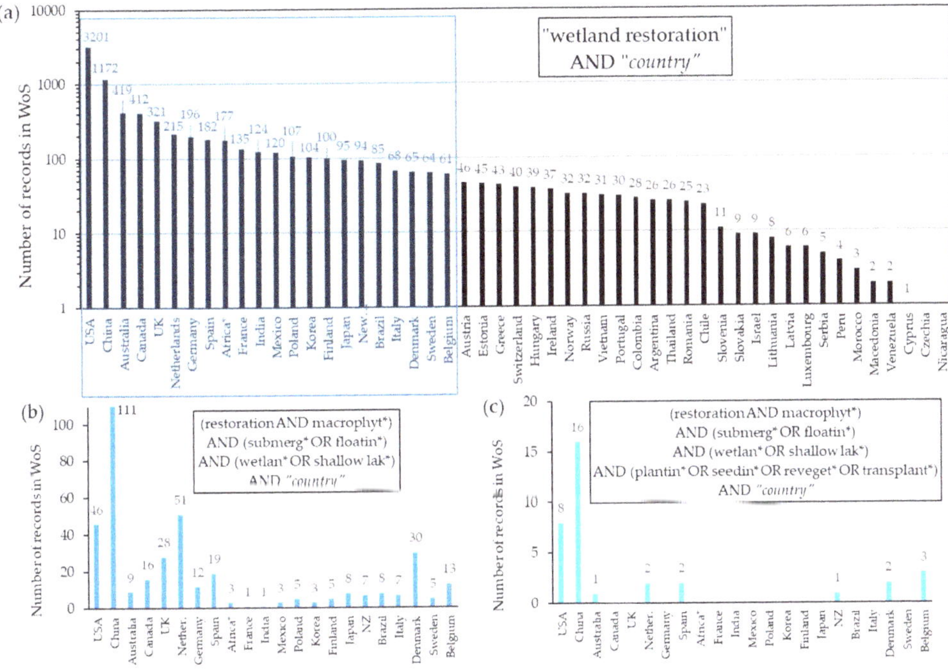

Figure 1. (**a**) The number of records obtained in the Web of Science (WoS) search for wetland restoration by countries (*except for the African continent) (notice the logarithmic scale). (**b**) The number of records in the WoS with the keywords: (restoration AND macrophyt*) AND (submerg* OR floatin*) AND (wetlan* OR shallow lak*) AND "country", for the countries that showed more than 60 records in the previous search (graph (**a**)). (**c**) Number of records for the countries of graph (**b**) now with the keywords: (restoration AND macrophyt*) AND (submerg* OR floatin*) AND (wetlan* OR shallow lak*) AND (plantin* OR seedin* OR reveget* OR transplant*) AND "country". Searches made in March 2021.

Firstly, among the 22 countries with more than 60 records, the highest number of records in the search in the WoS including submerged or floating macrophytes (using the string (restoration AND macrophyt*) AND (submerg* OR floatin*) AND (wetlan* OR shallow lak*) AND "country"), was for China with 111 records (Figure 1b), followed by The Netherlands, USA and Denmark. UK, Spain and Canada showed between 28 and 16 records. The rest of the countries produced less than 13 records. Secondly, focusing on revegetation approaches and adding to the above string "AND (plantin* OR seedin* OR reveget* OR transplant*)", the number of records was substantially reduced (Figure 1c), but China was again the country with the highest number of records (16), followed by the United States (8).

As seen above, both in Europe and the United States, but also in China, there has been a large tradition of aquatic ecosystems restoration [31–34]. For example, important native seed banking initiatives have been developed to improve the access to genetically diverse native wetland seeds for research, conservation, and restoration (European Native Seed Conservation Network (ENSCONET) and Seeds of Success (SOS)) (see references in [1]). China, with 68.5 million hectares of wetlands (36.2 and 32.3 million ha natural and constructed wetlands, respectively [26]), is one of the countries where many attempts at submerged macrophyte restoration and bioremediation have been made since the 1980s, despite the unfavorable results in some of the cases (references in [35]; reviewed in [36]). For example, from 2002 to 2006, nearly 60 programs have been developed to restore wetlands in this Asian country [37]. Their main results suggest that, combined with other technics (e.g., addition of filter-feeding aquatic animals in the proper biomass), submerged macrophyte restoration in wetlands might have a high success rate. The restoration or rehabilitation of wetlands around rivers and lakes has also been growing rapidly since the early 1990s in Japan including activities to recover lost or degraded vegetation and plant diversity [38]. In New Zealand, the most common action undertaken for restoration of wetlands is revegetation, involving removal of introduced weeds, and then the planting of native species appropriate to the habitat conditions and region [39].

4. The Scale of Revegetation

Some studies dealt with indoor aquaria or smaller outdoor mesocosm with the prospect of future wetland or shallow lake restoration program centered on hydrophyte replacement (see, for example, Ciurli et al. in Italy [40] or Fontanarrosa et al. in Argentina [41]). However, for successful revegetation, research using experimental on-site wetland mesocosms should be planned before starting larger-scale initiatives. Thus, knowledge of the biological and ecological requirements of the species, the choice of source material, the method of introduction and the selection of ecologically suitable translocation sites can reduce the failure rate [42]. Moreover, as revegetation represents a large proportion of the costs associated with restoration, developing cost-effective new planting methods would reduce the costs of large-scale restoration [43].

Table 1 shows some of the experiments which have been conducted related to revegetation with hydrophytes. Qiu et al. [44] performed in situ enclosure experiments in three parts of a eutrophic shallow lake in China with different trophic status, introducing both floating-leaved and submerged hydrophytes. All the introduced species grew well. The authors reported a monthly mean macrophyte biomass increase of 329 gWW/m^2. A large-scale experiment was conducted in the Danish shallow Lake Engelsholm, where three species were planted in three 25-m^2 exclosures with densities of 4–10 ramets/m^2 (with no roots) and 25 cm in length for *Stuckenia pectinata* and 40–50 cm in length for the other two *Potamogeton* species [45]. After two years following transplantation, the plant density development increased six-fold. Hilt et al. [46] described how Rott, in 2002, planted 200 m^2 of "macrophyte islands" with *Myriophyllum* spp. and *Chara contraria* in a 25-ha shallow lake in Southern Germany. One year later, the hydrophytes had already colonized 5.3 ha, which represented a monthly coverage increase rate of 0.4 ha. Ye et al. [47] performed an ecological restoration demonstration project in the shallow lake Taihu planting a total density of 105 plants/m^2 of four species of submerged hydrophytes in containers. After one year, *H. verticillata* dominated the composition of the communities, with only a few *P. malaianus*, *V. spiralis* and *N. marina* remaining, owing to the competitive exclusion from *H. verticillata*. Moore et al. [48] demonstrated with exclosure experiments that *Vallisneria americana* can be successfully restored in tidal freshwater areas of the Chesapeake Bay that were unvegetated for 60 years. These authors concluded that whole shoot transplants resulted in the most rapid cover, whereas direct dispersal of individual seeds or intact seed pods were also effective, but the recovery was slower. When protected from herbivory, approximately 3 years of growth were required for the transplants to reach 100% bottom cover at maximum densities of 100–150 shoots/m^2. Rodrigo et al. [16] set up 54 1 × 1 m-exclosures in two shallow lagoons within a newly created constructed wetland in Spain that were planted with cultures of two species of higher plants and two charophyte species (16 cultures of three specimens each/m^2). The higher plants developed better than the charophytes, but always when protected from biotic factors. Gao et al. [49] performed outdoor mesocosm experiments with four species of submerged macrophytes, planted at a density of 300 gWW/m^2 in the Gonghu Bay of Lake Taihu. They concluded that *H. verticillata* and *P. malaianus* are suitable submerged macrophyte species for restoration of eutrophicated shallow lakes. The relative growth rate of *H. verticillata* was maximum (0.03), and around 0.01 for the other three species. In Sweden, Nilsson et al. [50] planted *Elodea canadensis*, *Myriophyllum alterniflorum* and *Ceratophyllum demersum* in newly semi-natural constructed wetlands to intercept nitrogen from surface waters in an agricultural landscape and followed them for 12 years. Nitrogen removal increased with the ecosystem age, and the dominant submerged species was *Potamogeton natans*, which colonized naturally. Schad et al. [51] planted founder colonies of *Heteranthera dubia* and *Potamogeton nodosus* in a series of constructed floodway wetlands in the USA to analyze the influence of varying construction completion dates, water sources and ecosystem management stage on macrophyte development and its relationship with macroinvertebrate assemblages.

For macrophytes to maintain a clear water state, a minimum coverage of shallow water bodies is required [52]. As a rule of thumb, 30% coverage has been used as a minimum threshold, which is in the range of 10–40% reported by some authors, but lower than 50% indicated by others. In warm shallow lakes in tropical and subtropical regions, even a higher hydrophyte coverage may be needed as the grazing of zooplankton on phytoplankton is low due to high fish predation (see references in [52]). In Mediterranean regions, where high temperatures are reached in summer, a larger coverage would be necessary to outcompete the growth of phytoplankton and filamentous algae. Consequently, large-scale restoration efforts should be faced because they could potentially be more successful than smaller ones since large submerged aquatic vegetation beds are thought to be more stable and resilient to stress than small beds [53]. However, getting such high coverages requires a tremendous effort (high costs for material, installation and maintenance and solving difficulties, such as filamentous algal growth or high herbivory pressures, and interference with recreational use). Therefore, large-scale plantings of hydrophytes have not often been performed in wetlands, although there are some cases. Table 2 gathers study cases of hydrophyte revegetation at larger scales (from 0.4 ha or more). One of the most emblematic cases of a large-scale approach to submerged aquatic vegetation restoration is the Chesapeake Bay, the first estuary in the USA to apply an integrated watershed approach for restoration (13.4 ha/year were revegetated in 2003–2008 [43]). In New Zealand, Dugdale et al. [54] planted 1 ha of a shallow lake with charophytes protected from fish and it was successful in allowing founder colonies of charophytes to establish and expand ($\geq$75% cover within one year). At the Mediterranean, Sebastián et al. [55] restored two wetland areas in Spain by planting *Ceratophyllum demersum*, *Myriophyllum verticillatum*, *M. spicatum* and the floating-leaved species *Nymphaea alba* as well as two shallow lagoons at the end of a constructed wetland [16]. In China, Chen et al. [56] bordered an area of 10 ha of the littoral zone of lake Wuli (a bay of Taihu lake) with a waterproof fabric and planted four species of submerged macrophytes, three species of floating-leaved macrophytes and one species of a free-floating plant. The coverages of floating-leaved macrophytes, submerged macrophytes and free-floating macrophytes inside the enclosure were up to 9.7%, 8.1% and 2.9%, respectively, one month after plantation. One year later, the coverage area of aquatic macrophytes (including emergent species) expanded and increased to about 45.7%. Clarkson and Peters [39] revegetated New Zealand wetlands by planting *Potamogeton cheesemanii*, *Myriophyllum propinquum* and *Lemna minor*. Dick et al. [57] planted six species of submerged hydrophytes, one species of charophyte and four species of floating-leaved species in a chain of wetlands in the USA. Plant establishment continued with supplemental plantings 4–10 years later. Yu et al. [58] found that restoration by transplantation of six species of submerged macrophytes (at a density of 30–70 adult plants/m^2) after fish removal had major positive effects on water quality variables in three shallow lakes of China (an isolated 5-ha bay of shallow lake Taihu, lake Qinhu (8 ha) and South lake (0.4 ha)). Theÿsmeÿer et al. [59] planned the recovery of submerged and floating-leave species in the Canadian marsh area of the Great Lakes. More recently, Liu et al. [60] reported the successful restoration of a tropical shallow eutrophic lake combining fish removal with transplantation of submerged macrophytes. *Vallisneria denseserrulata* was planted at a density of 10–15 shoots/m^2 and *Hydrilla verticillata* of 20–30 shoots/m^2.

Table 1. Summary of outdoor experiments performed to plan larger-scale revegetation with hydrophytes. Plant species, some experiment features, and site (country) are indicated. References are ordered chronologically. N/C indicates if the wetland/shallow lake is natural (N) or constructed (C).

Planted Hydrophyte Species		Plant Origin	Experiment Features	Site (Country)	N/C	Ref.
Submerged	**Floating-Leaved**					
Vallisneria sp. *Hydrilla verticillata* *Potamogeton maackianus*	*Trapa bicornis* *Nelumbo nucifera*	Not indicated	800–3000 m^2 enclosures in three sublakes	The shallow lake Donghu (China)	N	[44]
Stuckenia pectinata *Potamogeton perfoliatus* *P. lucens*	–	Collected 80 km south of the lake in ditches and channels	25 m^2 protected and unprotected areas	The shallow lake Engelsholm (Denmark)	N	[45]
Myriophyllum spicatum *Chara contraria*	–	Not indicated	200 m^2 "macrophyte islands"	A shallow lake (Germany)	N	Rott (2005) in [46]
Hydrilla verticillata, *Potamogeton malaianus* *Vallisneria spiralis* *Najas marina*	–	Wuli Bay and East Taihu Bay (lake Taihu)	200 L containers in an outdoor green house	The shallow lake Taihu (China)	N	[47]
Vallisneria americana	–	From nursery grown stock in culture ponds at Virginia Inst. Marine Sci. campus; seed pods from beds in Nanjemoy Creek, Maryland; separated seeds	4 exclosures of 40 m^2 (with 2 × 2 m plots inside)	A tidal marsh area at James River (VA, USA)	N	[48]
M. spicatum *S. pectinata* *Chara hispida* *Nitella hyalina*	–	From cultures produced in indoors culture room	54 exclosures of 1 × 1 m	Tancat de la Pipa wetland (Spain)	C	[16]
P. malaianus *M. spicatum* *H. verticillata* *V. spinulosa*	–	From lake Taihu and cultivated in outdoor tanks (100 cm diam. 100 cm height)	15 680-L outdoor tanks (100 cm diam. 100 cm height)	Gonghu bay, lake Taihu (China)	N	[49]
Elodea canadensis *Myriophyllum alterniflorum* *Ceratophyllum demersum*	*Potamogeton natans* (but appear spontaneously)	Shoot fragments from nearby ponds	6 10 × 4 m surface-flow constructed semi-natural wetlands	Semi-natural wetlands in agricultural landscape (Sweden)	C	[50]
Heteranthera dubia	*Potamogeton nodosus*	Founder colonies from nearby sites	24 0.9-m diam. ring cages	Dallas Floodway Extension Lower Chain of Wetlands (USA)	C	[51]

Table 2. Summary of some larger (>0.4 ha) field revegetation with hydrophytes. Plant species used, surface treated, and site and country are indicated. References are ordered chronologically.

Planted Hydrophyte Species		Surface	Site (Country)	Reference
Submerged	**Floating-Leaved/Free Floating**	**(ha)**		
Chara australis	–	1	Shallow lake Rotoroa (NZ)	[54]
Ceratophyllum demersum *Myriophyllum verticillatum* *Myriophyllum spicatum*	*Nymphaea alba*	1.5 and 1.2	Almenara and Algemesí wetlands (Spain)	[55]
Potamogeton malaianus *Myriophyllum spicatum* *Potamogeton maackianus* *Hydrilla verticillata* *Vallisneria natans*	*Nymphoides peltata* *Nymphaea rubra* *Trapa bicornis* *non-native Alternanthera philoxeroides*	10	Large enclosure in Lake Wuli, northern bay of Lake Taihu (China)	[56]
Potamogeton cheesemanii *Myriophyllum propinquum*	*Lemna minor*	–	Several wetlands in New Zealand	[39]
Myriophyllum spicatum *Stuckenia pectinata* *Ceratophyllum submersum*	–	6 and 8	Educative and Reserve lagoons, Tancat de la Pipa wetland (Spain)	Sebastián and Peña in [16]
Ceratophyllum demersum *Chara vulgaris* *Heteranthera dubia* *Potamogeton illinoensis* *Potamogeton pusillus* *Vallisneria americana* *Zannichellia palustris*	*Potamogeton nodosus Nelumbo lutea* *Nymphaea mexicana* *Nymphaea odorata*	>10	Chain of wetlands at Dallas Floodway Extension (USA)	[57]
Hydrilla verticillata *Vallisneria spinulosa* *Potamogeton maackianus* *P. malaianus* *M. spicatum* *Ceratophyllum demersum*	–	5, 8 and 0.4	Shallow lakes Wuli (isolated bays of lake Taihu), Qinhu and South (China)	[58]
Vallisneria americana	*Potamogeton nodosus/natans* *Nymphaea odorota*	~18	Great Lakes wetland area (Canada)	[59]
P. malaianus *M. spicatum* *H. verticillata* *V. spinulosa*	–	0.4	Gonghu Bay, Lake Taihu (China)	[35]
Vallisneria denseserrulata *Hydrilla verticillata*	–	12	One basin of Huizhou West shallow lake (China)	[60]

Revegetation is essential, but whole-ecosystem, long-term interventions including most if not all ecosystem processes are desirable to be sure that the restoration result is the expected [61]. Furthermore, for large-scale hydrophyte restoration, the efforts should be in the framework of coordinated interagency programs, to develop, evaluate, and refine the suitable protocols and procedures. Maybe this is an issue not so easy to achieve. Guidelines should be published to help managers aiming to restore wetlands and shallow lakes, and critically assess and predict the potential development of submerged vegetation, taking into account the complex factors and interrelations that determine their occurrence, abundance and diversity. In very few countries (such as the USA in 2002 [62] and Germany in 2006 [46]) such guidelines were published.

5. Procedure Approaches in Revegetation with Hydrophytes

5.1. Seeding

The need to seed in wetland restoration has been widely and recently reviewed by Kettenring and Tarsa [1]. Employing strategic seed-based approaches in wetland restoration is a first step to more quickly and completely recover the underlying vegetation structure and composition that supports the vital functions and services of wetlands. Seed-based approaches are less expensive and more logistically feasible in treating larger areas than other wetland revegetation techniques (e.g., planting plugs, transplanting rhizomes and installing sod mats) despite the high cost of native seed (see [1] and references therein). However, frequently a deficiency of submerged macrophyte propagules is faced. This is why, on some occasions and for certain plants, in vitro propagation protocols have been developed [37].

Moreover, seeding results can be unpredictable and the mortality can be high [42,63]. The seed and seedling stages of plants are a demographic bottleneck, and often few seeds survive to become seedlings [64]. Hence, an effective seed-based approach should be driven by ecological, genetic and evolutionary principles, along with consideration for economics, logistics and other social constraints. Moreover, best practices for seed-based restoration must address limiting environmental factors and inform strategic management interventions for improving revegetation outcomes [1].

In many wetlands, recalcitrant seeds occur [1]. These species will require storage under high humidity or submerged in water. It has been shown that seeds kept under submerged storage conditions show higher seed longevity with aeration (this is the case of *Zostera marina* and *Potamogeton perfoliatus*), or under cool temperatures (e.g., *Z. marina, P. perfoliatus* and *Ruppia maritima*) or high salinity (e.g., *Z. marina, R. maritima*; [65–67]). Some species of submerged vegetation can also be stored under low humidity/low moisture for some temperatures. This is the case of the Hydrocharitaceae and Potamogetonaceae families [68,69], and thus may not be recalcitrant contrary to general predictions in aquatic plants. In the case of charophytes, there are also situations where stonewort seedlings do not develop in their original habitats, despite the restoration of optimal hydro-chemical conditions for their growth. These macroalgae produce extremely small propagules (oospores), and working with them is not an easy task. Recently, a protocol consisting of microencapsulating these oospores using sodium alginate has been published [70], and it is presented as a promising method for preserving charophyte oospores to support both laboratory and field experiments. The author proposes this procedure to greatly facilitate the conduct of both in situ and ex situ conditions' studies and experiments.

Seeds should be sown as soon as temperatures are within a species' optimal range, before the plant canopy has time to develop and inhibit light at the sediment surface. Regarding recommendations for seed sowing rates, data are quite limited and vary widely across wetland types. For tidal salt marshes, Broome et al. [71] suggested sowing 100 pure live seeds (PLS)/m^2, while Busch et al. [72] seeded *Zostera marina* at 37 PLS/m^2. In the large Chesapeake Bay restoration (41 ha), *Z. marina* seeding densities of 11 to 49 seeds/m^2 were used [43]. Current restoration techniques for seed sowing introduce the seeds by hand or with machines. They are designed to overcome dispersal limitations but do not necessarily mimic these natural dispersal mechanisms, which comprise water, wind, animals (particularly waterbirds) and gravity. One approach for keeping seeds in place is sowing seeds in burlap bags made with natural fibers. This has also been used with *Z. marina*, and it improves the recruitment outcomes in seagrass meadow restoration with high wave action [73]. To avoid the low seedling establishment rates (<10%) and seed loss through herbivory on seeds, mechanical devices for planting *Z. marina* seeds slightly beneath the sediment surface have been developed with improved seedling establishment rates [66,74,75]. Aerial seeding might be both economical and practically feasible to vegetate large wetland areas. It could accelerate restoration efforts by replacing the expensive hand-planting of vegetative clones (see below), but the unavailability of large quantities of viable seed is one of the major hindrances.

Seed pellets (also known as pods, seed balls and seed bombs), which are an aggregation of clay, soil, water, and multiple seeds, have been used in terrestrial restoration [76,77] and have been successfully used with emergent plants in wetlands (*Typha* seeds; Moreno L., pers. comm., Figure 2a). Thus, seed pellets are potentially a user-friendly way to establish wetland plants because they can be launched into the air in cases of hard-to-reach locations. No information has been found for the use of seed pellets in the case of hydrophytes. Therefore, this is a potential study field in the restoration of wetlands with hydrophytes, but it needs further research and testing. Moreover, making seed bombs has been a popular activity at garden centers, family events and student visits (Figure 2a), and it is potentially a viable method for engaging and educating the society, at the same time, it helps distribute native seeds across larger areas.

Mechanical planters designed for planting whole seagrass plants and sods have been developed in both the United States and Australia [43]. Despite their limitations in the operating procedures (e.g., weather, depth limits, donor bed proximity, need for SCUBA divers), they hold the potential for rapidly and cost-effectively planting larger areas of submerged aquatic vegetation than would be possible through manual means. For wetlands, SCUBA divers are not necessary, due to their shallowness, but large mechanical planters can have a great impact on the wetland fauna since planting should be done in spring when waterfowl is also breeding.

5.2. Planting: Translocations and Production of Hydrophyte Cultures

Translocations are among the techniques used in wetland restoration and plant conservation [9,78] (i.e., the human-mediated movement of living organisms for restoration and/or conservation benefit from one area, with release in another). If hydrophytes are not taken from existing areas, and to minimize potential impacts to wild populations, they must be cultivated previously to be planted. Hydrophytes can be cultured in aquaculture systems (as described in Tanner and Parham [79]). Zhou et al. [37] described how a 15 m^2 tissue culture room together with a 50 m^2 acclimation pond can produce 125,000 high-quality seedlings for *Myriophyllum spicatum* and *Potamogeton crispus* in 7 weeks, which could vegetate more than 2 ha sediments in shallow lake recovery programs, with the common density adopted in East China. Rodrigo et al. (unpublished results) prepared 2000 cultures of *M. spicatum*, *Stuckenia pectinata* and several species of charophytes in a 9-m^2 culture room in 8 weeks, ready to be planted in a Mediterranean wetland. Regarding other cultivation times, *Zostera marina* plants large enough for planting can be grown from seeds within 70–100 days under controlled conditions [79]. Rodrigo and Carabal [17] reported lengths of 80 cm for *Stuckenia pectinata* in one month planted from 10-cm cuts and cultured in an acclimated room; *M. spicatum* and *Chara vulgaris* grew up to 25 cm in one month starting from 5-cm apical parts [17]. Riis et al. [7] indicated that submerged plant shoots of 20–25 cm are adequate to be planted. Moreover, they recommend using shoots with an apical tip, since they have been previously shown to regenerate better than shoots without. For the case of constructed wetlands, the use of innovative tissue culture technologies allows isolation of plant clonal lines of single seed phenotype origin that can be screened for particular tolerances, such as cold temperature, high nitrate removal rates, etc. [80]. The sediment used for preparing the cultures should preferably be from the local site but if it is impossible to get enough top sediment from the selected wetland, the volume can be augmented with a mixture of commercial sand and sediment [16].

Figure 2. (**a**) Seed bombs with *Typha* seeds prepared by students with the supervision of the managers of Tancat de la Pipa wetland (Albufera de València Natural Park, Spain; photographs: Lourdes Ribera); (**b**) examples of *M. spicatum*, *S. pectinata* and the charophyte *Chara vulgaris* ready to be planted in the field without any kind of holder; (**c**) example of *M. spicatum* with the root-sediment system in a peat pot ready to be planted in the field; (**d**) fragments of hydrophytes with a stone to serve as "anchor" to be thrown inside enclosures (see Figure 2c) (*S. pectinata*, *M. spicatum*, *C. submersum* and *C. vulgaris*).

Planting of submerged hydrophytes can be done directly with the root-sediment "ball" (Figure 2b) or by using biodegradable holders. Rotting and non-rotting substrates (e.g., nets) that keep planted macrophytes on the waterbody bottom have been used in Germany [46]. Likewise, other types of substrate have been utilized on some occasions: wood cages [55] and peat pots [16] (Figure 2c) in Spanish wetlands, trays in rivers [7]. However, peat is an example of a non-renewable resource and its extraction could contribute to the degradation of wetland ecosystems. For this reason, a biodegradable material which, at the time is a waste, as is the case of rice straw in particular areas, could represent a good option to make holders for planting. For example, in the Albufera de València Natural Park (València, Spain), there are 15,000 ha of rice fields, which produce 80,000 tons of rice straw per year [81]. The elimination of this "waste" is a problem in that area. Part of this material could be used in the fabrication of biodegradable pots, after appropriate research, for restoration tasks. Rice straw substrates have already been used in floating beds planted with macrophytes for treating wastewater. This substrate enhanced bio-remediation efficiency and macrophyte growth in comparison to the inert palygorskite ceramsite which hindered the bio-remediation process [82]. On some occasions, planting has been performed using just a stone adhered to the fragments of plants (Figure 2d). This has been successfully used even with charophytes (Rodrigo et al., unpublished results; Figure 2d). Regarding charophytes, Blindow and Carlsson [83] (this issue) describe methods for oospore germination, cultivation and plantation of charophytes depending on the type of charophyte species (k-strategists vs. r-strategists) to support the existence of threatened species in Sweden. This knowledge might be used in a wider framework for restoration purposes.

Hydrophyte planting can be done by hand [16,58], etc., or mechanically [43]. Planting by hand is rather work-intensive and might ideally involve volunteers. Planting of hydrophytes is not a complicated task, and volunteers (e.g., undergraduate and master students, members of NGOs, etc.), following wetland managers and researchers' instructions, can do a great job. Furthermore, it may be worth doing to increase outreach and involve the local population in recovery efforts.

Plants cultivated indoors might be subjected to significant differences in ecological conditions between the cultivation site and recipient site; the larger the differences, the greater the potential negative impact on survival and fitness of the introduced plants. Therefore, to reduce environmentally mediated shocks, acclimatization techniques (hardening) must be adopted before locating the cultures to the recipient site. One method consists of placing the plant cultures in acclimation ponds close to the target wetland if available, or in tanks located in the wetland to be revegetated, for gradual acclimatization to external temperatures, decreased or increased shading, etc. (Figure 3a).

Once the restoration has been performed by planting, and if most of the plants have been produced by vegetative reproduction, seed pellets could be thrown near the hydrophyte stands to increase the genetic variability.

Figure 3. (**a**) Tanks of hydrophyte cultures produced indoors, now acclimatizing in the field before being planted in Tancat de la Pipa wetland (Albufera de València Natural Park, Spain; photographs: Ximo Fernández. (**b**) Exclosures to plant hydrophytes using the method shown in Figure 2d; the exclosure is almost 100% covered with the grown submerged macrophytes (the thin ropes in the upper part try to avoid grazing by herbivorous birds). (**c**) Exclosures to plant hydrophytes that are being enlarged successively to get wider coverages (photographs taken from [84]).

5.3. Sediment Transfer/Transplantation

Effective transplanting of wetland soil from small remnant wetlands in other areas of shallow marshes has been performed on several occasions [85–88]. However, most of the studies refer to wet meadows, *Sphagnum*-dominated peatlands, etc., and the study cases with sediment transfer involving submerged or floating species are few (Table 3). Brown and Bedford [86] observed little plant establishment at water depths greater than 45 cm, suggesting that transplantation of wetland soil should be concentrated in shallower zones; therefore, this method seems not to be the most adequate for submerged macrophytes that grow deeper. On the other hand, transfer of bulk soil has shown promising results in temporary wetlands, indicating that soil transfer may enhance the success of wetland restoration projects compared to natural colonization [89,90]. This technique could be the most efficient method for transferring a large number of temporary wetland plant species that have a short life cycle but can produce large quantities of seeds and rapidly form a large seed bank [91]. Muller et al. [90], in a temporary wetland restoration after rice cultivation in France (Table 3), found that soil transfer not only enhanced the target species introduction but also significantly reduced the establishment of undesired species emerging from the seed bank and from the surroundings, such as ricefield weeds (mostly exotic species introduced by rice cultivation). The pilot project of revegetation of lakeshore vegetation launched at Lake Kasumigaura in Japan by transferring lake bottom sediment achieved the recovery of 12 native submerged plants during the first year of restoration previously disappeared [89,92] (Table 3). Soil transfers also have the advantages of biotic interactions preservation by transferring soil microorganisms, important in structuring plant community and improving substrate conditions, and the transfer of zooplankton and macroinvertebrate egg bank [90]. However, nature protection aspects and the potential risk of transferring pollutants, or undesired species (such as fish or other animals' parasites, pathogens, etc.) must be considered.

Table 3. Some examples of practices with wetland sediment/soil transfers. References are ordered chronologically.

Recovered Species	Sediment Origin	Receptor	Site (Country)	Reference
Chara braunii, Nitella hyalina, Monochoria korsakowii, Nymphoides peltata, Limnophila sessiliflora, Vallisneria denseserrulata, Hydrilla vercillata, Ceratophyllum demersum and five species of *Potamogeton*	Seed banks from lake-bottom sediments	Lake shores ranging 5300–27,800 m^2 (width: 30–60 m). Sediments spread thinly (~10 cm)	Littoral areas of shallow Lake Kasumigaura (Japan)	[89,92]
(Mostly emergent plants) *Myriophyllum spicatum*	0–5 cm deep soil from 1 × 1 plots from different sites	Surface of 55 m^2	Yeyahu wetland natural reserve (China)	[88]
Callitriche sp., *Callitriche truncata, Chara aspera, C. canescens, C. globularis, Ranunculus peltatus, R. trichophyllus, Tolypella glomerata, T. hispanica, Zannichellia obtusifolia, Z. pedicellata*	40 L (from 45 × 45 cm, 3 cm deep) of soil per donor site (5 temporary wetlands)	50 L of soil on a 4 × 2 m plots at the bottom of each transfer mesocosm	Cassaïre site, Camargue area (France)	[90]

6. Selection of Species. Most Commonly Used Species

Different authors have indicated that the selection of hydrophytes for revegetation should be made based on (i) the type of wetlands considered, (ii) the former vegetation in the wetland and the species typically occurring in that type of water and region, (iii) the potential uses of the wetland (natural vs. constructed wetland), (iv) the suitability of the selected species for seeding and/or planting, (v) the habitat preferences of the selected species and (vi) the potential origin or source of the plants (or seeds). Moreover, the tools for the restoration of aquatic plant communities should consider the complex interactions between abiotic factors and aquatic plant requirements; otherwise, the objective of restoring such communities may be difficult to reach [93]. Recently, it has been stated that the efforts to build and maintain the resilience of an ecosystem after restoration by revegetation should be trait-based rather than merely focusing on vegetation abundance [94]. In addition, Song et al. [95] showed that the macrophyte effects on water quality vary by growth forms and that the growth forms which positively affect the water quality differ between the (sub)tropical and temperate areas. Dalla Vecchia et al. [96] stressed that root traits may explain important plant functions and need further research. Su et al. [94] suggested that plant height was one of the mechanisms underlying the positive feedbacks on water quality. Submerged plant species of taller-growing "rank", such as *M. spicatum* and *Stuckenia pectinata* have been suggested to be introduced initially in coastal eutrophic wetlands [17]. Choosing between r-selected and k-selected plants is also crucial. For example, Qiu et al. [44] attributed the failure of recovery of *P. maackianus*, a k-selected species, when it was used as initial species, to its poorly developed rhizome, weak regeneration capacity and relatively small seed bank. Pioneering species should have been used first for restoration and *P. maachianus* and other perennial plants could be re-introduced later to increase the biodiversity [44].

Hilt et al. [46] recommended the following submerged macrophytes species for potential successful use for artificial colonization in eutrophic shallow lakes in Germany: *Ceratophyllum demersum, Chara contraria, C. globularis, Nitella mucronata, Eleocharis acicularis, Myriophyllum spicatum, M. verticillatum, Najas marina, Potamogeton alpinus, P. berchtoldii, P. crispus, P. friesii, P. obtusifolius, P. pusillus, P. perfoliatus, Stuckenia pectinata, Ranunculus* subg. *Batrachium, Ranunculus trichophyllus* (only in alkaline lakes), *Zannichellia palustris* ssp. *palustris. H. verticillata* and *P. malaianus* have been described as suitable submerged macrophyte species for restoration of eutrophicated lakes and wetlands [49] when combined with filter-feeding aquatic animals. *Myriophyllum verticillatum, Potamogeton perfoliatus* and *Najas minor* yielded quite similar results when nutrient removal efficiencies were analyzed, although they were higher for *N. minor* and Zhou et al. [97] pointed at *N. minor* to be a promising plant for water purification. On some occasions, facilitation has been the proposed mechanism that may enhance the colonization of several submerged hydrophytes planted at the same time [98]. Thus, Dai et al. [99] proposed using the combination of *C. demersum* and *M. verticillatum* as the best choice for ecological restoration of eutrophic water bodies. The charophyte *Chara vulgaris* has also been used for replanting in eutrophic wetlands due to its high-nitrate concentrations tolerance and because it is a r-strategist that produces large amounts of oospores [100].

Among the criteria to select the hydrophyte species for revegetation is the availability of knowledge on each species. For example, methods for collecting, processing and storing large quantities of *Ruppia maritima* and *Potamogeton perfoliatus* seeds started in 2004 and protocols for using seeds of these species in restoration plantings are described in Ailstock et al. [65]. *Myriophyllum spicatum*, a perennial submerged macrophyte, is one of the species preferentially used in many restoration projects in lakes and wetlands [101] (see Tables 1 and 2 and Figure 4), mainly due to its strong resistance to pollution. Because of environmental disturbances, *M. spicatum* is easily broken to form apical fragments and then it is possible for them to develop into robust new plants and gradually settle to form colonies. *M. spicatum* can also tolerate both fresh and brackish water [102,103]. This tolerance range allows *M. spicatum* to live under a wide range of salinity and different oxidative stress conditions. This makes this species a good candidate for coastal wetlands affected by salinization [104]. Moreover, *M. spicatum* can secrete allelochemicals to inhibit the growth of microalgae [105]; the major components of these secondary metabolites secreted by plants are phenolic acids, fatty acids, alkaloids, terpenoids, flavonoids, etc. [106]. Other species which produce and secrete allelopathic compounds are, for example, *Vallisneria spiralis* [107], *Ceratophyllum demersum* [108], *Potamogeton malaianus* [109], and also charophyte species [110,111]. The allelopathy of macrophytes on microalgae growth is extremely promising due to its low cost, good algal inhibition effect and high environmental safety [112] and should be also a criterion to consider in the selection of hydrophytes for revegetation. The use of allelochemicals produced by macrophytes in the field of water ecological restoration has been recently reviewed by Li et al. [106]. These authors even propose searching in the micro-spheroidization technology as engineering applications of allelochemicals directly in water to prevent microalgal growth.

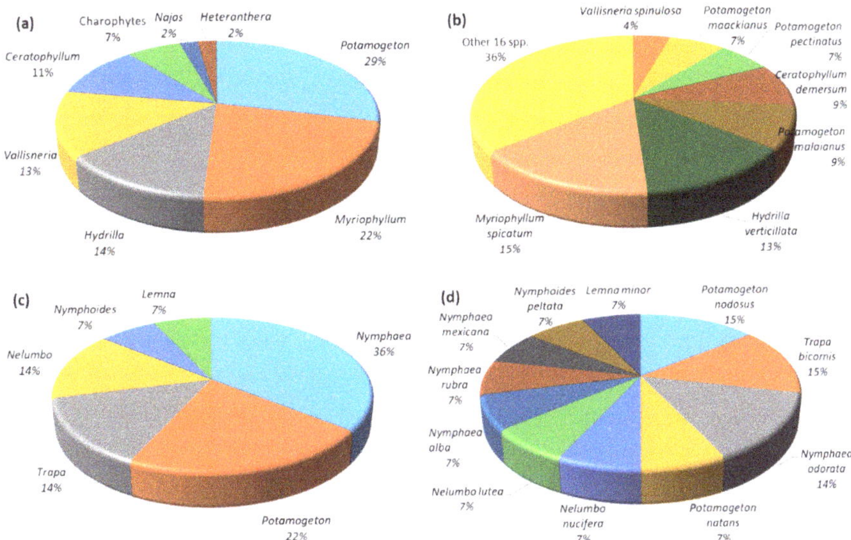

Figure 4. (a) Distribution of the native genera of submerged macrophytes (and charophytes) used for revegetation in wetlands; (b) the most used native species of submerged macrophytes for revegetation in wetlands. (c) Distribution of the native genera of floating-leaved and free-floating species used for revegetation in wetlands. (d) Percentage of the floating-leaved and free-floating native species used for revegetation in wetlands.

Regarding flow-surface constructed wetlands [113], *Ceratophyllum demersum*, *Hydrilla verticillata*, *Myriophyllum verticillatum*, *Vallisneria natans*, and *Potamogeton crispus* are commonly used among the submerged plants. The commonly used free-floating hydrophytes in CWs include *Lemna minor*, *Eichhornia crassipes*, *Salvinia natans*, and *Hyrocharis dubia*. Meanwhile, floating-leaved species in CWs are mainly *Nymphoides peltata*, *Trapa bispinosa*, *Nymphaea tetragona*, and *Marsilea quadrifolia* [114].

The use of resistant genotypes (to herbivores and salinity, for example) in hydrophyte restoration, as it is proposed for seagrasses [115], might be an approach for improving the extant genetic baselines of natural populations and for enhancing the resilience of the restored population to present and future stressors (e.g., climate change). The selection of more tolerant genotypes to improve restoration success could be performed by growing wild specimens under controlled conditions, but resistant genotypes can also be produced with a lower level of intervention through the use of priming/hardening methods [116]. Pre-exposing specimens to mild stress has the potential to induce stress memory, giving rise to genotypes with enhanced tolerance to subsequent stressful events. When stress memory is set by stress-induced epigenetic modifications, the acquired resistance can be passed to offspring leading to new generations with acquired resistance. Therefore, when dealing with clonal plants, such as most submerged and floating hydrophytes, restoration management should consider "epigenetic diversity" as an indicator of stability and functioning of the ecosystem equal to genetic diversity. However, this is a still unexplored issue in the field of wetlands restoration with hydrophytes.

7. Factors Affecting the Success of Restoration

7.1. Site Selection

Site selection is arguably one of the most critical steps in wetland restoration processes [43]. Many restoration projects have failed due to inadequate site selection. Among the factors to be considered are the historical presence or absence of submerged aquatic vegetation, water depth and light availability, water column nutrient concentrations, sediment quality, wave exposure, etc. Interannual variability in climate and water quality conditions (see below) also play a critical role in the initial establishment and survival of planted submerged aquatic vegetation. This is why planting efforts may need to be repeated over multiple years to achieve great success [43]. Regarding climate variability, site selection should consider the foreseen changes that will occur due to climate change in the near future that will surely affect the success of revegetation [117].

7.2. Time Selection

Introduction of hydrophytes should be carried out early in the favorable season in eutrophic wetlands, before the development of microalgae and/or filamentous algae (see below). However, the period for plant introduction should not affect other important phenological events such as the breeding of waterfowl [118]. Rodrigo and Segura [119] reported unsuccessful revegetation in 2020, due to an inappropriate time for it in a Mediterranean wetland. The revegetation was planned for mid-March 2020, and all cultures were prepared, but then the lockdown of the whole society was declared due to the pandemic caused by the SARS-CoV-2 virus. The planting was finally performed in mid-June 2020, when the "normal" people mobility situation was restored. However, the hydrophyte recovery failed due to the intensive growth of filamentous green algae, which had already developed at this time of the year; hydrophytes could not outcompete with the filamentous algae. In the planning for planting, the acclimation time of the hydrophytes in the field has to be also considered.

7.3. Herbivory

Herbivory can be performed mainly by waterfowl, fish, crayfish and turtles. Moore et al. [48] pointed at herbivory as the main factor for the lack of success in the restoration with *Vallisneria americana*. The growth of both adult transplants and plants developing from seeds was good, but using mesh exclosures to protect the plants from herbivory proved to be critical to the restoration success. Similar results were obtained by Rodrigo et al. [16] in Mediterranean wetlands. Studies in the United Kingdom [120,121], Denmark [45] and Germany [46] showed higher survival and number of plants and longer total shoot length in enclosures that prevented bird access. Thus, long-term protection with exclosures may be required to establish large founder colonies that are of sufficient size to withstand initial grazing pressures (Figure 3b). The size of the exclosures can be progressively enlarged to obtain wider surface coverages each time [84] (Figure 3c). However, the use of such protective exclosures as a restoration tool for very large-scale use is difficult, due to high costs for material, installation and maintenance and difficulties, such as filamentous algal growth (see below) and interference with fauna and recreational use.

Herbivore species may have preferences on particular species. For example, Yu et al. [122] described how grass carp preferred *Vallisneria spinulosa* and *Ceratophyllum demersum* to *Myriophyllum spicatum*. Waterfowl have been documented to graze selectively on *Stuckenia pectinata* (herbivores electively removed *S. pectinata* specimens in favor of charophytes in Estonia [123], and waterfowl suppressed dominance of *S. pectinata* in favor of subordinate *Zannichellia palustris* and *Potamogeton pusillus* in The Netherlands [124]). Invasive red swamp crayfish preferentially fed on charophytes [125]. Therefore, having a good knowledge of the herbivores present in the wetland is also essential for planning the species selection. Finally, hydrophyte palatability and disturbance tests should be carried out directly in the field to allow determining the likelihood of consumption/resistance to disturbances of the different macrophyte species.

7.4. Massive Filamentous Algal Development

Since the wetlands to be restored are, in most cases, eutrophic systems, submerged macrophyte recovery is often accompanied by an excessive proliferation of filamentous green algae [126]. Filamentous algae compete with submerged macrophytes for space, light, nutrients and other resources; they also mechanically damage hydrophyte stems and leaves by twining around them, negatively influencing their normal growth. Moreover, the response of regeneration ability of apical fragments to decaying green filamentous algae is negatively affected (see, for example, the adverse influence of *Cladophora oligoclona* on *Hydrilla verticillata* seed germination and seedling growth [127] or on *Myriophyllum spicatum* formation of buds and roots [126]. Furthermore, high growth rates, as high as 0.7-0.8 d^{-1} [128], have been described for several species of *Cladophora* and also can grow from their internal nutrient storages. This confers a large advantage over hydrophytes. Since filamentous green algae grow adhered to substrates in their benthic stage [129], planted hydrophytes (as well as the nets of the exclosures, see above), can be used by them as suitable substrates. All this can lead to the recession or even the disappearance of the hydrophytes in the restored system. Zhang et al. [130] suggested the importance of appropriately selecting macrophyte species to prevent filamentous algal bloom in shallow water bodies restoration. They recommended avoiding planting of *H. verticillata* and *C. oryzetorum* because these species promoted the growth of filamentous algae in the early spring, while *P. malaianus* might inhibit filamentous algae and this species was recommended as a pioneer species. Therefore, the excessive growth of filamentous green algae should be regulated during revegetation, although this is a real challenge in the restoration of eutrophic wetlands. Bearing climate change in mind [131], the increase of temperature foreseen for areas such as the Mediterranean would favor the early development of green filamentous algae in wetlands located in that region. The removal of the dense mats formed by the filamentous algae and the use of this waste in collaboration with biotechnological companies could be a solution, since the use, for example, of *Cladophora glomerata* removed from sites where it forms green tides [132] is described for the production of highly crystalline cellulose [133]. In this way, similar to what is done with the withdrawal of emergent vegetation biomass in constructed wetlands, three goals can be achieved: (i) elimination of large quantities of nutrients from water which are now retained in the filamentous algal biomass, (ii) the use of waste which represents an important environmental problem, and (iii) help the revegetation with hydrophytes.

7.5. Water and Sediment Quality Conditions

Wetlands to be restored are frequently rich in water and sediment nutrients but also in contaminants, such as metals or organic compounds. These concentrations should be reduced below the tolerance thresholds of the hydrophyte species prior to being reintroduced. Experimental mesocosm studies performed in Denmark indicated that the threshold concentrations above which is likely to lose submerged macrophytes in shallow systems are 1.2–2 mg/L of TN and 0.13–0.20 mg/L of TP [134,135]. Wang et al. [136] found the TP thresholds for the shift from clear-water state to turbid-water state at 0.08–0.12 mg/L. Submerged macrophytes cannot tolerate high ammonia concentrations, and may cause damage to and loss of macrophytes in wetlands and shallow lakes [137]. For example, the threshold value of ammonia for *Potagometon crispus* is 4 mg/L [138], but it has been accepted that ammonia tolerance differs greatly among wetland plant species [139]. Wang et al. [140] found that the increase of TN removal efficiency in *Myriophyllum aquaticum* was hindered when treated with high levels of NH_4^+ (26–36 mmol/L), suggesting this as the threshold for its tolerance to NH_4^+. Regarding nitrogen and charophytes [141,142], Lambert et al. [142] predicted a transition from charophyte presence to absence in aquatic ecosystems at a concentration of approximately 2 mg NO_3-N/L. However, Rodrigo et al. [100] found *C. hispida* and *C. vulgaris* forming meadows with nitrate concentrations higher than 2 mg NO_3-N/L in water bodies affected by seepage from agricultural runoff. Moreover, in laboratory experiments, these species grew well up to 30 NO_3-N/L. Performed research related to the use of submerged macrophytes in constructed wetlands has provided a wide knowledge in terms of thresholds for water quality conditions for particular species (e.g., [139,140]). Some wetland hydrophytes are being used as phytoremediating plants capable of taking up heavy metals and other pollutants from water and sediments. For example, *Ceratophyllum demersum* can remove cadmium from sediments by phytoextraction by means of the production of phytochelatin for metal binding in shoots [143,144], *Potamogeton pectinatus* and *P. malaianus* has also been attributed a high capability to remove heavy metals and other pollutants directly from the contaminated water [145,146]. Among charophytes, *Chara vulgaris* has been lately proposed to be used in phytoremediation [147–149]. As the tolerance to nutrients and different pollutants varies among the hydrophyte species, this is also an important aspect to take into account when selecting plants species for wetland restoration according to the state of water and sediments in each particular wetland.

8. Evaluation of the Success of Revegetation

Revegetation should be evaluated at different scales in both spatial (from the community up to the landscape) and temporal (from seasonal dynamics up to long-term changes) dimensions [43]. This requires approaches that are, at the same time, effective and feasible in the long-term. Some revegetation projects have been followed in the long-term [90]; however, other programs have been abandoned at relatively early stages because meaningful follow-up is a monumental undertaking, and scientists often lack the necessary opportunities and funding, while developers probably lack interest.

The monitoring can be done by using control plots and aerial photography surveys and other remote sensing methods, when possible. Unmanned (or Unoccupied) Aerial Vehicles (UAVs), known by the popularized name of drones, have been utilized for algal bloom and submerged aquatic vegetation detection for nearly two decades [150]. This type of high-resolution aerial imagery offers a cost-effective and rapid method to assess primary producer assemblages in aquatic environments, and provides great spatial resolutions for imaging [151]. Moreover, UAVs have advantages over manned vehicles for remote sensing: (i) flying UAVs is less expensive, (ii) is more flexible in scheduling, (iii) enables lower altitudes, (iv) uses lower speeds, (v) and the already cited provision of better image spatial resolution. Mistch et al. [152] used color aerial photography followed by ground-truth verifications (normalized maps and a grid system marked with permanent, numbered white poles to facilitate identification of the locations of plant communities in the wetland during ground-truthing and aerial photography). However, permissions are required in many countries for the use of UAVs [153]. Reflectance and transmittance spectra of floating-leaved plants can be measured, to know their influence on light availability in the water column which can alter the environmental conditions underneath the water surface [154]. With the data obtained at the ground level, macrophyte community diversity indexes should be applied to examine if the desired goals in terms of hydrophyte biodiversity have been achieved (see for example [152]).

9. Final Remarks and Conclusions

Restoration of wetlands by revegetation with native hydrophytes is a challenging task. Several countries have developed pilot case studies and field-scale projects with more or less success. The number of large field-scale cases are less due to all the needed issues that have to be solved (not only biological but financial, staff resources, etc.). Most published papers (more than 90%) only refer to successful results, but study cases in which failure in revegetation has been the outcome and that analyze the reasons for such a result, should be published as well, to learn from "what not to do". Some of the shortcomings of experimental designs which could significantly limit the interpretation of hydrophyte reintroduction projects are: (i) inadequate previous information and documentation, (ii) lack of understanding of the underlying reasons for the decline in existing plant populations, (iii) poorly defined success criteria for revegetation projects, (iv) insufficient monitoring following reintroduction, which can drive to an overly optimistic evaluation of success based on short-term results. Clearly, successful revegetation needs to be accompanied (in advance) by other management actions, such as external and internal nutrient load reduction, food web biomanipulation, increasing light availability by water level drawdowns in spring, etc. (for detailed information see Hilt et al. [46]). Moreover, before starting revegetation, the existence of any legal restrictions should be checked, because they can be different in each country.

It can be concluded that the value of hydrophyte revegetation as a restoration tool could be improved by:

(i)   Performing research in advance. Experimental out-site (culture room) and on-site (wetland mesocosms) should be planned before starting larger-scale initiatives.

(ii)   Selecting suitable wetlands with ecologically suitable revegetation sites. It is very important to consider the clear negative factors which prevent the success of revegetation (herbivory, microalgae and filamentous green algae, etc.). If revegetation is performed in sites with high nutrient and pollutant concentrations, high density of herbivorous fish, very low water transparency, etc., the result will be a total failure [46].

(iii)   An increased focus on species biology (including genetics) and ecology. Selecting and obtaining native (and typically occurring in the wetland previous to its degradation) suitable hydrophyte species is fundamental. In the studies reviewed here, the use of floating hydrophyte species has been less generalized than the submerged species. A total of 45 different species of submerged hydrophytes and 14 floating-leaved and free-floating species have been used for revegetation in wetlands (Figure 4). The genus *Potamogeton* has been used the most among the submerged hydrophytes (in 29% of the occasions), but *Myriophyllum spicatum* and *Hydrilla verticilata* have been the two most used species (15% and 13%). The genus *Nymphaea* has been the most used as a free-floating hydrophyte (36% of occasions), followed by the floating-leaved species of *Potamogeton* (22%). Introducing highly competitive species (r-strategists) has the risk that they outcompete part of the original vegetation including rare species. However, if the initial aim is to have a large cover of hydrophytes to prevent the growth of phytoplankton, resuspension of the sediment, etc., they can be chosen, and, in a second step, other species, specifically rare species, could be reintroduced in particular sites suitable for them. Although other management actions had been applied (i.e., nutrient and pollutant reductions), species or ecotypes/genotypes with high capacity to tolerate stress conditions should be initially chosen. *Potamogeton pectinatus*, *P. malaianus*, and *Ceratophyllum demersum* can live in contaminated water with heavy metals and other pollutants and remove them [143–146]. Among charophytes, *Chara vulgaris* is maybe the best candidate [100,147–149]. The selection of species with high allelopathic capacity against phytoplankton and periphyton is a complementary issue (e.g., *Myriophyllum spicatum* [105], *Vallisneria spiralis* [107], *Ceratophyllum demersum* [108], *Potamogeton malaianus* [109]). *P. malaianus* also inhibits filamentous algae growth.

(iv)   Deciding the appropriate wetland surface area to be potentially planted with hydrophytes. To increase light availability and be sure that clear-water conditions will be maintained, this area should be at least 30–40% of the wetland surface where hydrophytes could grow (this has to be determined in advance, based on wetland morphometry, water column light attenuation, light requirements for growth of the selected species according to their type, such as caulescent or rosette-type angiosperms, charophytes, etc.).

(v)   Selecting the appropriated revegetation techniques, considering the seed production and recruitment. The studies reviewed here suggest that sediment transfer is more adequate for temporary wetlands. However, in the cases of transferences from other sites to the target wetland, nature protection aspects and the potential risk of transferring pollutants, fish parasites, pathogens or other undesired species must be considered. Samples of this sediment have to be chemically analyzed to dismiss the presence of different kinds of pollutants and also carefully observed by experts to be sure that no unwanted propagules are present. If nutrient or pollutant contents are high, experimental tests of the sediment suitability by planting test species are recommended.

(vi) Choosing the suitable propagation technique. Seed-based approaches are less expensive and more logistically feasible in treating larger areas than other wetland revegetation techniques. For seeding, densities varying from 11 to 100 seeds/m^2 have been used for coastal wetlands. A high number of "transplants" and of adequate length should be selected: around 10 ramets/m^2 with lengths of 20–30 cm seem to be the most adequate to be planted (with apical parts) [7,17,45]. The use (or not) of a substrate to plant the prepared cultures in the wetland will depend on the type of the radicular system the hydrophyte develops and the features of the receptor sediment. Hydrophytes, such as *M. spicatum, S. pectinata* or *C. vulgaris,* for example, do not need any kind of support substrate. If the sediment is unconsolidated with low cohesive strength—typical for waterbodies with previous phytoplankton dominance—degradable substrates should be used. Planting by hand, although work-intensive, can be achieved by involving volunteers. Mechanical planters might have a great impact on the wetland fauna. When a moderate herbivory pressure on hydrophytes is suspected, protective exclosures should be used in initial trials to determine if the magnitude of this pressure will cause the failure of the revegetation. Protective exclosures can be also used, progressively enlarging them until established hydrophyte stands resistant to herbivory are formed to facilitate submerged macrophyte growth and dispersal.

(vii) Performing long-term monitoring programs to assess the performance and the variability of the restored populations over time. Whole-ecosystem, long-term interventions including most if not all ecosystem processes are desirable to be sure that the restoration result is the expected [61]. Furthermore, for large-scale hydrophyte restoration, the efforts should be in the framework of coordinated interagency programs, to develop, evaluate, and refine the suitable protocols and procedures. All this information will allow modeling the transition to an alternative stable clear macrophyte-dominated state and its future resilience [155].

It is necessary to encourage countries to publish scientifically sound guidelines to help managers aiming to restore wetlands and shallow lakes, and critically assess and predict the potential development of submerged vegetation, taking into account the complex factors and interrelations that determine their occurrence, abundance and diversity. Despite all the information already found in the published documents regarding revegetation with hydrophytes (approaches and experiments, manipulations in the field, etc.), further research is needed to key issues, such as target recruitment bottlenecks, interactive factors, foreseen climate change, etc., specific to many species and wetland types, which can yield insights into environmental manipulations or species selection that maximizes recruitment and ensures restoration. Not only ecology but also microbiology, soil and genetic sciences are necessary to improve the success of revegetation with hydrophytes, because they can provide new insights into why revegetation fails. The inclusion of an "epigenetic restoration and conservation" perspective together with a genetic one is also desirable as has been suggested for seagrass restoration [115]. Many papers lack precise data on the speed and efficiency of colonization of the wetlands by the different species, and this information is very valuable for wetland restoration practices with hydrophytes elsewhere.

Finally, the revegetation with hydrophytes must be performed in the context of broader wetland habitat restoration projects to have a greater chance of success. Restoration needs a continued effort (in terms of time and economic and personal resources) of research and implementation. It is clear that research so far has been very productive, but the results obtained should be more effectively integrated with policy-making, general wetland restoration practices and with a landscape perspective [156], particularly under future climatic scenarios.

Funding: The writing of this paper received no external funding.

Acknowledgments: All Integrative Ecology Group members, master students, Flora Unit staff of CIP El Palmar-GV and Tancat de la Pipa owner and managers (CHJ, AE-Agró, SEO-BirdLife) are acknowledged for their support and help. I also thank William Colom for his comments of improvement on a previous version of the manuscript. I acknowledge the comments and suggestions of three anonymous reviewers as well.

Conflicts of Interest: The author declares no conflict of interest.

References

1. Kettenring, K.M.; Tarsa, E.E. Need to Seed? Ecological, Genetic, and Evolutionary Keys to Seed-Based Wetland Restoration. *Front. Environ. Sci.* **2020**, *8*, 109. [CrossRef]
2. Brander, L.; Brouwer, R.; Wagtendonk, A. Economic valuation of regulating services provided by wetlands in agricultural landscapes: A meta-analysis. *Ecol. Eng.* **2013**, *56*, 89–96. [CrossRef]
3. Duarte, C.M.; Losada, I.J.; Hendriks, I.E.; Mazarrasa, I.; Marbà, N. The role of coastal plant communities for climate change mitigation and adaptation. *Nat. Clim. Chang.* **2013**, *3*, 961–968. [CrossRef]
4. Espeland, E.K.; Kettenring, K.M. Strategic plant choices can alleviate climate change impacts: A review. *J. Environ. Manag.* **2018**, *222*, 316–324. [CrossRef] [PubMed]
5. Endter-Wada, J.; Kettenring, K.M.; Sutton-Grier, A. Protecting wetlands for people: Strategic policy action can help wetlands mitigate risks and enhance resilience. *Environ. Sci. Policy* **2020**, *108*, 37–44. [CrossRef]
6. Eisele, F.; Hwang, B.S. New UN Decade on Ecosystem Restoration Offers Unparalleled Opportunity for Job Creation, Food Security and Addressing Climate Change. 2019. Available online: https://www.unep.org/news-and-stories/press-release/new-un-decade-ecosystem-restoration-offers-unparalleled-opportunity (accessed on 8 March 2021).
7. Riis, T.; Schultz, R.; Olsen, H.-M.; Katborg, C.K. Transplanting macrophytes to rehabilitate streams: Experience and recommendations. *Aquat. Ecol.* **2009**, *43*, 935–942. [CrossRef]
8. Wu, J.; Cheng, S.; Li, Z.; Guo, W.; Zhong, F.; Yin, D. Case study on rehabilitation of a polluted urban water body in Yangtze River Basin. *Environ. Sci. Pollut. Res.* **2013**, *20*, 7038–7045. [CrossRef]
9. Orsenigo, S. Editorial: How to halt the extinction of wetland-dependent plant species? The role of translocations and restoration ecology. *Aquat. Conserv. Mar. Freshw. Ecosyst.* **2018**, *28*, 772–775. [CrossRef]
10. Sparks, E.L.; Cebrian, J.; Biber, P.D.; Sheehan, K.L.; Tobias, C.R. Cost-effectiveness of two small-scale salt marsh restoration designs. *Ecol. Eng.* **2013**, *53*, 250–256. [CrossRef]
11. Sloey, T.M.; Willis, J.M.; Hester, M.W. Hydrologic and edaphic constraints on *Schoenoplectus acutus, Schoenoplectus californicus*, and *Typha latifolia* in tidal marsh restoration. *Restor. Ecol.* **2015**, *23*, 430–438. [CrossRef]
12. Rodrigo, M.A.; Valentín, A.; Claros, J.; Moreno, L.; Segura, M.; Lassalle, M.; Vera, P. Assessing the effect of emergent vegetation in a surface-flow constructed wetland on eutrophication reversion and biodiversity enhancement. *Ecol. Eng.* **2018**, *113*, 74–87. [CrossRef]
13. Land, M.; Granéli, W.; Grimvall, A.; Hoffmann, C.C.; Mitsch, W.J.; Tonderski, K.S.; Verhoeven, J.T.A. How effective are created or restored freshwater wetlands for nitrogen and phosphorus removal? A systematic review. *Environ. Évid.* **2016**, *5*, 9. [CrossRef]
14. Becerra-Jurado, G.; Harrington, R.; Kelly-Quinn, M. A review of the potential of surface flow constructed wetlands to enhance macroinvertebrate diversity in agricultural landscapes with particular reference to Integrated Constructed Wetlands (ICWs). *Hydrobiologia* **2012**, *692*, 121–130. [CrossRef]
15. Mitsch, W.J.; Zhang, L.; Waletzko, E.; Bernal, B. Validation of the ecosystem services of created wetlands: Two decades of plant succession, nutrient retention, and carbon sequestration in experimental riverine marshes. *Ecol. Eng.* **2014**, *72*, 11–24. [CrossRef]
16. Rodrigo, M.A.; Rojo, C.; Alonso-Guillén, J.L.; Vera, P. Restoration of two small Mediterranean lagoons: The dynamics of submerged macrophytes and factors that affect the success of revegetation. *Ecol. Eng.* **2013**, *54*, 1–15. [CrossRef]
17. Rodrigo, M.A.; Carabal, N. Selecting submerged macrophyte species for replanting in Mediterranean eutrophic wetlands. *Glob. Ecol. Conserv.* **2020**, *24*, e01349. [CrossRef]
18. Kaiser, J. WETLANDS RESTORATION: Recreated Wetlands No Match for Original. *Science* **2001**, *293*, 25. [CrossRef]
19. Scholz, M.; Harrington, R.; Carroll, P.; Mustafa, A. The Integrated Constructed Wetlands (ICW) concept. *Wetlands* **2007**, *27*, 337–354. [CrossRef]
20. Sartori, L.; Canobbio, S.; Cabrini, R.; Fornaroli, R.; Mezzanotte, V. Macroinvertebrate assemblages and biodiversity levels: Ecological role of constructed wetlands and artificial ponds in a natural park. *J. Limnol.* **2014**, *73*, 335–345. [CrossRef]
21. De Martis, G.; Mulas, B.; Malavasi, V.; Marignani, M. Can Artificial Ecosystems Enhance Local Biodiversity? The Case of a Constructed Wetland in a Mediterranean Urban Context. *Environ. Manag.* **2016**, *57*, 1088–1097. [CrossRef]
22. Sánchez-Ramos, D. Design and Modelling of Wetlands for the Treatment of Wastewater Treatment Plant Effluents. Application in the Surroundings of the Tablas de Daimiel Natural Park. Ph.D. Thesis, Universidad de Castilla-La Mancha, Ciudad Real, Spain, (In Spanish). 2013. Available online: http://hdl.handle.net/10578/3487 (accessed on 9 March 2021).

23. Jóźwiakowski, K.; Marzec, M.; Kowalczyk-Juśko, A.; Gizińska-Górna, M.; Pytka-Woszczyło, A.; Malik, A.; Listosz, A.; Gajewska, M. 25 years of research and experiences about the application of constructed wetlands in southeastern Poland. *Ecol. Eng.* **2019**, *127*, 440–453. [CrossRef]

24. Swartz, L.K.; Hossack, B.R.; Muths, E.; Newell, R.L.; Lowe, W.H. Aquatic macroinvertebrate community responses to wetland mitigation in the Greater Yellowstone Ecosystem. *Freshw. Biol.* **2019**, *64*, 942–953. [CrossRef]

25. Kennedy, G.; Mayer, T. Natural and Constructed Wetlands in Canada: An Overview. *Water Qual. Res. J.* **2002**, *37*, 295–325. [CrossRef]

26. An, S.; Tian, Z.; Cai, Y.; Wen, T.; Xu, D.; Jiang, H.; Yao, Z.; Guan, B.; Sheng, S.; Ouyang, Y.; et al. Wetlands of Northeast Asia and High Asia: An overview. *Aquat. Sci.* **2012**, *75*, 63–71. [CrossRef]

27. Osaliya, R.; Kansiime, F.; Oryem-Origa, H.; Kateyo, E. The potential use of storm water and effluent from a constructed wetland for re-vegetating a degraded pyrite trail in Queen Elizabeth National Park, Uganda. *Phys. Chem. Earth Parts A B C* **2011**, *36*, 842–852. [CrossRef]

28. Waltham, N.J.; Barry, M.; McAlister, T.; Weber, T.; Groth, D. Protecting the Green Behind the Gold: Catchment-Wide Restoration Efforts Necessary to Achieve Nutrient and Sediment Load Reduction Targets in Gold Coast City, Australia. *Environ. Manag.* **2014**, *54*, 840–851. [CrossRef] [PubMed]

29. Almeida, B.A.; Sebastián-González, E.; Dos Anjos, L.; Green, A.J. Comparing the diversity and composition of waterbird functional traits between natural, restored, and artificial wetlands. *Freshw. Biol.* **2020**, *65*, 2196–2210. [CrossRef]

30. Hsu, C.-B.; Hsieh, H.-L.; Yang, L.; Wu, S.-H.; Chang, J.-S.; Hsiao, S.-C.; Su, H.-C.; Yeh, C.-H.; Ho, Y.-S.; Lin, H.-J. Biodiversity of constructed wetlands for wastewater treatment. *Ecol. Eng.* **2011**, *37*, 1533–1545. [CrossRef]

31. Stable states, buffers and switches: An ecosystem approach to the restoration and management of shallow lakes in the Netherlands. *Water Sci. Technol.* **1998**, *37*, 151–164. [CrossRef]

32. Jeppesen, E.; Søndergaard, M.; Kronvang, B.; Jensen, J.P.; Svendsen, L.M.; Lauridsen, T.L. Lake and catchment management in Denmark. *Hydrobiol.* **1999**, *395/396*, 419–432. [CrossRef]

33. Annadotter, H.; Cronberg, G.; Aagren, R.; Lundstedt, B.; Nilsson, P.Å.; Ströbeck, S. Multiple techniques for lake restoration. *Hydrobiol.* **1999**, *395/396*, 77–85. [CrossRef]

34. Smart, M.; Dick, G.O. *Propagation and Establishment of Aquatic Plants: A Hand-Book for Ecosystems Restoration Projects*; Technical Report A-99-4, February199; US Army Corps of Engineers, Waterways Experimental Station: Vicksburg, MS, USA, 1999.

35. Gao, H.; Qian, X.; Wu, H.; Li, H.; Pan, H.; Han, C. Combined effects of submerged macrophytes and aquatic animals on the restoration of a eutrophic water body—A case study of Gonghu Bay, Lake Taihu. *Ecol. Eng.* **2017**, *102*, 15–23. [CrossRef]

36. Wu, J.; Dai, Y.; Cheng, S. General trends in freshwater ecological restoration practice in China over the past two decades: The driving factors and the evaluation of restoration outcome. *Environ. Sci. Eur.* **2020**, *32*, 1–12. [CrossRef]

37. Zhou, C.; An, S.; Jiang, J.; Yin, D.; Wang, Z.; Fang, C.; Sun, Z.; Qian, C. An in vitro propagation protocol of two submerged macrophytes for lake revegetation in east China. *Aquat. Bot.* **2006**, *85*, 44–52. [CrossRef]

38. Nakamura, K.; Tockner, K.; Amano, K. River and Wetland Restoration: Lessons from Japan. *BioScience* **2006**, *56*, 419–429. [CrossRef]

39. Clarkson, B.; Peters, M. (Eds.) Chapter 10: Revegetation. In *Wetland Restoration: A Handbook for NZ Freshwater Systems*; Manaaki Whenua Press: Lincoln, New Zealand, 2010; pp. 156–184.

40. Ciurli, A.; Zuccarini, P.; Alpi, A. Growth and nutrient absorption of two submerged aquatic macrophytes in mesocosms, for reinsertion in a eutrophicated shallow lake. *Wetl. Ecol. Manag.* **2008**, *17*, 107–115. [CrossRef]

41. Fontanarrosa, M.S.; Allende, L.; Rennella, A.M.; Boveri, M.B.; Sinistro, R. A novel device with macrophytes and bio balls as a rehabilitation tool for small eutrophic urban ponds: A mesocosm approximation. *Limnologica* **2019**, *74*, 61–72. [CrossRef]

42. Godefroid, S.; Piazza, C.; Rossi, G.; Buord, S.; Stevens, A.-D.; Aguraiuja, R.; Cowell, C.; Weekley, C.W.; Vogg, G.; Iriondo, J.M.; et al. How successful are plant species reintroductions? *Biol. Conserv.* **2011**, *144*, 672–682. [CrossRef]

43. Shafer, D.; Bergstrom, P. An Introduction to a Special Issue on Large-Scale Submerged Aquatic Vegetation Restoration Research in the Chesapeake Bay: 2003–2008. *Restor. Ecol.* **2008**, *18*, 481–489. [CrossRef]

44. Qiu, D.; Wu, Z.; Liu, B.; Deng, J.; Fu, G.; He, F. The restoration of aquatic macrophytes for improving water quality in a hypertrophic shallow lake in Hubei Province, China. *Ecol. Eng.* **2001**, *18*, 147–156. [CrossRef]

45. Lauridsen, T.L.; Sandsten, H.; Møller, P.H. The restoration of a shallow lake by introducing Potamogeton spp.: The impact of waterfowl grazing. *Lakes Reserv. Res. Manag.* **2003**, *8*, 177–187. [CrossRef]

46. Hilt, S.; Gross, E.M.; Hupfer, M.; Morscheid, H.; Mählmann, J.; Melzer, A.; Poltz, J.; Sandrock, S.; Scharf, E.-M.; Schneider, S.; et al. Restoration of submerged vegetation in shallow eutrophic lakes—A guideline and state of the art in Germany. *Limnologica* **2006**, *36*, 155–171. [CrossRef]

47. Ye, C.; Yu, H.-C.; Kong, H.-N.; Song, X.-F.; Zou, G.-Y.; Xu, Q.-J.; Liu, J. Community collocation of four submerged macrophytes on two kinds of sediments in Lake Taihu, China. *Ecol. Eng.* **2009**, *35*, 1656–1663. [CrossRef]

48. Moore, K.A.; Shields, E.C.; Jarvis, J.C. The Role of Habitat and Herbivory on the Restoration of Tidal Freshwater Submerged Aquatic Vegetation Populations. *Restor. Ecol.* **2008**, *18*, 596–604. [CrossRef]

49. Gao, H.; Shi, Q.; Qian, X. A multi-species modelling approach to select appropriate submerged macrophyte species for ecological restoration in Gonghu Bay, Lake Taihu, China. *Ecol. Model.* **2017**, *360*, 179–188. [CrossRef]

50. Nilsson, J.E.; Liess, A.; Ehde, P.M.; Weisner, S.E. Mature wetland ecosystems remove nitrogen equally well regardless of initial planting. *Sci. Total. Environ.* **2020**, *716*, 137002. [CrossRef]

51. Schad, A.N.; Kennedy, J.H.; Dick, G.O.; Dodd, L. Aquatic macroinvertebrate richness and diversity associated with native submerged aquatic vegetation plantings increases in longer-managed and wetland-channeled effluent constructed urban wetlands. *Wetl. Ecol. Manag.* **2020**, *28*, 461–477. [CrossRef]

52. Bakker, E.S.; Sarneel, J.M.; Gulati, R.D.; Liu, Z.; Van Donk, E. Restoring macrophyte diversity in shallow temperate lakes: Biotic versus abiotic constraints. *Hydrobiologia* **2012**, *710*, 23–37. [CrossRef]

53. Wilcox, D.J.; Harwell, M.C.; Orth, R.J. Modeling Dynamic Polygon Objects in Space and Time: A New Graph-based Technique. *Cartogr. Geogr. Inf. Sci.* **2000**, *27*, 153–164. [CrossRef]

54. Dugdale, T.M.; Hicks, B.J.; De Winton, M.; Taumoepeau, A. Fish exclosures versus intensive fishing to restore charophytes in a shallow New Zealand lake. *Aquat. Conserv. Mar. Freshw. Ecosyst.* **2006**, *16*, 193–202. [CrossRef]

55. Sebastián, A.; Peña, C.; Laguna, E. Experiencias de restauración de balsas temporales y otras zonas húmedas en el territorio valenciano. In *Conservació, Problemàtiques i Gestió de les Llacunes Temporànies Mediterrànies*; Vila, X., Campos, M., Feo, C., Eds.; Consorci de l'Estany: Girona, Spain, 2008. (In Catalan)

56. Chen, K.-N.; Bao, C.-H.; Zhou, W.-P. Ecological restoration in eutrophic Lake Wuli: A large enclosure experiment. *Ecol. Eng.* **2009**, *35*, 1646–1655. [CrossRef]

57. Dick, G.O.; Dodd, L.L.; Schad, A.N.; Smith, D.H.; Owens, C.S.; Dallas Floodway Extension Lower Chain of Wetlands and Grasslands Ecological Management and Monitoring. Status Report to the USACE SWF. 2015. Available online: https://www.swf.usace.army.mil/Portals/47/docs/PAO/DFE/PDF/DFE_March_2015_Status_Report.pdf (accessed on 18 May 2021).

58. Yu, J.; Liu, Z.; Li, K.; Chen, F.; Guan, B.; Hu, Y.; Zhong, P.; Tang, Y.; Zhao, X.; He, H.; et al. Restoration of Shallow Lakes in Subtropical and Tropical China: Response of Nutrients and Water Clarity to Biomanipulation by Fish Removal and Submerged Plant Transplantation. *Water* **2016**, *8*, 438. [CrossRef]

59. Theÿsmeÿer, T.; Bowman, J.; Court, A.; Richer, S. *Wetlands Conservation Plan 2016–2021*; Royal Botanical Gardens: Burlington, ON, Canada, 2016.

60. Liu, Z.; Hu, J.; Zhong, P.; Zhang, X.; Ning, J.; Larsen, S.E.; Chen, D.; Gao, Y.; He, H.; Jeppesen, E. Successful restoration of a tropical shallow eutrophic lake: Strong bottom-up but weak top-down effects recorded. *Water Res.* **2018**, *146*, 88–97. [CrossRef]

61. Mitsch, W.J.; Day, J.W. Thinking big with whole-ecosystem studies and ecosystem restoration—a legacy of H.T. Odum. *Ecol. Model.* **2004**, *178*, 133–155. [CrossRef]

62. Orth, R.J.; Batiuk, R.A.; Bergstrom, P.W.; Moore, K.A. A perspective on two decades of policies and regulations influencing the protection and restoration of submerged aquatic vegetation in Chesapeake Bay, USA. *Bull. Mar. Sci.* **2002**, *71*, 1391–1403.

63. Greet, J.; Ede, F.; Robertson, D.; McKendrick, S. Should I plant or should I sow? Restoration outcomes compared across seven riparian revegetation projects. *Ecol. Manag. Restor.* **2019**, *21*, 58–65. [CrossRef]

64. Palma, A.C.; Laurance, S.G. A review of the use of direct seeding and seedling plantings in restoration: What do we know and where should we go? *Appl. Veg. Sci.* **2015**, *18*, 561–568. [CrossRef]

65. Ailstock, M.S.; Shafer, D.J.; Magoun, A.D. Protocols for Use of *Potamogeton perfoliatus* and *Ruppia maritima* Seeds in Large-Scale Restoration. *Restor. Ecol.* **2008**, *18*, 560–573. [CrossRef]

66. Marion, S.R.; Orth, R.J. Innovative Techniques for Large-scale Seagrass Restoration Using *Zostera marina* (eelgrass) Seeds. *Restor. Ecol.* **2008**, *18*, 514–526. [CrossRef]

67. Baskin, C.C.; Baskin, J.M. *Seeds: Ecology, Biogeography, and Evolution of Dormancy and Germination*, 2nd ed.; Academic Press: San Diego, CA, USA, 2014; 1600p.

68. Kauth, P.J.; Biber, P.D. Moisture content, temperature, and relative humidity influence seed storage and subsequent sur-vival and germination of *Vallisneria americana* seeds. *Aquat. Bot.* **2015**, *120*, 297–303. [CrossRef]

69. Zhao, S.; Zhang, R.; Liu, Y.; Yin, L.; Wang, C.; Li, W. The effect of storage condition on seed germination of six Hydrocharitaceae and Potamogetonaceae species. *Aquat. Bot.* **2017**, *143*, 49–53. [CrossRef]

70. Rybak, A.S. Microencapsulation with the usage of sodium alginate: A promising method for preserving stonewort (Characeae, Charophyta) oospores to support laboratory and field experiments. *Algal Res.* **2021**, *54*, 102236. [CrossRef]

71. Broome, S.W.; Seneca, E.D.; Woodhouse, W.W. Tidal salt marsh restoration. *Aquat. Bot.* **1988**, *32*, 1–22. [CrossRef]

72. Busch, K.E.; Golden, R.R.; Parham, T.A.; Karrh, L.P.; Lewandowski, M.J.; Naylor, M.D. Large-Scale *Zostera marina* (eelgrass) Restoration in Chesapeake Bay, Maryland, USA. Part I: A Comparison of Techniques and Associated Costs. *Restor. Ecol.* **2008**, *18*, 490–500. [CrossRef]

73. Unsworth, R.K.F.; Bertelli, C.M.; Cullen-Unsworth, L.C.; Esteban, N.; Jones, B.L.; Lilley, R.; Lowe, C.; Nuuttila, H.K.; Rees, S.C. Sowing the Seeds of Seagrass Recovery Using Hessian Bags. *Front. Ecol. Evol.* **2019**, *7*, 311. [CrossRef]

74. Traber, M.; Granger, S.; Nixon, S. Mechanical seeder provides alternative method for restoring eelgrass habitat (Rhode Island). *Ecol. Restor.* **2003**, *21*, 213–214.

75. Golden, R.R.; Busch, K.E.; Karrh, L.P.; Parham, T.A.; Lewandowski, M.J.; Naylor, M.D. Large-Scale *Zostera marina* (eelgrass) Restoration in Chesapeake Bay, Maryland, USA. Part II: A Comparison of Restoration Methods in the Patuxent and Potomac Rivers. *Restor. Ecol.* **2010**, *18*, 501–513. [CrossRef]

76. Gornish, E.; Arnold, H.; Fehmi, J. Review of seed pelletizing strategies for arid land restoration. *Restor. Ecol.* **2019**, *27*, 1206–1211. [CrossRef]

77. Anderson, E.C.; Minor, E.S. Assessing four methods for establishing native plants on urban vacant land. *Ambio* **2021**, *50*, 695–705. [CrossRef]
78. Abeli, T.; Dixon, K. Translocation ecology: The role of ecological sciences in plant translocation. *Plant Ecol.* **2016**, *217*, 123–125. [CrossRef]
79. Tanner, C.E.; Parham, T. Growing Zostera marina (eelgrass) from Seeds in Land-Based Culture Systems for Use in Restoration Projects. *Restor. Ecol.* **2010**, *18*, 527–537. [CrossRef]
80. Lin, D.; Hu, L.; You, H.; Sarkar, D.; Xing, B.; Shetty, K. Plant clonal systems as a strategy for nitrate pollution removal in cold latitudes. In *Molecular Environmental Soil Science at the Interfaces in the Earth's Critical Zone*; Xu, J.M., Huang, P.M., Eds.; Springer: Berlin/Heidelberg, Germany, 2010; pp. 75–77.
81. Carricondo-Antón, J.M.; González-Romero, J.A.; Mengual-Cuquerella, J.; Turegano-Pastor, J.V.; Oliver-Villanueva, J.V. Alternative use of rice straw ash as natural fertilizer to reduce phosphorus pollution in protected wetland ecosystems. *Int. J. Recycl. Org. Waste Agric.* **2020**, *9*, 61–74. [CrossRef]
82. Shen, X.X.; Cao, W.P.; Huang, D.C.; Zhang, C.Z.; Jiang, J.L. Removal of nutrient compounds from eutrophic water using two hybrid floating beds (HFBS) with different substrates. *Fresenius Environ. Bull.* **2017**, *26*, 661–665.
83. Blindow, I.; Carlsson, M. Establishment of charophytes—A method to support threatened charophytes. *Plants* **2021**, *10*. in press.
84. Guillem, A.; Management of Vegetation for the Improvement of Water Quality and Habitat. Technical Manuals for the Management of Artificial Wetlands in Natural Environments (in Spanish). LIFE Albufera. Spain. 2016. Available online: https://www.antoniogiraldez.es/index.php/es/noticias/novedades (accessed on 18 May 2021).
85. McKnight, S.K. Transplanted seed bank response to drawdown time in a created wetland in East Texas. *Wetlands* **1992**, *12*, 79–90. [CrossRef]
86. Brown, S.C.; Bedford, B.L. Restoration of wetland vegetation with transplanted wetland soil: An experimental study. *Wetlands* **1997**, *17*, 424–437. [CrossRef]
87. Fahselt, D. Is transplanting an effective means of preserving vegetation? *Can. J. Bot.* **2007**, *85*, 1007–1017. [CrossRef]
88. Hong, J.; Liu, S.; Shi, G.; Zhang, Y. Soil seed bank techniques for restoring wetland vegetation diversity in Yeyahu Wetland, Beijing. *Ecol. Eng.* **2012**, *42*, 192–202. [CrossRef]
89. Nishihiro, J.; Nishihiro, M.A.; Washitani, I. Restoration of wetland vegetation using soil seed banks: Lessons from a project in Lake Kasumigaura, Japan. *Landsc. Ecol. Eng.* **2006**, *2*, 171–176. [CrossRef]
90. Muller, I.; Buisson, E.; Mouronval, J.-B.; Mesléard, F. Temporary wetland restoration after rice cultivation: Is soil transfer required for aquatic plant colonization? *Knowl. Manag. Aquat. Ecosyst.* **2013**, *411*, 03. [CrossRef]
91. Mouronval, J.-B.; Baudouin, S. *Plantes aquatiques de Camargue et de Crau*; Office National de la Chasse et de la Faune Sauvage: Paris, France, 2010; 120p.
92. Nishihiro, J.; Nishihiro, M.A.; Washitani, I. Assessing the potential for recovery of lakeshore vegetation: Species richness of sediment propagule banks. *Ecol. Res.* **2005**, *21*, 436–445. [CrossRef]
93. Bornette, G.; Puijalon, S. Response of aquatic plants to abiotic factors: A review. *Aquat. Sci.* **2011**, *73*, 1–14. [CrossRef]
94. Su, H.; Chen, J.; Wu, Y.; Chen, J.; Guo, X.; Yan, Z.; Tian, D.; Fang, J.; Xie, P. Morphological traits of submerged macrophytes reveal specific positive feedbacks to water clarity in freshwater ecosystems. *Sci. Total. Environ.* **2019**, *684*, 578–586. [CrossRef] [PubMed]
95. Song, Y.; Liew, J.H.; Sim, D.Z.H.; Mowe, M.A.D.; Mitrovic, S.M.; Tan, H.T.W.; Yeo, D.C.J. Effects of macrophytes on lake-water quality across latitudes: A meta-analysis. *Oikos* **2018**, *128*, 468–481. [CrossRef]
96. Vecchia, A.D.; Villa, P.; Bolpagni, R. Functional traits in macrophyte studies: Current trends and future research agenda. *Aquat. Bot.* **2020**, *167*, 103290. [CrossRef]
97. Zhou, X.; Li, Z.; Zhao, R.; Gao, R.; Yun, Y.; Saino, M.; Wang, X. Experimental comparisons of three submerged plants for reclaimed water purification through nutrient removal. *Desalination Water Treat.* **2015**, *57*, 1–10. [CrossRef]
98. Hao, B.; Wu, H.; Shi, Q.; Liu, G.; Xing, W. Facilitation and competition among foundation species of submerged macrophytes threatened by severe eutrophication and implications for restoration. *Ecol. Eng.* **2013**, *60*, 76–80. [CrossRef]
99. Dai, Y.; Tang, H.; Chang, J.; Wu, Z.; Liang, W. What's better, *Ceratophyllum demersum* L. or *Myriophyllum verticillatum* L., individual or combined? *Ecol. Eng.* **2014**, *70*, 397–401. [CrossRef]
100. Rodrigo, M.A.; Puche, E.; Rojo, C. On the tolerance of charophytes to high-nitrate concentrations. *Chem. Ecol.* **2018**, *34*, 22–42. [CrossRef]
101. Wang, C.; Liu, S.; Jahan, T.E.; Liu, B.; He, F.; Zhou, Q.; Wu, Z. Short term succession of artificially restored submerged macrophytes and their impact on the sediment microbial community. *Ecol. Eng.* **2017**, *103*, 50–58. [CrossRef]
102. Hillmann, E.R.; La Peyre, M.K. Effects of salinity and light on growth and interspecific interactions between *Myriophyllum spicatum* L. and *Ruppia maritima* L. *Aquat. Bot.* **2019**, *155*, 25–31. [CrossRef]
103. Kaijser, W.; Kosten, S.; Hering, D. Salinity tolerance of aquatic plants indicated by monitoring data from the Netherlands. *Aquat. Bot.* **2019**, *158*, 103129. [CrossRef]
104. Gil, L.; Capó, X.; Tejada, S.; Mateu-Vicens, G.; Ferriol, P.; Pinya, S.; Sureda, A. Salt variation induces oxidative stress response in aquatic macrophytes: The case of the Eurasian water-milfoil *Myriophyllum spicatum* L. (Saxifragales: Haloragaceae). *Estuarine Coast. Shelf Sci.* **2020**, *239*, 106756. [CrossRef]

105. Nakai, S.; Zou, G.; Okuda, T.; Nishijima, W.; Hosomi, M.; Okada, M. Polyphenols and fatty acids responsible for anti-cyanobacterial allelopathic effects of submerged macrophyte *Myriophyllum spicatum*. *Water Sci. Technol.* **2012**, *66*, 993–999. [CrossRef] [PubMed]
106. Li, B.; Yin, Y.; Kang, L.; Feng, L.; Liu, Y.; Du, Z.; Tian, Y.; Zhang, L. A review: Application of allelochemicals in water ecological restoration—Algal inhibition. *Chemosphere* **2021**, *267*, 128869. [CrossRef] [PubMed]
107. Xian, Q.; Chen, H.; Liu, H.; Zou, H.; Yin, D. Isolation and Identification of Antialgal Compounds from the Leaves of *Vallisneria spiralis* L. by Activity-Guided Fractionation (5 pp). *Environ. Sci. Pollut. Res.* **2006**, *13*, 233–237. [CrossRef] [PubMed]
108. Dong, J.; Chang, M.; Li, C.; Dai, D.; Gao, Y. Allelopathic effects and potential active substances of *Ceratophyllum demersum* L. on *Chlorella vulgaris* Beij. *Aquat. Ecol.* **2019**, *53*, 651–663. [CrossRef]
109. Zuo, S.; Wang, H.; Gan, L.D.; Shao, M. Allelopathy appraisal of worm metabolites in the synergistic effect between *Limnodrilus hoffmeisteri* and *Potamogeton malaianus* on algal suppression. *Ecotoxicol. Environ. Saf.* **2019**, *182*, 109482. [CrossRef] [PubMed]
110. Mulderij, G.; Van Donk, E.; Roelofs, J.G.M. Differential sensitivity of green algae to allelopathic substances from Chara. *Hydrobiologia* **2003**, *491*, 261–271. [CrossRef]
111. Rojo, C.; Segura, M.; Rodrigo, M. The allelopathic capacity of submerged macrophytes shapes the microalgal assemblages from a recently restored coastal wetland. *Ecol. Eng.* **2013**, *58*, 149–155. [CrossRef]
112. Xie, S.L.; Wang, J.; Liu, Q.; Feng, J.; Lv, J.P.; Shi, Y.; Li, Z. Research progresses on plant allelopathic effects for algal control. *J. Shanxi Univ.* **2017**, 652–660. [CrossRef]
113. Li, J.; Zheng, B.; Chen, X.; Li, Z.; Xia, Q.; Wang, H.; Yang, Y.; Zhou, Y.; Yang, H. The Use of Constructed Wetland for Mitigating Nitrogen and Phosphorus from Agricultural Runoff: A Review. *Water* **2021**, *13*, 476. [CrossRef]
114. Nan, X.; Lavrnić, S.; Toscano, A. Potential of constructed wetland treatment systems for agricultural wastewater reuse under the EU framework. *J. Environ. Manag.* **2020**, *275*, 111219. [CrossRef]
115. Pazzaglia, J.; Nguyen, H.; Santillán-Sarmiento, A.; Ruocco, M.; Dattolo, E.; Marín-Guirao, L.; Procaccini, G. The Genetic Component of Seagrass Restoration: What We Know and the Way Forwards. *Water* **2021**, *13*, 829. [CrossRef]
116. Jisha, K.C.; Vijayakumari, K.; Puthur, J.T. Seed priming for abiotic stress tolerance: An overview. *Acta Physiol. Plant.* **2013**, *35*, 1381–1396. [CrossRef]
117. Sadat-Noori, M.; Rankin, C.; Rayner, D.; Heimhuber, V.; Gaston, T.; Drummond, C.; Chalmers, A.; Khojasteh, D.; Glamore, W. Coastal wetlands can be saved from sea level rise by recreating past tidal regimes. *Sci. Rep.* **2021**, *11*, 1–10. [CrossRef]
118. Paillisson, J.-M.; Reeber, S.; Carpentier, A.; Marion, L. Plant-water regime management in a wetland: Consequences for a floating vegetation-nesting bird, whiskered tern Chlidonias hybridus. *Biodivers. Conserv.* **2006**, *15*, 3469–3480. [CrossRef]
119. Rodrigo, M.A.; Segura, M. *Study of Phytoplankton, Zooplankton and Submerged Vegetation for the Project "Protection and Enhancing Biodiversity and Monitoring of Biological Variables in Tancat de la Pipa"*; Technical Report; University of Valencia: Valencia, Spain, 2020; 43p. (In Spanish)
120. Chaichana, R.; Leah, R.; Moss, B. Seasonal impact of waterfowl on communities of macrophytes in a shallow lake. *Aquat. Bot.* **2011**, *95*, 39–44. [CrossRef]
121. Irfanullah, H.M.; Moss, B. Factors influencing the return of submerged plants to a clear-water, shallow temperate lake. *Aquat. Bot.* **2004**, *80*, 177–191. [CrossRef]
122. Yu, J.; Zhen, W.; Guan, B.; Zhong, P.; Jeppesen, E.; Liu, Z. Dominance of *Myriophyllum spicatum* in submerged macrophyte communities associated with grass carp. *Knowl. Manag. Aquat. Ecosyst.* **2016**, *417*, 24. [CrossRef]
123. Hidding, B.; Bakker, E.; Keuper, F.; De Boer, T.; De Vries, P.; Nolet, B.A. Differences in tolerance of pondweeds and charophytes to vertebrate herbivores in a shallow Baltic estuary. *Aquat. Bot.* **2010**, *93*, 123–128. [CrossRef]
124. Hidding, B.; Nolet, B.A.; De Boer, T.; De Vries, P.P.; Klaassen, M. Above- and below-ground vertebrate herbivory may each favour a different subordinate species in an aquatic plant community. *Oecologia* **2009**, *162*, 199–208. [CrossRef] [PubMed]
125. Cirujano, S.; Camargo, J.A.; Gómez-Cordovés, C. Feeding Preference of the Red Swamp Crayfish *Procambarus clarkii* (Girard) on Living Macrophytes in a Spanish Wetland. *J. Freshw. Ecol.* **2004**, *19*, 219–226. [CrossRef]
126. Zhang, L.; Huang, S.; Peng, X.; Liu, B.; Zhang, Y.; Zhou, Q.; Wu, Z. The Response of Regeneration Ability of Myriophyllum spicatum Apical Fragments to Decaying *Cladophora oligoclona*. *Water* **2019**, *11*, 1014. [CrossRef]
127. Zhang, L.; Peng, X.; Liu, B.; Zhang, Y.; Zhou, Q.; Wu, Z. Effects of the decomposing liquid of *Cladophora oligoclona* on *Hydrilla verticillata* turion germination and seedling growth. *Ecotoxicol. Environ. Saf.* **2018**, *157*, 81–88. [CrossRef] [PubMed]
128. Higgins, S.N.; Malkin, S.Y.; Howell, E.T.; Guildford, S.J.; Campbell, L.; Hiriart-Baer, V.; Hecky, R.E. An ecological review of cladophora glomerata (chlorophyta) in the laurentian great lakes. *J. Phycol.* **2008**, *44*, 839–854. [CrossRef]
129. Harris, V.A. Cladophora confounds coastal communities—Public perceptions and management dilemmas. In Proceedings of the Workshop "Cladophora Research and Management in the Great Lakes", Milwaukee, WI, USA, 8 December 2004; Boostma, H., Jenson, E., Young, E., Berges, J., Eds.; University of Wisconsin-Milwaukeel: Milwaukee, WI, USA, 2005. no. 2005-01.
130. Zhang, W.; Shen, H.; Zhang, J.; Yu, J.; Xie, P.; Chen, J. Physiological differences between free-floating and periphytic filamentous algae, and specific submerged macrophytes induce proliferation of filamentous algae: A novel implication for lake restoration. *Chemosphere* **2020**, *239*, 124702. [CrossRef]
131. Short, F.T.; Kosten, S.; Morgan, P.A.; Malone, S.; Moore, G.E. Impacts of climate change on submerged and emergent wetland plants. *Aquat. Bot.* **2016**, *135*, 3–17. [CrossRef]

132. Gladyshev, M.I.; Gubelit, Y.I. Green Tides: New Consequences of the Eutrophication of Natural Waters (Invited Review). *Contemp. Probl. Ecol.* **2019**, *12*, 109–125. [CrossRef]
133. Mihranyan, A. Cellulose from cladophorales green algae: From environmental problem to high-tech composite materials. *J. Appl. Polym. Sci.* **2011**, *119*, 2449–2460. [CrossRef]
134. Sagrario, M.A.G.; Jeppesen, E.; Goma, J.; Søndergaard, M.; Jensen, J.P.; Lauridsen, T.; Landkildehus, F. Does high nitrogen loading prevent clear-water conditions in shallow lakes at moderately high phosphorus concentrations? *Freshw. Biol.* **2005**, *50*, 27–41. [CrossRef]
135. Jeppesen, E.; Søndergaard, M.; Meerhoff, M.; Lauridsen, T.L.; Jensen, J.P. Shallow lake restoration by nutrient loading reduction—some recent findings and challenges ahead. *Hydrobiologia* **2007**, *584*, 239–252. [CrossRef]
136. Wang, H.-J.; Wang, H.-Z.; Liang, X.-M.; Wu, S.-K. Total phosphorus thresholds for regime shifts are nearly equal in subtropical and temperate shallow lakes with moderate depths and areas. *Freshw. Biol.* **2014**, *59*, 1659–1671. [CrossRef]
137. Dong, B.; Zhou, Y.; Jeppesen, E.; Shi, K.; Qin, B. Response of community composition and biomass of submerged macrophytes to variation in underwater light, wind and trophic status in a large eutrophic shallow lake. *J. Environ. Sci.* **2021**, *103*, 298–310. [CrossRef]
138. Yin, X.; Zhang, J.; Guo, Y.; Fan, J.; Hu, Z. Physiological Responses of Potamogeton crispus to Different Levels of Ammonia Nitrogen in Constructed Wetland. *Water Air Soil Pollut.* **2016**, *227*, 1–9. [CrossRef]
139. Li, C.; Zhang, B.; Zhang, J.; Wu, H.; Xie, H.; Xu, J.; Qi, P. Physiological responses of three plant species exposed to excess ammonia in constructed wetland. *Desalination Water Treat.* **2011**, *32*, 271–276. [CrossRef]
140. Wang, R.; Bai, N.; Xu, S.; Zhuang, G.; Bai, Z.; Zhao, Z.; Zhuang, X. The adaptability of a wetland plant species *Myriophyllum aquaticum* to different nitrogen forms and nitrogen removal efficiency in constructed wetlands. *Environ. Sci. Pollut. Res.* **2018**, *25*, 7785–7795. [CrossRef] [PubMed]
141. Simons, J.; Ohm, M.; Daalder, R.; Boers, P.; Rip, W. Restoration of Botshol (The Netherlands) by reduction of external nutrient load: Recovery of a characean community, dominated by *Chara connivens*. *Hydrobiologia* **1994**, *275*, 243–253. [CrossRef]
142. Lambert, S.J.; Davy, A.J. Water quality as a threat to aquatic plants: Discriminating between the effects of nitrate, phosphate, boron and heavy metals on charophytes. *New Phytol.* **2011**, *189*, 1051–1059. [CrossRef]
143. Mishra, S.; Tripathi, R.; Srivastava, S.; Dwivedi, S.; Trivedi, P.K.; Dhankher, O.; Khare, A. Thiol metabolism play significant role during cadmium detoxification by *Ceratophyllum demersum* L. *Bioresour. Technol.* **2009**, *100*, 2155–2161. [CrossRef]
144. Andresen, E.; Mattusch, J.; Wellenreuther, G.; Thomas, G.; Abad, U.A.; Küpper, H. Different strategies of cadmium detoxification in the submerged macrophyte *Ceratophyllum demersum* L. *Metallomics* **2013**, *5*, 1377–1386. [CrossRef]
145. Singh, N.; Pandey, G.; Rai, U.; Tripathi, R.; Singh, H.; Gupta, D. Metal Accumulation and Ecophysiological Effects of Distillery Effluent on Potamogeton pectinatus L. *Bull. Environ. Contam. Toxicol.* **2005**, *74*, 857–863. [CrossRef]
146. Peng, K.; Luo, C.; Lou, L.; Li, X.; Shen, Z. Bioaccumulation of heavy metals by the aquatic plants Potamogeton pectinatus L. and Potamogeton malaianus Miq. and their potential use for contamination indicators and in wastewater treatment. *Sci. Total. Environ.* **2008**, *392*, 22–29. [CrossRef]
147. Laffont-Schwob, I.; Triboit, F.; Prudent, P.; Soulié-Märsche, I.; Rabier, J.; Despréaux, M.; Thiéry, A. Trace metal extraction and biomass production by spontaneous vegetation in temporary Mediterranean stormwater highway retention ponds: Freshwater macroalgae (*Chara spp.*) vs. cattails (*Typha spp.*). *Ecol. Eng.* **2015**, *81*, 173–181. [CrossRef]
148. Mahajan, P.; Kaushal, J.; Upmanyu, A.; Bhatti, J. Assessment of Phytoremediation Potential of Chara vulgaris to Treat Toxic Pollutants of Textile Effluent. *J. Toxicol.* **2019**, *2019*, 1–11. [CrossRef]
149. Mahajan, P.; Kaushal, J. Phytoremediation of azo dye methyl red by macroalgae *Chara vulgaris* L.: Kinetic and equilibrium studies. *Environ. Sci. Pollut. Res.* **2020**, *27*, 26406–26418. [CrossRef] [PubMed]
150. Nowak, M.M.; Dziób, K.; Bogawski, P. Unmanned Aerial Vehicles (UAVs) in environmental biology: A review. *Eur. J. Ecol.* **2019**, *4*, 56–74. [CrossRef]
151. Kislik, C.; Genzoli, L.; Lyons, A.; Kelly, M. Application of UAV Imagery to Detect and Quantify Submerged Filamentous Algae and Rooted Macrophytes in a Non-Wadeable River. *Remote. Sens.* **2020**, *12*, 3332. [CrossRef]
152. Mitsch, W.J.; Zhang, L.; Stefanik, K.C.; Nahlik, A.M.; Anderson, C.J.; Bernal, B.; Hernandez, M.; Song, K. Creating Wetlands: Primary Succession, Water Quality Changes, and Self-Design over 15 Years. *Bioscience* **2012**, *62*, 237–250. [CrossRef]
153. EASA. *Easy Access Rules for Unmanned Aircraft Systems (Regulation (EU) 2019/947 and Regulation (EU) 2019/945)*; European Union Aviation Safety Agency: Cologne, Germany, 2021.
154. Klančnik, K.; Iskra, I.; Gradinjan, D.; Gaberščik, A. The quality and quantity of light in the water column are altered by the optical properties of natant plant species. *Hydrobiologia* **2017**, *812*, 203–212. [CrossRef]
155. Kim, D.; Yang, C.; Parsons, C.T.; Bowman, J.; Theÿsmeÿer, T.; Arhonditsis, G.B. Eutrophication management in a Great Lakes wetland: Examination of the existence of alternative ecological states. *Ecosphere* **2021**, *12*, e03339. [CrossRef]
156. Von Holle, B.; Yelenik, S.; Gornish, E.S. Restoration at the landscape scale as a means of mitigation and adaptation to climate change. *Curr. Landsc. Ecol. Rep.* **2020**, *5*, 1–13. [CrossRef]

MDPI
St. Alban-Anlage 66
4052 Basel
Switzerland
Tel. +41 61 683 77 34
Fax +41 61 302 89 18
www.mdpi.com

Plants Editorial Office
E-mail: plants@mdpi.com
www.mdpi.com/journal/plants